Bayesian Networks for Reliability Engineering

Baoping Cai · Yonghong Liu ·
Zengkai Liu · Yuanjiang Chang ·
Lei Jiang

Bayesian Networks for Reliability Engineering

Springer

Baoping Cai
China University of Petroleum
Qingdao, Shandong, China

Yonghong Liu
China University of Petroleum
Qingdao, Shandong, China

Zengkai Liu
China University of Petroleum
Qingdao, Shandong, China

Yuanjiang Chang
China University of Petroleum
Qingdao, Shandong, China

Lei Jiang
Southwest Jiaotong University
Chengdu, Sichuan, China

ISBN 978-981-13-6515-7 ISBN 978-981-13-6516-4 (eBook)
https://doi.org/10.1007/978-981-13-6516-4

Library of Congress Control Number: 2019932691

This Springer imprint is published by the registered company Springer Nature Singapore Pte Ltd.
The registered company address is: 152 Beach Road, #21-01/04 Gateway East, Singapore 189721, Singapore

Contents

Application of Bayesian Networks in Reliability Evaluation

Abstract The Bayesian network (BN) is a powerful model for probabilistic knowledge representation and inference and is increasingly used in the field of reliability evaluation. This paper presents a bibliographic review of BNs that have been proposed for reliability evaluation in the last decades. Studies are classified from the perspective of the objects of reliability evaluation, i.e., hardware, structures, software, and humans. For each classification, the construction and validation of a BN-based reliability model are emphasized. The general procedural steps for BN-based reliability evaluation, including BN structure modeling, BN parameter modeling, BN inference, and model verification and validation, are investigated. Current gaps and challenges in reliability evaluation with BNs are explored, and a few upcoming research directions that are of interest to reliability researchers are identified.

Keywords Bayesian network (BN) · Reliability · Hardware · Structure · Software · Human

1 Introduction

RELIABILITY is an item's probability that it performs its required function under given conditions for a stated time interval. This characteristic is intrinsically uncertain and a stochastic variable of an object, which can be hardware, structures, software, or humans. Evaluating the reliability of these objects is a challenging problem for reliability engineers and researchers.

Reliability can be evaluated using appropriate statistical inference techniques. For example, hardware reliability focuses on the research of systems and hardware and can be researched using fault tree analysis, event tree analysis, reliability block diagrams, Markov and semi-Markov models, and Petri nets. Structures are subsumed under hardware; however, we categorize them separately because they are closely related to structural mechanics principles. Structural reliability has been researched using response surface methods, first-order reliability methods, and second-order fourth-moment methods. Software reliability is defined by IEEE as the probability of failure-free software operation for a specified period of time in a specified environ-

B. Cai et al., *Bayesian Networks for Reliability Engineering*,
https://doi.org/10.1007/978-981-13-6516-4_1

ment. It has been researched using relevance vector regression, Gaussian processes, and the Markov-modulated Poisson process. Human reliability is the probability that an individual conducts system-required activities correctly for a specified period of time. It has been researched using ATHEANA, CREAM, and SPAR-H. Each reliability evaluation technique has its advantages and inherent disadvantages. Representing the uncertainties in the dependencies between different components or factors of the evaluated objects with many reliability evaluation methods, such as fault tree and reliability block diagram, is difficult because of the binary variable restriction [1]. Other techniques, such as Markov models and Petri nets, suffer from state space explosion problems [2].

Bayesian networks (BNs) are important probabilistic directed acyclic graphical models that can effectively characterize and analyze uncertainty, which is a problem commonly encountered in real-world domains, and handle state space explosion problems [3]. The applications of BNs have been extended to many fields involving uncertainty [4], from risk analysis [5, 6], safety engineering [7], resilience engineering [8], and fault diagnosis [9–11] to current reliability engineering, which is mainly discussed in the present work. BN-based reliability evaluation is conducted by forward (or predictive) analysis of BNs with various inference algorithms. That is, the probability of occurrence of the node that denotes the state of the evaluated object is calculated on the basis of the prior probabilities of the root nodes that denote the components or factors of the evaluated object and the conditional dependence of each node.

Reliability evaluation with BNs in the hardware, structure, software, and human domains is a particularly active research area that is attracting considerable attention from reliability engineers and researchers. Several review articles have summarized previous related studies. Langseth and Portinale [12] provided a thorough literature survey on BNs applied to reliability engineering, focusing on modeling framework, including BN model construction, causal interpretation, and BN inference. Tosun et al. [13] provided a systematic review of BNs applied to software quality prediction, also focusing on BN modeling steps, namely structure learning, parameter learning, use of tools, data characteristics, and validation. Mkrtchyan et al. [14] reviewed the use of BNs in human reliability analysis, analyzed five groups of BN applications, and identified the process of constructing BNs. In another work, Mkrtchyan et al. [15] reviewed five approaches to creating conditional probability tables and evaluated the performance of each approach.

This study aims to summarize and review recent studies on BNs used for reliability from the perspective of the objects of reliability evaluation, namely hardware, structures, software, and humans, because these categories of objects cover nearly the entire BN-based reliability literature. The general BN-based reliability evaluation procedural steps are investigated and compared between evaluation objects. Moreover, the potential challenging problems of BNs in reliability evaluation are identified, and upcoming research directions that are of interest to reliability engineers and researchers are presented. The remainder of this paper is structured as follows. Section 2 presents the BN-based reliability evaluation methodology. Section 3 summarizes and analyzes the applications of BNs in evaluating the reliability of hardware, structures, software, and humans. Section 4 suggests research directions. Section 5 summarizes the study.

2 BN-Based Reliability Evaluation Methodology

A. *Overview of BNs*

BNs, also known as static BNs, are probabilistic directed acyclic graphical models. They use nodes to represent variables, arcs to signify direct dependencies between the linked nodes, and conditional probabilities to quantify the dependencies. Static BNs are widely used in reliability evaluation, and many monographs have introduced BNs in detail [16–18].

For n random variables $X_1, X_2, \ldots, X_n$ and a directed acyclic graph with n nodes, among which node j ($1 \leq j \leq n$) is associated with X_j variable, the graph is the BN representing the variables $X_1, X_2, \ldots, X_n$ in the following equation:

$$P(X_1,\ X_2,\ \ldots,\ Xn) = \prod_{j=1}^{n} P\big(X_j|\text{parent}\big(X_j\big)\big), \tag{1}$$

where the parents (X_j) denote the set of all variables X_i and an arc connects node i to node j in the graph.

According to conditional independence assumptions and chain rules, the joint probability of variables $U = \{X_1, X_2, \ldots, X_\text{n}\}$ can be calculated as follows:

$$P(U) = \prod_{i=1}^{n} P(X_i|\text{Pa}(X_i)), \tag{2}$$

where Pa (X_i) is the parent node of X_i in the BN.

BNs can perform backward or diagnostic analyses with various inference algorithms based on *Bayes' theorem*, which is expressed as follows:

$$P(U|E) = \frac{P(E|U)P(U)}{P(E)} = \frac{P(E, U)}{\sum_U P(E, U)}. \tag{3}$$

Several limitations may be observed when BNs are adopted to evaluate the reliability of dynamic or complex systems [16] because BNs are static models and do not involve classes and objects. By contrast, dynamic BNs (DBNs) can represent the dynamic behavior of systems [18], and object-oriented BNs (OOBNs) can model complex systems with identical or similar components [17].

Evaluating the reliability of objects at the present moment does not involve temporal features; thus, BNs in such an evaluation scenario are appropriate. By contrast, in predicting reliability in the future, temporal features are involved; thus, DBNs are required. DBNs are extended BNs that relate variables to each other between adjacent time steps; that is, DBNs include multiple copies of the same variables, and the different copies represent different states of the variables over time [18]. DBNs are powerful tools in representing dynamic systems and therefore widely used in the reliability prediction of objects.

Establishing a BN-based reliability evaluation model is difficult and tedious when the evaluated object is too complex to be modeled with BNs, especially when the object is composed of collections of identical or similar components. Object-oriented methods are integrated into BNs to form OOBNs. An OOBN is a BN that contains not only the usual nodes but also instance nodes, which represent an instance of another generic BN fragment termed as class. An object, which is the fundamental unit of an OOBN, is produced by instantiating the class. OOBNs allow for simple model reuse, submodel encapsulation, and model construction in a top-down fashion, a bottom-up fashion, or a mixture of the two [17]. OOBNs are suitable tools for evaluating the reliability of objects with large, complex, and hierarchical structures.

Complex dynamic systems can be modeled by dynamic OOBNs (DOOBNs), which are an integration of DBNs and OOBNs [19]. A DOOBN-based model was used in the reliability evaluation of a water heater process in a previous work, from implementation to operation [20].

B. *Procedure for Reliability Evaluation with BNs*

For each category of evaluated objects (i.e., hardware, structures, software, and humans), BN-based reliability evaluation mainly includes four steps, namely BN structure modeling, BN parameter modeling, BN inference, and verification and validation. A detailed flowchart of this procedure is given in Fig. 1.

The BN structure is the model's qualitative part and corresponds to the directed acyclic graph. BN structure modeling consists of determining nodes and specifying linking arcs. BN structure modeling can be completed using the following methods: knowledge representation (for hardware, structures, software, and humans), mapping (for hardware, software, and humans), and structure learning (for hardware).

BN parameter modeling includes assigning prior probability and specifying conditional probability. Prior probability refers to the probability distribution of variables before any evidence is considered. Conditional probability refers to the posterior probability of variables when evidence is observed. A conditional probability table is used for discrete variables, whereas a conditional probability distribution is used for continuous variables. BN parameter modeling can be completed using the following methods: expert elicitation (for hardware, structures, software, and humans), mapping (for hardware), and parameter learning (for hardware, structures, software, and humans).

In BN-based reliability evaluation, BN inference is used to update the probability evaluation of networks given new observations or information. BN inference can be conducted using exact inference (for hardware and structures) and approximate inference (for hardware and software). Many BN tools, such as Netica and Hugin, are widely used for BN inference (for hardware, structures, software, and humans).

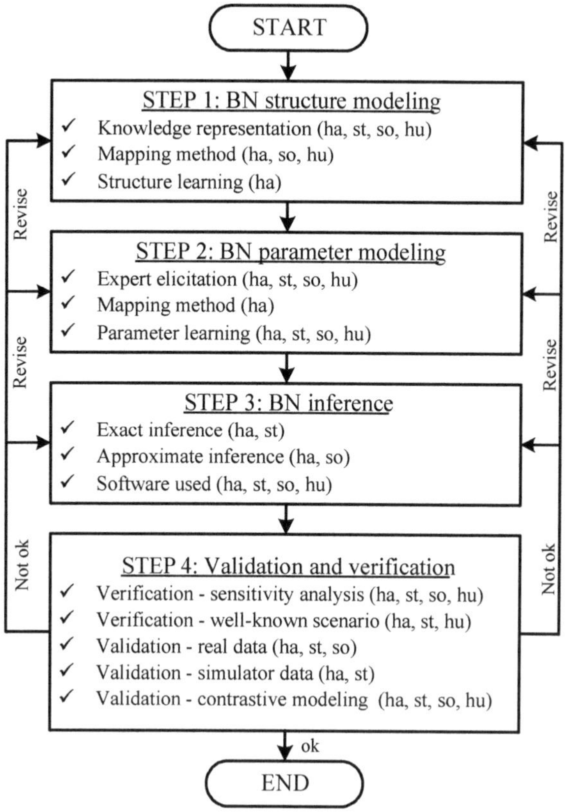

Fig. 1 Detailed flowchart of BN-based reliability evaluation (ha: hardware; st: structures; so: software; hu: humans)

The verification and validation of BN models are of great significance to reliability evaluation because they provide reasonable confidence to the assessment results. Verification is determining whether a model accurately represents the corresponding description and specifications. It can be completed using sensitivity analysis (for hardware, structures, software, and humans) and well-known scenarios (for hardware, structures, and humans). Validation is determining whether a model accurately reflects reality and can be accomplished using real data (for hardware, structures, and software), simulated data (for hardware and structures), and contrastive modeling (for hardware, structures, software, and humans).

3 Applications of BNs in Reliability Evaluation

A. *Hardware Reliability Evaluation with BNs*

Figure 2 shows a typical BN for the reliability evaluation of a series hardware system composed of a series system and a parallel system. Each root node represents the

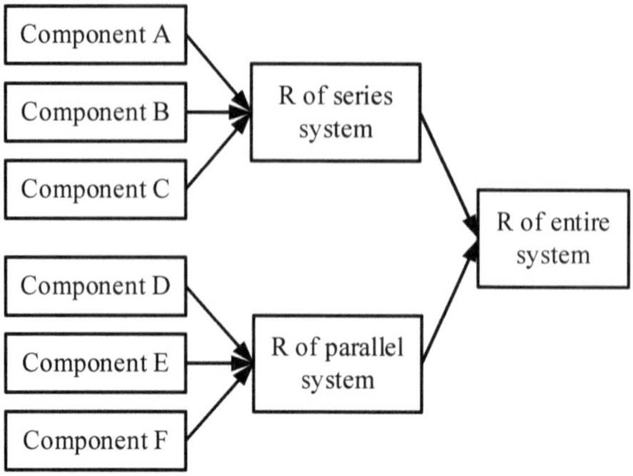

Fig. 2 Typical BN for hardware reliability evaluation

status of each component in the systems, and the final leaf node represents the reliability of the entire system. A review of the procedural steps for hardware reliability evaluation with BNs follows.

(1) *BN Structure Modeling*: In the literature, three main types of methods have been identified for constructing BNs for the reliability evaluation of hardware, namely knowledge representation, mapping, and structure learning.

Knowledge representation techniques are also known as cause-and-effect relationship methods. The knowledge about the evaluated hardware is collected by experts and captured into BNs. For example, Sättele et al. [21] developed a BN-based reliability evaluation method for alarm systems for natural hazards; the structure of the BNs was constructed on the basis of the influence relationships of the components in the monitoring, data interpretation, and information dissemination units. Honari et al. [22] used BNs to evaluate the reliability of an (r, s)-out-of-(m, n): F distributed communication system; the structure of the BNs was constructed on the basis of the logical relationship of the (r, s)-out-of-(m, n): F system. Eliassi et al. [23, 24] proposed a BN-based composite power system reliability modeling and evaluation methodology; the structure of the BNs was extracted by a minimal cutset-based approach, and the nodes at the three main levels of the structure were linked together according to the logical relationships of the components, cutsets, and system state. Cai et al. [25–27] adopted BNs and DBNs to evaluate the reliability and availability of subsea blowout preventer systems while considering common cause failure, imperfect repair, and preventive maintenance [28, 29]; the structures of the BNs and DBNs were constructed using the logical relationships of the components and knowledge of experts. These knowledge representation methods are largely subjective and may produce inaccurate reliability evaluation models. In addition, BN structures constructed by knowledge representation methods are large and pose substantial computational demand on inference algorithms.

Mapping techniques are also termed translating or transforming methods. The structure of the BNs is transformed from other reliability evaluation models. Fault tree analysis is a popular technique for hardware reliability evaluation and a widely used mapping model. Conversion algorithms from fault tree to BNs have been thoroughly researched. Bobbio et al. [1] proposed a mapping algorithm from a fault tree to BNs and evaluated the reliability of a redundant multiprocessor system. Huang et al. [30] analyzed the reliability of the electrical system of a CNC machine tool by using BNs; the structure of the BNs was also mapped from fault trees. Mi et al. [31] proposed a reliability evaluation methodology for electromechanical systems; the structure of the BNs was translated from dynamic fault trees. Montani et al. [32] and Portinale et al. [33] developed DBN-based reliability analysis tools; the structure of their DBNs was translated from corresponding dynamic fault trees. Boudali and Dugan [34] proposed a discrete-time BN-based reliability modeling and analysis methodology for critical systems; the structure of the discrete-time BNs was translated from dynamic fault trees. Liang et al. [35] proposed a reliability assessment methodology that is based on the integration of DBN and numerical simulation for warships; the structure of the DBNs was transformed from fault trees. Cai et al. [36] constructed BNs by transforming fault trees and assessed the reliability and availability of a subsea blowout preventer system by considering imperfect repair. Reliability models aside from fault trees can also be mapped into BNs. For example, Liu et al. [37] proposed a reliability assessment method for auxiliary feedwater systems by translating GO-FLOW models to BNs. Mapping methods are less subjective, and the accuracy of reliability evaluation by mapping methods is higher than that by knowledge representation methods. The superiority of the former can be ascribed to the use of mapping models, such as fault trees, instead of the knowledge of experts.

Structure learning techniques are machine learning methods. The structure of BNs for reliability evaluation is learned from massive reliability data related to the failure of each component and the entire hardware system. Structure learning methods are completely objective. The advantage of this type of method is the higher accuracy of the reliability evaluation results compared with those obtained by knowledge representation and mapping methods. The apparent disadvantage of structure learning methods is the difficulty or infeasibility of collecting sufficient data for structure learning. An alternative is to use simulation methods to generate sufficient data to fill the gap. For example, Daemi et al. [38] proposed a BN-based reliability evaluation methodology for composite power systems by focusing on the importance of components; they used state sampling using Monte Carlo simulation to generate sufficient training data and used common structure learning algorithms to establish the structure of the BNs with the data. Doguc and Ramirez-Marquez [39, 40] developed a generic BN-based evaluation methodology for system reliability; they used the $K2$ algorithm to learn the structure of the BNs from historical data of the system. Only few studies have used structure learning methods in constructing BNs for reliability evaluation mainly because the difficulty in obtaining training data limits the application of these methods.

(2) *BN Parameter Modeling*: This process is similar to BN structure modeling. Three main types of methods for BN parameter modeling for the reliability evaluation of hardware have been identified, namely expert elicitation, mapping, and parameter learning.

In expert elicitation methods, the prior and conditional probabilities of a BN for reliability evaluation are specified by domain experts on the basis of their knowledge, experience, and statistical reliability data. For example, Daemi et al. [38] assigned the parameters of component nodes artificially from forced outage rates to establish a BN-based reliability evaluation model for composite power systems. Yontay and Pan [41] developed a BN-based tool for evaluating the reliability dependencies between components and systems; the prior distributions of the model parameters were obtained by converting knowledge of experts into corresponding statistical expression. Zhang et al. [42] proposed a BN-based reliability evaluation method for a power system and evaluated the probabilities of successful cyberattacks on the system; the conditional probabilities of the nodes were determined using simple "AND" and "OR" relationships. Ruijun et al. [43] used interval-valued triangular fuzzy BNs to evaluate the reliability of multistate systems; the conditional probability tables of the networks were established by combining the knowledge of experts and practical experience. A major issue that arises when modeling BN parameters is the potentially large size of conditional probability tables. Micromodels such as noisy-OR and noisy-MAX have been proposed to solve this problem. These models present a local conditional probability distribution that depends on fewer parameters than complete ones. Therefore, these micromodels can be used for local structures. However, they may be inaccurate in real-world scenarios.

If the structures of BNs are translated from other kinds of reliability evaluation models by mapping methods, then the parameters of the BNs can also be translated from those models. With a fault tree as an example, the prior probabilities of the root nodes in BNs are identical to the corresponding probabilities of the leaf nodes in the fault tree. The conditional probabilities of the non-root nodes in the BNs are assigned on the basis of the logical relationships of the nodes in the fault tree, such as AND gate, OR gate, R/N gate, PAND gate, SEQ gate, and SP gate [1, 30–37]. The gate logic in a fault tree is usually binary, but mapping the two-state relationship in a fault tree to a multistate relationship in BNs is difficult. Therefore, inaccurate reliability evaluation models will be established when multistate relationships are considered in reality.

Parameter learning in BNs is known as parameter estimation. It is the task of estimating prior and conditional probabilities corresponding to the network structure. Like structure learning methods, parameter learning methods have the strength of highly accurate reliability models but also the weakness of difficulty in obtaining sufficient training data. Many algorithms have been developed for parameter learning, such as the expectation maximization and the penalized expectation maximization algorithms. Several parameter learning methods have been researched and used in the field of BN-based reliability evaluation. For example, Doguc and Ramirez-Marquez [39, 40] used the unsupervised construction algorithm $K2$ to calculate the conditional probabilities of BNs for system reliability evaluation with the help of Bayes' theorem. Daemi et al. [38] specified the parameters of load point nodes using a maximum likelihood method for the BN-based reliability evaluation model for composite power systems. Liu et al. [44] established a BN-based reliability evaluation model by considering common cause failures and studied the influence of extreme averse weather on the reliability of composite power systems; the conditional probability distributions of the BNs were obtained using random sampling, a parameter learning algorithm.

(3) *BN Inference*: Exact and approximate inference algorithms are used to update the probability evaluation of networks given new observations or information.

Exact algorithms, such as variable elimination, junction tree, and conditioning algorithms, guarantee correct answers but tend to be computationally demanding. Tien and Der Kiureghian [45] proposed a BN-based reliability evaluation methodology for infrastructure systems, presented a BN parameter compression approach, and developed an exact inference algorithm that is based on variable elimination, which used compressed conditional probability tables without decompressing them, for system reliability assessment. Tong and Tien [46] extended the research from binary systems to multistate systems, which are also based on an exact inference algorithm.

By contrast, approximate algorithms relax the demand for exact answers to ease the computational demand. These algorithms are usually based on sampling or optimization. This category includes stochastic sampling, importance sampling, Markov chain Monte Carlo, and belief propagation. Marquez et al. [47] proposed a novel hybrid BN-based framework containing a mixture of discrete and continuous variables for system reliability analysis and developed an approximate inference algorithm that is based on dynamic discretization for the hybrid BNs by combining an iterative discretization scheme with a junction tree inference algorithm. Zhong et al. [48] applied a time-to-failure modeling and reliability evaluation method that is based on a continuous BN for complex mechatronic systems; a revised nonparametric belief propagation inference algorithm, which is an approximation inference algorithm, was developed to perform the reliability analysis.

Notably, research on BN inference algorithms in the field of reliability engineering has received limited attention. Conversely, inference is performed using various commercial or free software, including Netica [26, 27, 36, 37, 49], AgenaRisk [31], MSBNs [25], GeNIe [21, 50], and BayesiaLab [51].

(4) *Verification and Validation*: The verification methods for BN-based hardware reliability evaluation are based on either sensitivity analysis or well-known scenarios. For the first type of method, Cai et al. [49] used a three-axiom sensitivity analysis method to verify the DBN-based reliability evaluation methodology for grid-connected photovoltaic systems. For the second type of method, Daemi et al. [38] verified the efficiency of a BN-based reliability evaluation model for composite power systems by applying it to the IEEE Reliability Test System. Similarly, Zhang et al. [42] and Liu et al. [44] verified their BN-based

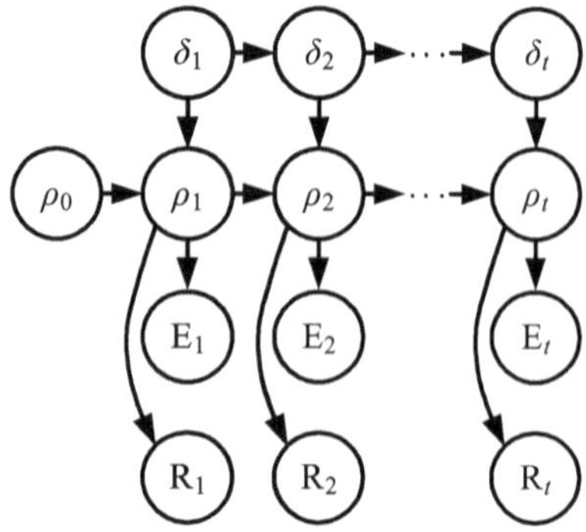

Fig. 3 Typical BN for structural reliability evaluation

reliability analysis methods for composite power systems by applying them to IEEE Reliability Test System 79 and the modified IEEE Reliability Test System, respectively.

The validation methods for BN-based hardware reliability evaluation are based on real data, simulated data, and contrastive modeling. For the first type of method, Doguc and Ramirez-Marquez [39, 40] validated the reliability evaluation results obtained by BNs by comparing them with actual values obtained by other researchers. For the second type of method, Zhong et al. [48] validated their reliability evaluation method on the basis of a continuous BN for complex mechatronic systems by using simulated data obtained from an active vehicle suspension simulation that was based on MATLAB's Kernel Density Estimation Toolbox. For the last type of method, Eliassi et al. [23] validated their BN-based reliability evaluation method for composite power systems by comparing it with the state enumeration approach and Monte Carlo simulation. Similarly, Marquez et al. [47] and Simon et al. [51] used analytical solutions, exact probability theorem, and Markov chain approach to validate their respective BN-based reliability analysis and inference methods.

B. *Structural Reliability Evaluation with BNs*

A structure is considered a specific kind of hardware because the reliability of a structure is closely related to the mechanics principles but not to the configurations or relationships of components. The DBN is the main modeling tool for structural reliability evaluation because structural reliability is closely related to the degradation behavior of the structure. Figure 3 depicts a typical DBN for the reliability evaluation of a degraded structure. In general, nodes δ and ρ are the parameters in the structure degradation model, node E represents the observed evidence, and node R represents the reliability of the structure. A detailed review of the procedural steps for structural reliability evaluation with this type of DBN follows.

(1) *BN Structure Modeling:* In the literature, only the first type of method for the BN structure modeling of hardware, which is knowledge representation, is used in structural reliability research. For example, Groden and Collette [52] proposed

a BN-based framework for updating the structural performance and reliability of marine structures; fatigue, probabilistic, and permanent set models were integrated into single BNs artificially, and the possible values of various distributions in these models were represented by the nodes of the BNs. Hackl and Kohler [53] proposed a DBN-based evaluation methodology for the structural reliability of reinforced concrete structures, focusing on degradation caused by corrosion; the structure of the BNs was constructed by transforming the parameters in corrosion and structural models into the nodes of the BNs and the causal relationships between the parameters of the models into the edges of the BNs. Mokhtar et al. [54] developed an evaluation methodology for the structural reliability of corroded interdependent pipe networks by using BNs, whose structure was constructed using the structural relationships of the pipelines and segments. Lee and Achenbach [55] analyzed the reliability of a jet engine compressor rotor blade with a fatigue crack by using DBNs; the structure of the DBNs was based on a fatigue crack growth model, and the parameters in the model were represented by the nodes. In [56], a similar method was used for the structural reliability analysis of stress corrosion crack growth. Straub [57] established a generic stochastic structural reliability analysis model for deterioration processes by using DBNs; the nodes represented the parameters in structural models, such as the fatigue crack growth model, and the structure of DBNs was constructed artificially. The literature shows that the structure of BNs is mainly constructed from structural performance models or the artificial integration of these models, such as the integration of a corrosion model and a mechanical model. The major feature of BNs is that the nodes represent the parameters in these models.

Mapping methods have not been used for the BN structure modeling for structural reliability evaluation because of the absence of one-to-one correspondence between the original reliability evaluation models, such as response surface models, and the BNs. In addition, structure learning methods have not been reported because the available structural reliability data are insufficient, especially those on degradation.

(2) *BN Parameter Modeling:* The prior probabilities and conditional probabilities of BNs for structural reliability evaluation are specified using expert elicitation methods or parameter learning methods.

Using an expert elicitation method, Mokhtar et al. [54] established the prior probability table of BNs for a structural reliability evaluation methodology for corroded interdependent pipe networks by using the failure probabilities obtained from a first-order reliability method; the authors specified the conditional probability on the basis of series or parallel relationships. Mahadevan et al. [58] proposed a BN-based structural system reliability evaluation methodology and used a branch-and-bound method to construct the conditional probability tables.

Similar to BN structure modeling, parameter learning methods suffer from insufficiency of structural reliability data, especially those on the degradation of structures. Therefore, parameter learning with complete data is impossible. Certain parameter learning algorithms with incomplete data have been developed and used in the field of BN-based reliability engineering. For example, Lee and Achenbach [55] calibrated the parameters of DBNs for the reliability analysis of the rotor blade of a jet engine compressor by using an expectation maximization algorithm for the incomplete available inspection data. A similar method was used in [56] for the structural reliability analysis of stress corrosion crack growth.

(3) *BN Inference:* Given that structural reliability modeling focuses on degradation behavior, the variables in structural reliability models are represented as a continuous space in DBNs. The convergence rate of the approximate inference algorithms for the models involving continuous variables is extremely slow, and in reality, the algorithms may not even converge. In addition, the application of simulation-based approximate inference algorithms in the case of rejection or likelihood sampling is also limited due to the computational inefficiency of such algorithms [57, 59]. A review of the literature shows that only exact inference algorithms have been used in the field of BN-based reliability evaluation. For example, Mokhtar et al. [54] used a junction tree algorithm to assess the structural failure probability of corroded interdependent pipe networks. Lee and Achenbach [55] used a forward–backward algorithm to achieve the exact inference of DBNs for the reliability analysis of a jet engine compressor rotor blade. A similar method was used in the structural reliability analysis of stress corrosion crack growth in [56]. In [57] and [60], a forward–backward algorithm was adopted to perform the exact inference of DBNs for a stochastic structural reliability analysis model for deterioration processes, and Straub and Kiureghian [59, 61] applied exact inference algorithms to BNs for their proposed structural reliability framework by minimizing the enhanced BNs into reduced BNs.

Few commercial BN tools have been used in structural reliability evaluation; alternatively, researchers have written their own BN codes for structural reliability evaluation by using MATLAB [54–57, 59, 61–63].

(4) *Verification and Validation:* Verification and validation methods for structural reliability evaluation are identical to those for hardware reliability evaluation. The verification methods are based on sensitivity analysis and well-known scenarios. For example, Mokhtar et al. [54] used both types of method as they conducted a sensitivity analysis and behavior tests by simulating known scenarios to verify the proposed BN-based structural reliability evaluation methodology. Using the second type of method, Zwirglmaier and Straub [62] used two application examples to verify the discretization procedures for rare events in discrete BNs for structural reliability assessment.

Three types of validation methods for BN-based structural reliability evaluation are used, namely validations based on real data, simulated data, and contrastive modeling. Regarding the first type of method, Lee and Achenbach [55] validated a DBN-based model for the reliability analysis of a jet engine compressor rotor blade by using a Bayes factor and filed inspection data. For the second type of method, Lee et al. [56] validated a structural reliability analysis methodology for stress corrosion

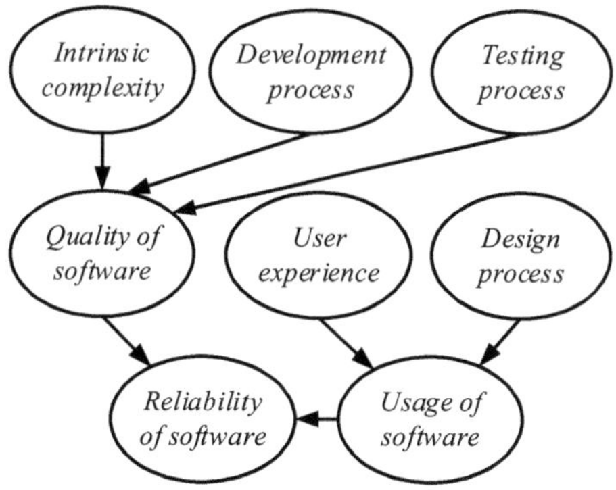

Fig. 4 Typical BN for software reliability evaluation

crack growth by using Bayesian hypothesis testing and simulated inspection data. Groden and Collette [52] used synthetic inspection data generated by Monte Carlo simulation to validate a BN-based framework for updating structural performance and reliability of marine structures. Regarding the last type of method, Mahadevan et al. [58] validated the proposed BN-based structural system reliability evaluation methodology by comparing its results with those of traditional reliability methods. Zhu and Collette [63] proposed a dynamic discretization method for DBN inference for structural reliability analysis; the robustness and efficiency of the method were validated by comparing it with the existing ones by using crack growth examples. Straub [57] validated a DBN-based stochastic model for structural reliability analysis by comparing its results with those of a second-order reliability method and Monte Carlo simulation. Luque and Straub [60] validated the computational efficiency of a DBN-based reliability analysis method for deteriorating systems by comparing its results with those of a standard Markov chain Monte Carlo method.

C. *Software Reliability Evaluation with BNs*

Software is different from hardware and structure in that it is not subject to degradation. Any operational failure is caused by faults inherent to the software. Software failures are caused by random input data, maintenance activities, or changing environments over time. BNs are used to combine various information sources to evaluate the reliability of software systems. Figure 4 shows a typical BN for the reliability evaluation of software. In general, the root nodes represent the factors affecting reliability, and the final leaf node represents the reliability of the software. A review of the procedural steps of software reliability evaluation with BNs follows.

(1) *BN Structure Modeling*: Knowledge representation and mapping are the two main types of methods of constructing BN structures.

In knowledge representation methods, the information sources related to software reliability and the interdependent relationships are determined mainly by the developers and users of the software. For example, Fenton et al. [64–66] proposed a BN-based model for software reliability and defect prediction and constructed the BN structure artificially by considering the factors affecting the reliability of a software product, such as the experience of staff, the use of formal methods, and the complexity of the problem. Neil et al. [67] adopted OOBNs to construct large-scale models for predicting software reliability and safety; the structure of the BN fragment was constructed on the basis of the experience of experts. Dahll and Gran [68] proposed a BN-based reliability and safety evaluation approach for the software of programmable safety systems; all available relevant information was integrated into the BNs, and the structure was constructed gradually by integrating the target nodes with observable and intermediate ones. Gran [69] and Dahll [70] used BNs to combine disparate sources of information in the safety and reliability evaluation of software-based systems and adopted a causal direction approach to establish the structure of the BNs on the basis of the experience and judgment of experts. Mohanta et al. [71] proposed a BN-based bottom-up method for the early prediction of software reliability from product metrics and established the topology of the BNs by using the faults and corresponding design metrics. Si et al. [72] developed a BN-based dependability assessment method for Internet-scale software and established the structure of BNs by analyzing the software architecture on the basis of its characteristics.

In mapping methods, the structures of the BNs are translated from other software reliability evaluation models or from the logic of the software. For example, Zou et al. [73] integrated a flow network model, BNs, and the proposed contribution method to evaluate the reliability of a digital instrumentation and control software system; they analyzed the sensitive edges by mapping the flow network model into the BNs automatically. Roshandel et al. [74] applied a DBN-based software reliability prediction approach at the architectural level, constructed the structure of the BNs from the corresponding global behavioral model, and then extended the model to the DBNs; however, they did not perform a strict model transformation. Jiang et al. [75] proposed a BN-based reliability analysis model for programmable logic controller systems; the proposed hybrid relation model, which was identified as a BN, was constructed by mapping on the basis of the execution logic of the embedded software.

Software is not subject to aging; thus, the tools used are nearly all BNs but not DBNs. In addition, similar to the structural reliability literature, the software reliability literature has not reported the use of structure learning methods for BN-based software reliability evaluation.

(2) *BN Parameter Modeling*: Establishing a parameter model for BNs by using parameter learning methods is difficult because complete data are required by collecting sufficient software reliability data. In BN parameter modeling for software reliability evaluation, expert elicitation methods and parameter learning methods with incomplete data are mainly used. In the BN-based software defect and reliability evaluation models proposed by Fenton et al. [64] and Neil et al. [67], the prior probabilities and conditional probabilities of the BNs were determined using expert elicitation and data. Gran and Helminen [69, 76] proposed a method for merging a BN for a software safety assessment with a BN for the reliability evaluation of software-based digital systems; the conditional probability tables of the BNs were elicited through brainstorming exercises with

all the project participants, who shared their general knowledge and experience in software development and evaluation. In their BN-based software reliability and safety evaluation models, Dahll and Gran [68, 70] assigned the prior and conditional probabilities of the BNs by using expert judgment. In their BN-based software reliability prediction method, Mohanta et al. [71] obtained the conditional probability distributions by using parametric or functional form and conducting multiple linear regression analysis.

For the second type of method, Bai et al. [77] proposed a software reliability prediction approach based on Markov BNs and used an expectation maximization algorithm to estimate the unknown parameters in the distribution from incomplete data.

(3) *BN Inference*: No major study has focused on BN inference algorithms for software reliability evaluation. To our knowledge, only one approximate inference algorithm related to dynamic discretization has been reported. Fenton et al. [66] proposed a software defect and reliability prediction model based on BNs; an approximate inference method using dynamic discretization algorithm was adopted to perform the prediction, and the method exhibited higher accuracy and required less storage space than did a static one. Alternatively, many researchers have used BN tools, such as Hugin [64, 65, 68–70, 76], Netica [71, 78], and AgenaRisk [66], to conduct BN inference for software reliability evaluation. This situation is similar to that of hardware reliability research and totally different from that of structure reliability research.
(4) *Verification and Validation*: Verification based on sensitivity analysis is the main verification method for BN-based software reliability evaluation models. Roshandel et al. [74] used sensitivity analysis to demonstrate that the proposed DBN-based software reliability prediction approach was effective at the architectural level, and its results were helpful in making architectural decisions. Liu et al. [78] partially used sensitivity analysis to validate that the proposed BN-based software reliability evaluation method for subsea blowout preventers was accurate and rational. Verification based on well-known scenarios is not widely used possibly because no well-known software scenario similar to the IEEE Reliability Test System for hardware has been established for such verification.

Two validation methods for software reliability evaluation are used, namely validation based on real data and validation based on contrastive modeling. Real data are easier and more likely to be obtained from experiments on software than from those on hardware and structure. Therefore, many researchers have opted for validation based on real data. Neil et al. [67] used real test data and expert opinion that were not used to derive the BNs to validate the proposed OOBN-based software reliability and safety prediction approach. Si et al. [72] performed experiments on a real enterprise e-commerce application system to validate the effectiveness of their BN-based dependability assessment method for Internet-scale software. Jiang et al. [75] used experimental results to demonstrate the accuracy of the proposed BN-based reliability analysis model for programmable logic controller systems. Performing

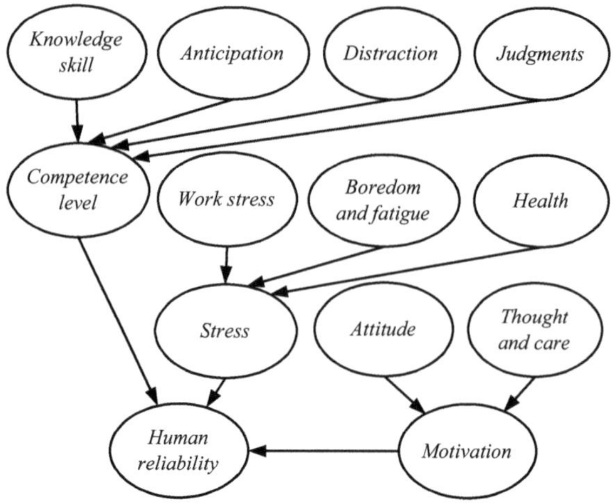

Fig. 5 Typical BN for human reliability evaluation

validation through contrastive modeling, Bai [79] evaluated the performance of a Markov BN-based software reliability prediction model with an operational profile by comparing the results of the proposed model with those of the Kaaniche–Kanoun model. In addition, Mohanta et al. [71] validated their BN-based software reliability early prediction method by using real data obtained from a set of experiments and investigated the accuracy of the method by comparing its results with those of the Rome Air Forces Development Centre method proposed by McCall [80].

D. *Human Reliability Evaluation with BNs*

Human reliability analysis, which is an important research field in reliability engineering, aims to identify and analyze the causes, consequences, and contributions of human failures in various industrial systems [81]. BNs are increasingly used in this field due to their capability to describe the complex influencing relationships of human factors. Figure 5 illustrates an example of a BN used in human reliability evaluation. In general, the root nodes represent the factors affecting reliability, and the final leaf node represents human reliability. A review of the procedural steps for human reliability evaluation with BNs follows.

(1) *BN Structure Modeling*: With regard to the structure of BNs for human reliability evaluation, knowledge representation method is still the primary method. The structure can be derived from the judgment and experience of experts, with limited human performance data. For example, Li et al. [82] studied organizational influences quantitatively by using a fuzzy BN-based human reliability analysis method and established the structure of the BNs by using the cause-and-effect relationship method on the basis of the experiences of maintenance and human factor experts. Aalipour et al. [83] used BNs to research the causes of human errors in the maintenance activities in a production process and constructed the structure of BNs on the basis of the causal dependencies between the variables. Zwirglmaier et al. [84] developed a BN-based methodology to capture cognitive

causal paths in human reliability analysis and adopted a two-level method (i.e., causal path identification and model reduction) to construct the structure of the BNs.

Numerous human reliability analysis methods have been developed, and some of these methods have been mapped to BNs in the study of human reliability. Groth and Swiler [85] proposed a BN version of SPAR-H for human reliability analysis by directly translating the SPAR-H model to BNs. Sundaramurthi and Smidts [86] proposed a BN-based human reliability modeling methodology for the Next Generation System Code, and the structure of the BNs was transformed from the simplified causal graph.

Cai et al. [87] used BNs to study the influence of human factors on offshore blowouts, and the structure of the BNs was mapped from pseudo-fault tree models. Martins and Maturana [88] proposed a BN-based framework for human reliability analysis for collision accidents during oil tanker operation and transformed the collision fault tree into BNs that represented the same domain.

Similar to software, human factors are not subject to degradation and aging; therefore, BNs and not DBNs are mainly used for human reliability analysis.

(2) *BN Parameter Modeling*: The uncertainties in human factors are more numerous and complex than those in hardware, structures, and software. Expert elicitation methods may be the most appropriate method for parameter modeling. Li et al. [82] estimated the prior and conditional probabilities of fuzzy BNs for human reliability analysis on the basis of the engineering judgments of domain experts. Aalipour et al. [83] assigned the marginal probability tables of BNs for human error analysis in the cable manufacturing industry by using direct elicitation and expert judgment. Zwirglmaier et al. [84] quantified a BN for human reliability analysis by combining human performance data and expert elicitation results as information sources. Martins and Maturana [88] obtained the conditional probability tables through an iterative search and linear interpolation approach in the context of the lack of data and expert opinion. Baraldi et al. [89] derived the conditional probability tables of BNs for human reliability analysis from the elicitation of a limited number of relationships provided by experts in the form of rules. Musharraf et al. [90] developed a collection method for human performance data by virtual experiments and assigned the prior and conditional probabilities of BNs for human reliability analysis in the event of an offshore emergency evacuation. Li et al. [91] used situational awareness data measured by simulation experiments to determine the conditional probability distribution of a BN-based model for operators' situational awareness reliability.

Collecting sufficient human performance data for general industrial applications is difficult. However, for certain industries, some data have already been collected and sorted, and parameter learning methods can be used to determine the prior and conditional probabilities. For example, Sundaramurthi and Smidts [86] calculated the conditional probabilities and node probabilities by using the parameter learning function of the software GeNIe and data collected from aviation and nuclear accidents.

(3) *BN Inference*: No BN inference algorithms have been studied for human reliability analysis. However, many BN tools have been used directly to perform BN inference, including MSBNx [82, 92, 93], AgenaRisk [83], Hugin [85, 94], GeNIe [86, 90, 95], and Netica [87, 88].
(4) *Verification and Validation*: Validation techniques for human reliability evaluation methodologies were researched and classified by Kirwan [96, 97], but he focused on all evaluation methodologies instead of only BN-based ones. In BN-based human reliability analysis, verification and validation are distinct. For verification, sensitivity analysis is the primary method. Cai et al. [87] performed a sensitivity analysis to verify the proposed BN models for human reliability analysis of offshore blowouts. Gregoriades and Sutcliffe [98] used BNs to model the cause-and-effect relationships of performance-shaping factors and assessed the agent's operational reliability; the models were validated using data mining techniques, including relevance analysis, association rules, and classification. Yang et al. [94] developed a human reliability quantification method in marine engineering by integrating fuzzy logic, evidential reasoning, and BNs and used sensitivity analysis to validate the proposed method. In addition, verification based on well-known scenarios has also been reported. Lee and Seong [93] developed a BN-based computational model to evaluate the situation of nuclear power plant operators and verified the proposed method by comparing the results of expert mental model and less-skilled operator mental model.

Acquiring real and even simulated data is difficult; consequently, only validation based on contrastive modeling has been used. For instance, Groth and Swiler [85] validated the proposed BN version of SPAR-H for human reliability analysis by comparing it with the SPAR-H method. Musharraf et al. [90] validated their BN-based human reliability analysis method for offshore emergency evacuation by comparing its results with those of the Bradbury-Squires method [99] by using the same data. Musharraf et al. [95] used a success likelihood index methodology to validate another BN-based human reliability assessment model for offshore emergency conditions. Kim and Seong [92] proposed a BN-based analytic model to evaluate the situation of nuclear power plant operators and validated this method by comparing it with the situation evaluation model proposed by Miao et al. [100]. This comparison of methods validates existing human reliability evaluation methodologies partially but not completely. A complete validation should be performed further in real industrial applications with real data.

E. *Discussion*

The applications of BNs in reliability evaluation are reviewed in the previous section. The important issues mentioned can be summarized as follows.

(1) Most of the studies on BN-based reliability evaluation focus on hardware, whereas only few studies explore the BN-based reliability evaluation of structures, software, and humans. In the reliability evaluation of hardware, structures, software, and humans, most studies focus on BN structure modeling and BN parameter modeling, whereas only few studies include BN inference, verification, and validation.

(2) Knowledge representation and expert elicitation methods are the predominantly used techniques for structure and parameter modeling because they offer the simplest solution to determining the uncertainty between the nodes of BNs, that is, by using the knowledge of experts. However, these methods are highly subjective and may produce inaccurate reliability evaluation models.
(3) Mapping methods are less subjective, and the accuracy of reliability evaluation by this type of method is higher than that by knowledge representation methods. Therefore, if the structure modeling is using mapping method, the parameter modeling is not always using mapping method, just like software and human.
(4) Structure and parameter learning are the most accurate methods for constructing BNs; however, in practice, obtaining sufficient and valuable data for training structure and parameter models is nearly impossible.
(5) BN inference algorithms, including exact and approximate inference, have been seldom investigated in studies on BN-based reliability evaluation methodologies. Instead, inference is mainly performed using various commercial or free software programs because they can help solve computational complexity and reduce memory usage considerably.
(6) Verification and validation are important to BN-based reliability evaluation methodologies; however, many proposed methodologies have not been verified and validated using suitable methods. Real data have only seldom been applied to the verification and validation of BN-based reliability methodologies. Instead, contrastive modeling and simulated data are widely used.
(7) DBNs are extensively used to study the reliability of hardware and structures. However, their application in software and human reliability evaluation is limited mainly because only hardware and structures are subject to degradation and aging, which can be well described by DBNs.

4 Research Directions

In view of the literature review of BN-based reliability evaluation methodologies, a few upcoming research directions in this field that are of interest to reliability researchers and practitioners are presented in this section.

A. *BN Modeling Methods Considering Cascading Failures*

Failure dependency remarkably affects the reliability of systems, particularly hardware and structures. Common cause failure and cascading failure are two typical examples of failure dependency. BN modeling methods considering common cause failure have been extensively researched [25, 101]. A cascading failure is a failure in an interconnected system, in which the failure of a part can trigger the failure of successive parts. Few studies on the BN-based reliability evaluation methodology have considered cascading failures [102]. For hardware and structure reliability evaluation, constructing the structure and parameter models of BNs for reliability evaluation by considering the cascading failure of components, especially when temporal and dynamic features are involved, is a challenging problem.

B. *DBN-Based Reliability Prediction for Software and Humans*

Software and humans are not subject to degradation and aging when they are modeled for reliability evaluation. Software behavior changes with time because maintenance activities occur or the environment changes over time. Human errors are more complex than software errors because human reliability is influenced by intrinsic factors (e.g., skill) and external factors (e.g., weather). Reliability can be predicted well if the dynamic changes in the environmental factors related to software and human reliability can be modeled using DBNs.

C. *Integration of Big Data and BN Reliability Evaluation Methodology*

Big data is a popular research topic and has three attributes: volume, variety, and velocity. Massive amounts of data on hardware, structures, software, and humans are collected or recorded by sensors or humans. Numerous pieces of useful information, especially those about system or component failures, are contained by these data. The structure learning and parameter learning of BNs need these useful data for reliability modeling. The integration of big data and BN reliability evaluation presents a definite orientation for interdisciplinary research. However, extracting useful information from big data and constructing reliability evaluation, analysis, and prediction models by using BNs are challenging problems.

D. *Rapid Approximate Inference Algorithms for DBNs for Reliability Evaluation*

For reliability prediction, DBN inference is mainly accomplished using various commercial or free software programs. In reality, a complex reliability model leads to high computational cost of the inference with software. A time slice with a long period of time should be set, thereby leading to inaccurate prediction results. By contrast, rapid approximate inference algorithms may achieve good prediction results. Therefore, research on rapid approximate inference algorithms for DBNs is a worthwhile future direction.

5 Conclusion

Since the introduction of BNs by Pearl in the early 1980s, their application in reliability engineering has been widely researched and obtained favorable achievements. This work provides a literature review of BN-based reliability evaluation methodologies from the perspective of the objects of evaluation, focusing on the general procedure of reliability modeling with BNs. For each evaluated object (i.e., hardware, structures, software, and humans), the various methods for each procedural step (i.e., BN structure modeling, BN parameter modeling, BN inference, and model verification and validation) are reviewed and analyzed in detail. The potential problems and current gaps in applying BNs to reliability evaluation are discussed, and upcoming research directions that are of interest to reliability researchers are presented. We hope that the current paper can provide researchers and practitioners a helpful guide for BN-based reliability evaluation methodology with BNs.

References

1. A. Bobbio, L. Portinale, M. Minichino, E. Ciancamerla, Improving the analysis of dependable systems by mapping fault trees into Bayesian networks. Reliab. Eng. Syst. Saf. **71**(3), 249–260, (2001). (2001-01-01)
2. IEC, Electric/Electronic/Programmable Electronic safety-related systems, parts 1–7. *Technical Report, International Electrotechnical Commission* (2010)
3. B. Cai, L. Huang, M. Xie, Bayesian networks in fault diagnosis. IEEE Trans. Ind. Inf. **13**(5), 2227–2240 (2017)
4. H.A. Khorshidi, I. Gunawan, M.Y. Ibrahim, Data-Driven system reliability and failure behavior modeling using FMECA. IEEE Trans. Ind. Inf. **12**(3), 1253–1260 (2016)
5. Q. Zhang, C. Zhou, Y. Tian, N. Xiong, Y. Qin, B. Hu, A fuzzy probability Bayesian network approach for dynamic cybersecurity risk assessment in industrial control systems. IEEE Trans. Ind. Inf. (2018), https://doi.org/10.1109/tii.2017.2768998
6. Z. Liu, Y. Liu, X.L. Wu, B. Cai, Risk analysis of subsea blowout preventer by mapping GO models into Bayesian networks. J. Loss Prev. Process Ind. **52**, 54–65 (2018)
7. B. Cai, Y. Liu, Q. Fan, A multiphase dynamic Bayesian networks methodology for the determination of safety integrity levels. Reliab. Eng. Syst. Saf. **150**, 105–115 (2016)
8. B. Cai, M. Xie, Y. Liu, Y. Liu, Q. Feng, Availability-based engineering resilience metric and its corresponding evaluation methodology. Reliab. Eng. Syst. Saf. **172**, 216–224 (2018)
9. B. Cai, Y. Liu, M. Xie, A dynamic-bayesian-network-based fault diagnosis methodology considering transient and intermittent faults. IEEE Trans. Autom. Sci. Eng. **1**(14), 276–285 (2017)
10. Y. Luo, L. Kaicheng, Y. Li, D. Cai, C. Zhao, Q. Meng, Three layer Bayesian network for classification of complex power quality disturbances. IEEE Trans. Ind. Inf. (2018), https://doi.org/10.1109/tii.2017.2785321
11. Z. Wang, Z. Wang, X. Gu, S. He, Z. Yan, Feature selection based on Bayesian network for chiller fault diagnosis from the perspective of field applications. Appl. Therm. Eng. **129**, 674–683 (2018)
12. H. Langseth, L. Portinale, Bayesian networks in reliability. Reliab. Eng. Syst. Saf. **92**(1), 92–108 (2007)
13. A. Tosun, A.B. Bener, S. Akbarinasaji, A systematic literature review on the applications of Bayesian networks to predict software quality. Softw. Qual. J. **25**(1), 273–305 (2017)
14. L. Mkrtchyan, L. Podofillini, V.N. Dang, Bayesian belief networks for human reliability analysis: A review of applications and gaps. Reliab. Eng. Syst. Saf. **139**, 1–16 (2015)
15. L. Mkrtchyan, L. Podofillini, V.N. Dang, Methods for building conditional probability tables of Bayesian belief networks from limited judgment: An evaluation for human reliability application. Reliab. Eng. Syst. Saf. **151**, 93–112 (2016)
16. F.V. Jensen, T.D. Nielsen, *Bayesian Networks and Decision Graphs* (Springer Science and Business Media, 2007)
17. U.B. Kjærulff, A.L. Madsen, *Bayesian Networks and Influence Diagrams: A Guide to Construction and Analysis* (Springer Science and Business Media, 2013)
18. A. Darwiche, *Modeling and Reasoning with Bayesian Networks* (Cambridge University Press, 2009)
19. X. Yuan, B. Cai, Y. Ma, J. Zhang, K. Mulenga, Y. Liu, G. Chen, Reliability evaluation methodology of complex systems based on dynamic object-oriented Bayesian networks. IEEE Access **6**, 11289–11300 (2018)

20. P. Weber, L. Jouffe, Complex system reliability modelling with dynamic object oriented Bayesian networks (DOOBN). Reliab. Eng. Syst. Saf. **91**(2), 149–162 (2006)
21. M. Sättele, M. Bründl, D. Straub, Reliability and effectiveness of early warning systems for natural hazards: Concept and application to debris flow warning. Reliab. Eng. Syst. Saf. **142**, 192–202 (2015)
22. B. Honari, J. Donovan, E. Murphy, Using Bayesian networks in reliability evaluation for an (r, s)-out-of-(m, n): F distributed communication system. J. Stat. Plann. Infer. **139**(5), 1756–1765 (2009)
23. M. Eliassi, A.K. Dashtaki, H. Seifi, M. Haghifam, C. Singh, Application of Bayesian networks in composite power system reliability assessment and reliability-based analysis. IET Gener. Transm. Distrib. **9**(13), 1755–1764 (2015). (2015-10-01)
24. H. Seifi, M. Eliassi,M. Haghifam, Incorporation of protection system failures into bulk power system reliability assessment by Bayesian networks. IET Gener. Transm. Distrib. **9**(11), 1226–1234 (2015). (2015-08-06)
25. B. Cai, Y. Liu, Z. Liu, X. Tian, X. Dong, S. Yu, Using Bayesian networks in reliability evaluation for subsea blowout preventer control system. Reliab. Eng. Syst. Saf. **108**, 32–41 (2012)
26. B. Cai, Y. Liu, Q. Fan, Y. Zhang, S. Yu, Z. Liu, X. Dong, Performance evaluation of subsea BOP control systems using dynamic Bayesian networks with imperfect repair and preventive maintenance. Eng. Appl. Artif. Intell. **26**(10), 2661–2672 (2013)
27. B. Cai, Y. Liu, Y. Ma, Z. Liu, Y. Zhou, J. Sun, Real-time reliability evaluation methodology based on dynamic Bayesian networks: A case study of a subsea pipe ram BOP system. ISA Trans. **58**, 595–604 (2015)
28. Q. Feng, X. Bi, X. Zhao, Y. Chen, B. Sun, Heuristic hybrid game approach for fleet condition-based maintenance planning. Reliab. Eng. Syst. Saf. **157**, 166–176 (2017)
29. Q. Feng, W. Bi, Y. Chen, Y. Ren, D. Yang, Cooperative game approach based on agent learning for fleet maintenance oriented to mission reliability. Comput. Ind. Eng. **112**, 221–230 (2017)
30. T. Huang, J. Yan, M. Jiang, W. Peng, H. Huang, Reliability analysis of electrical system of computer numerical control machine tool based on bayesian networks. J. Shanghai Jiaotong Univ. (Science) **21**(5), 635–640 (2016)
31. J. Mi, Y. Li, Y. Yang, W. Peng, H. Huang, Reliability assessment of complex electromechanical systems under epistemic uncertainty. Reliab. Eng. Syst. Saf. **152**, 1–15 (2016)
32. S. Montani, L. Portinale, A. Bobbio, D. Codetta-Raiteri, Radyban: A tool for reliability analysis of dynamic fault trees through conversion into dynamic Bayesian networks. Reliab. Eng. Syst. Saf. **93**(7), 922–932 (2008)
33. L. Portinale, D.C. Raiteri, S. Montani, Supporting reliability engineers in exploiting the power of dynamic Bayesian networks. Int. J. Approximate Reasoning **51**(2), 179–195 (2010)
34. H. Boudali, J.B. Dugan, A discrete-time Bayesian network reliability modeling and analysis framework. Reliab. Eng. Syst. Saf. **87**(3), 337–349 (2005)
35. X.F. Liang, H.D. Wang, H. Yi, D. Li, Warship reliability evaluation based on dynamic bayesian networks and numerical simulation. Ocean Eng. **136**, 129–140 (2017)
36. B. Cai, Y. Liu, Y. Zhang, Q. Fan, S. Yu, Dynamic Bayesian networks based performance evaluation of subsea blowout preventers in presence of imperfect repair. Expert Syst. Appl. **40**(18), 7544–7554 (2013)
37. Z. Liu, Y. Liu, X. Wu, D. Yang, B. Cai, C. Zheng, Reliability evaluation of auxiliary feedwater system by mapping GO-FLOW models into Bayesian networks. ISA Trans. **64**, 174–183 (2016)
38. T. Daemi, A. Ebrahimi, M. Fotuhi-Firuzabad, Constructing the Bayesian network for components reliability importance ranking in composite power systems. Int. J. Electr. Power Energy Syst. **43**(1), 474–480 (2012)
39. O. Doguc, J.E. Ramirez-Marquez, A generic method for estimating system reliability using Bayesian networks. Reliab. Eng. Syst. Saf. **94**(2), 542–550 (2009)
40. O. Doguc, J.E. Ramirez-Marquez, An automated method for estimating reliability of grid systems using Bayesian networks. Reliab. Eng. Syst. Saf. **104**, 96–105 (2012)

41. P. Yontay, R. Pan, A computational Bayesian approach to dependency assessment in system reliability. Reliab. Eng. Syst. Saf. **152**, 104–114 (2016)
42. Y. Zhang, L. Wang, Y. Xiang, C. Ten, Power system reliability evaluation with SCADA cybersecurity considerations. IEEE Trans. Smart Grid **6**(4), 1707–1721 (2015)
43. Z. Ruijun, Z. Lulu, W. Nannan, W. Xiaowei, Reliability evaluation of a multi-state system based on interval-valued triangular fuzzy Bayesian networks. Int. J. Syst. Assur. Eng. Manag. **7**(1), 16–24 (2016)
44. Y. Liu, C. Singh, Evaluation of hurricane impact on composite power system reliability considering common-cause failures. Int. J. Syst. Assur. Eng. Manag. **1**(2), 135–145 (2010)
45. I. Tien, A. Der Kiureghian, Algorithms for Bayesian network modeling and reliability assessment of infrastructure systems. Reliab. Eng. Syst. Saf. **156**, 134–147 (2016)
46. Y. Tong, I. Tien, Algorithms for Bayesian network modeling, inference, and reliability assessment for multistate flow networks. J. Comput. Civil Eng. **5**(31), 04017051 (2017)
47. D. Marquez, M. Neil, N. Fenton, Improved reliability modeling using Bayesian networks and dynamic discretization. Reliab. Eng. Syst. Saf. **95**(4), 412–425 (2010)
48. X. Zhong, M. Ichchou, A. Saidi, Reliability assessment of complex mechatronic systems using a modified nonparametric belief propagation algorithm. Reliab. Eng. Syst. Saf. **95**(11), 1174–1185 (2010)
49. B. Cai, Y. Liu, Y. Ma, L. Huang, Z. Liu, A framework for the reliability evaluation of grid-connected photovoltaic systems in the presence of intermittent faults. Energy **93**, 1308–1320 (2015)
50. A.O. Connor, A. Mosleh, A general cause based methodology for analysis of common cause and dependent failures in system risk and reliability assessments. Reliab. Eng. Syst. Saf. **145**, 341–350 (2016)
51. C. Simon, P. Weber, A. Evsukoff, Bayesian networks inference algorithm to implement Dempster Shafer theory in reliability analysis. Reliab. Eng. Syst. Saf. **93**(7), 950–963 (2008)
52. M. Groden, M. Collette, Fusing fleet in-service measurements using Bayesian networks. Mar. Struct. **54**, 38–49 (2017)
53. J. Hackl, J. Kohler, Reliability assessment of deteriorating reinforced concrete structures by representing the coupled effect of corrosion initiation and progression by Bayesian networks. Struct. Saf. **62**, 12–23 (2016)
54. E.H. Ait Mokhtar, A. Chateauneuf, R. Laggoune, Bayesian approach for the reliability assessment of corroded interdependent pipe networks. Int. J. Press. Vessels Pip. **148**, 46–58 (2016)
55. D. Lee, J.D. Achenbach, Analysis of the reliability of a jet engine compressor rotor blade containing a fatigue crack. J. Appl. Mech. **4**(83), 041004 (2016)
56. D. Lee, Y. Huang, J.D. Achenbach, Probabilistic analysis of stress corrosion crack growth and related structural reliability considerations. J. Appl. Mech. **2**(83), 021003 (2016)
57. D. Straub, Stochastic modeling of deterioration processes through dynamic Bayesian networks. J. Eng. Mech. **10**(135), 1089–1099 (2009)
58. S. Mahadevan, R. Zhang, N. Smith, Bayesian networks for system reliability reassessment. Struct. Saf. **23**(3), 231–251 (2001). (2001-01-01)
59. D. Straub, A.D. Kiureghian, Bayesian Network enhanced with structural reliability methods: Methodology. J. Eng. Mech. **10**(136), 1248–1258 (2009)
60. J. Luque, D. Straub, Reliability analysis and updating of deteriorating systems with dynamic Bayesian networks. Struct. Saf. **62**, 34–46 (2016)
61. D. Straub, A.D. Kiureghian, Bayesian network enhanced with structural reliability methods: Application. J. Eng. Mech. **10**(136), 1259–1270 (2010)
62. K. Zwirglmaier, D. Straub, A discretization procedure for rare events in Bayesian networks. Reliab. Eng. Syst. Saf. **153**, 96–109 (2016)
63. J. Zhu, M. Collette, A dynamic discretization method for reliability inference in dynamic Bayesian networks. Reliab. Eng. Syst. Saf. **138**, 242–252 (2015)
64. N. Fenton, B. Littlewood, M. Neil, L. Strigini, A. Sutcliffe, D. Wright, Assessing dependability of safety critical systems using diverse evidence. IEE Proc.-Softw. **1**(145), 34–46 (1998)

65. N.E. Fenton, M. Neil, A critique of software defect prediction models. EEE Trans. Softw. Eng. **5**(25), 675–689 (1999)
66. N. Fenton, M. Neil, D. Marquez, Using Bayesian networks to predict software defects and reliability. Proc. Inst. Mech. Eng. Part O J. Risk Reliab. **222**(4), 701–712 (2008)
67. M. Neil, N. Fenton, L. Nielson, Building large-scale Bayesian networks. Knowl. Eng. Rev. **3**(15), 257–284 (2000)
68. G. Dahll, B.A. Gran, The use of Bayesian belief nets in safety assessment of software based systems. Int. J. Gen. Syst. **2**(29), 205–229 (2001)
69. B.A. Gran, Use of Bayesian belief networks when combining disparate sources of information in the safety assessment of software-based systems. Int. J. Syst. Sci. **6**(33), 529–542 (2002)
70. G. Dahll, Combining disparate sources of information in the safety assessment of software-based systems. Nucl. Eng. Des. **6**(33), 529–542 (2000)
71. S. Mohanta, G. Vinod, R. Mall, A technique for early prediction of software reliability based on design metrics. Int. J. Syst. Assur. Eng. Manag. **2**(4), 261–281 (2011)
72. G. Si, J. Xu, J. Yang, S. Wen, An evaluation model for dependability of Internet-scale software on basis of Bayesian networks and trustworthiness. J. Syst. Softw. **89**, 63–75 (2014)
73. B. Zou, M. Yang, E. Benjamin, H. Yoshikawa, Reliability analysis of digital instrumentation and control software system. Prog. Nucl. Energy **98**, 85–93 (2017)
74. R. Roshandel, N. Medvidovic, L. Golubchik, A Bayesian model for predicting reliability of software systems at the architectural level. QoSA 2007 LNCS **4880**, 108–126 (2007)
75. Y. Jiang, H. Zhang, X. Song, X. Jiao, W.N.N. Hung, M. Gu, J. Sun, Bayesian-network-based reliability analysis of PLC systems. IEEE Trans. Ind. Electron. **60**(11), 5325–5336 (2013)
76. B.A. Gran, A. Helminen, A Bayesian belief network for reliability assessment. SAFECOMP 2001 LNCS **2187**, 35–45 (2001)
77. C.G. Bai, Q.P. Hu, M. Xie, S.H. Ng, Software failure prediction based on a Markov Bayesian network model. J. Syst. Softw. **74**(3), 275–282 (2005)
78. Y. Liu, B. Cai, R. Ji, Z. Liu, Y. Zhang, *Reliability Modeling and Evaluation of Subsea Blowout Preventer Systems* (Science Press, 2015)
79. C. Bai, Bayesian network based software reliability prediction with an operational profile. J. Syst. Softw. **77**(2), 103–112 (2005)
80. J.A. McCall, W. Randell, J. Dunham, L. Lauterbach, Software reliability, measurement, and testing software reliability and test integration. *Tech. Rep. Final Technical Report RL-TR-92-52* (Rome Laboratory, Rome) (1992)
81. Q. Zhou, Y.D. Wong, H.S. Loh, K.F. Yuen, A fuzzy and Bayesian network CREAM model for human reliability analysis—The case of tanker shipping. Saf. Sci. **105**, 149–157 (2018)
82. P. Li, G. Chen, L. Dai, L. Zhang, A fuzzy Bayesian network approach to improve the quantification of organizational influences in HRA frameworks. Saf. Sci. **50**(7), 1569–1583 (2012)
83. M. Aalipour, Y.Z. Ayele, A. Barabadi, Human reliability assessment (HRA) in maintenance of production process: A case study. Int. J. Syst. Assur. Eng. Manag. **7**(2), 229–238 (2016)
84. K. Zwirglmaier, D. Straub, K.M. Groth, Capturing cognitive causal paths in human reliability analysis with Bayesian network models. Reliab. Eng. Syst. Saf. **158**, 117–129 (2017)
85. K.M. Groth, L.P. Swiler, Bridging the gap between HRA research and HRA practice: A Bayesian network version of SPAR-H. Reliab. Eng. Syst. Saf. **115**, 33–42 (2013)
86. R. Sundaramurthi, C. Smidts, Human reliability modeling for the next generation system code. Ann. Nucl. Energy **52**, 137–156 (2013)
87. B. Cai, Y. Liu, Y. Zhang, Q. Fan, Z. Liu, X. Tian, A dynamic Bayesian networks modeling of human factors on offshore blowouts. J. Loss Prev. Process Ind. **26**(4), 639–649 (2013)
88. M.R. Martins, M.C. Maturana, Application of Bayesian belief networks to the human reliability analysis of an oil tanker operation focusing on collision accidents. Reliab. Eng. Syst. Saf. **110**, 89–109 (2013)
89. P. Baraldi, L. Podofillini, L. Mkrtchyan, E. Zio, V.N. Dang, Comparing the treatment of uncertainty in Bayesian networks and fuzzy expert systems used for a human reliability analysis application. Reliab. Eng. Syst. Saf. **138**, 176–193 (2015)

90. M. Musharraf, D. Bradbury-Squires, F. Khan, B. Veitch, S. MacKinnon, S. Imtiaz, A virtual experimental technique for data collection for a Bayesian network approach to human reliability analysis. Reliab. Eng. Syst. Saf. **132**, 1–8 (2014)
91. P. Li, L. Zhang, L. Dai, X. Li, Study on operator's SA reliability in digital NPPs. Part 3: A quantitative assessment method. Ann. Nucl. Energy **109**, 82–91 (2017)
92. M.C. Kim, P.H. Seong, An analytic model for situation assessment of nuclear power plant operators based on Bayesian inference. Reliab. Eng. Syst. Saf. **91**(3), 270–282 (2006)
93. H. Lee, P. Seong, A computational model for evaluating the effects of attention, memory, and mental models on situation assessment of nuclear power plant operators. Reliab. Eng. Syst. Saf. **94**(11), 1796–1805 (2009)
94. Z.L. Yang, S. Bonsall, A. Wall, J. Wang, M. Usman, A modified CREAM to human reliability quantification in marine engineering. Ocean Eng. **58**, 293–303 (2013)
95. M. Musharraf, J. Hassan, F. Khan, B. Veitch, S. MacKinnon, S. Imtiaz, Human reliability assessment during offshore emergency conditions. Saf. Sci. **59**, 19–27 (2013)
96. B. Kirwan, Validation of human reliability assessment techniques: Part 1—Validation issues. Saf. Sci. **27**(6), 359–73 (1996). (1996-12-01)
97. B. Kirwan, Validation of human reliability assessment techniques: Part 2—Validation results. Saf. Sci. **1**(27), 43–75 (1997)
98. A. Gregoriades, A. Sutcliffe, Workload prediction for improved design and reliability of complex systems. Reliab. Eng. Syst. Saf. **93**(4), 530–549 (2008)
99. D.J. Bradbury-Squires, *Simulation Training in a Virtual Environment of an Offshore Oil Installation* (2013)
100. A.X. Miao, G.L. Zacharias, S. Kao, A Computational situation assessment model for nuclear power plant operations. IEEE Trans. Syst. Man. Cybern.-Part A: Systems and humans **6**(27), 728–742 (1997)
101. Z. Liu, Y. Liu, B. Cai, D. Zhang, C. Zheng, Dynamic Bayesian network modeling of reliability of subsea blowout preventer stack in presence of common cause failures. J. Loss Prev. Process Ind. **38**, 58–66 (2015)
102. M. Li, J. Liu, J. Li, B.U. Kim, Bayesian modeling of multi-state hierarchical systems with multi-level information aggregation. Reliab. Eng. Syst. Saf. **124**, 158–164 (2014)

A Framework for the Reliability Evaluation of Grid-Connected Photovoltaic Systems in the Presence of Intermittent Faults

Abstract A framework for the reliability evaluation of grid-connected photovoltaic (PV) systems with intermittent faults is proposed using dynamic Bayesian networks (DBNs). A three-state Markov model is constructed to represent the state transition relationship of no faults, intermittent faults, and permanent faults for PV components. The model is subsequently fused into the DBNs. The reliability and availability of three simple PV systems with centralized, string, and multistring configurations, as well as a complex PV system, are analyzed through the proposed framework. The sequence of the degree of importance of PV components is investigated using mutual information. The effects of intermittent fault parameters, including the coefficients of intermittent fault, permanent fault, and intermittent repair, on the reliability and availability are explored. Results show that the reliability and availability of the PV system with centralized configuration rapidly decrease, compared with those of the PV systems with string and multistring configurations. The sequence of the degree of importance of PV components is DC/AC inverter, DC/DC converter, DC combiner, and PV module arranged from the largest to the smallest. The finding indicates that the DC/AC inverter should be given considerable attention to improve the reliability and availability and to prevent their possible failures.

Keywords Dynamic bayesian networks · Reliability evaluation · Intermittent faults · Photovoltaic systems

1 Introduction

Given the concerns about increasing environmental problems, the development and application of grid-connected photovoltaic (PV) power systems to reduce fossil fuel consumption and greenhouse gas emissions have aroused a great deal of interest [1–4]. However, PV systems usually work in extreme conditions, (e.g., desert), and the modules and balance-of-system components of these systems deteriorate because of environmental and operational stresses [5–7]. Therefore, the reliability and availability of these systems need to be quantitatively predicted.

B. Cai et al., *Bayesian Networks for Reliability Engineering*,
https://doi.org/10.1007/978-981-13-6516-4_2

A few studies have evaluated the reliability and availability of PV systems and their components using general reliability analysis methods, such as fault tree, Monte Carlo simulation, Petri nets, and Markov models. Each of these methods has its advantages and disadvantages in terms of reliability evaluation. Gautam and Kaushika investigate the operational lifetime of large solar PV arrays for reliability evaluation with probability theory [8]. Urbina et al. analyze the reliability of a rechargeable battery in a PV power supply system; they construct the model by integrating artificial neural network to simulate the damage that occurs in deep discharge cycles [9]. For the simple configurations of PV systems, reliability function is derived for quantitative analysis to obtain failure rate, probability density function, and average useful life [10, 11]. Chan et al. propose a method for optimizing the reliability of inverters in grid-connected PV systems using the design-of-experiments technique [12]. Monte Carlo simulation was also used to evaluate the reliability of a small isolated power system with solar photovoltaic and the customers' nodal reliability and reserve deployment with high PV power penetration [13–15]. Harb and Balog introduce a stress-factor reliability method to calculate the mean time between failures of a PV module-integrated inverter using a usage model approach [16]. Zhang et al. present a systematic technique for assessing the reliability of grid-connected PV power systems with a state enumeration method by considering the variations of failure rate and the input power of components [17]. Katsigiannis et al. propose a fluid stochastic Petri net-based method for the reliability assessment of small isolated power systems, including wind turbines, PV, and diesel generators [18]. Zini et al. present a method based on fault tree to evaluate the reliability evaluation of large-scale grid-connected PV systems; these researchers look into the effects of PV components on system reliability to improve the performance of diagnosis and maintenance [19, 20]. Markov method was also used to evaluate the reliability and performance of standalone PV systems, grid-connected PV systems, and multiphase DC/DC converters deployed in PV applications [21–23].

An intermittent fault is a recurrent event that appears and disappears with the changes in operation conditions. Some functions or performance characteristics fail for a while, but they subsequently recover [24]. Intermittent fault strongly affects the reliability of electronic products, such as DC combiner, DC/DC converter, and DC/AC inverter, deployed in PV applications. In the past three decades, few studies have reported on reliability modeling and evaluation in the presence of intermittent faults. Prasad develops a Markov model for the reliability assessment of digital system subject to intermittent and permanent faults [25, 26]. Considering the influences of permanent and intermittent faults, Cheng et al. present a reliability evaluation method using an improved neural network training algorithm and architecture [27]. In view of the influences of permanent and intermittent faults, Habib et al. introduce a neural network-based Markov and fault-tolerant model for the reliability assessment of a consecutive r-out-of-n: F system a consecutive r-out-of-n: F system [28]. The influences of the high occurrence rates of transient and intermittent faults on a microprocessor have also been examined using the generalized stochastic Petri net modeling [29].

Bayesian network (BN) and dynamic BNs (DBNs) are probabilistic graphical models that represent a set of random variables, including their conditional dependencies through directed acyclic graphs [30]. Since these models were first proposed by Pearl [31], they have been considered powerful tools for handling uncertainty information and have therefore received increasing attention in the field of reliability evaluation. Our previous work systematically investigates the reliability modeling and evaluation methodology using BNs and DBNs taking into account common cause failure, imperfect coverage, imperfect repair, and preventive maintenance [32–34]. However, to the best of our knowledge, the application of either BNs or DBNs to the evaluation of the reliability of grid-connected PV systems, especially in the presence of intermittent faults, has not been reported. Correspondingly, many critical issues on this subject should be investigated.

This study focuses on the reliability and availability evaluation of grid-connected PV systems in the presence of intermittent faults using the DBNs method. The rest of the paper organized into five sections. Following the introduction, Sect. 2 describes the configuration of grid-connected PV systems. Section 3 develops the DBN-based framework for the reliability evaluation of PV systems. Section 4 presents the reliability evaluation results and discussions. Finally, Sect. 5 summarizes the whole study.

2 Description of Grid-Connected PV Systems

A grid-connected PV system consists of PV modules and balance-of-system components. The PV modules can be arranged in different configurations that directly affect the structure and topology of the balance-of-system electronic components [35–37]. Different configurations of PV modules have been proposed during in the past, such as centralized, string, multistring, and modular configurations [38–40]. The balance-of-system components of PV systems include string protection, DC combiner, DC/DC converter, DC/AC inverter, DC disconnect, AC disconnect, grid protection, and others [6, 19, 41–44].

In this study, three PV system configurations, i.e., centralized, string, and multistring configurations, are analyzed to compare their respective system reliabilities in the context of intermittent faults of electronic components. For simplicity, only a few electronic devices are considered, including PV module, DC combiner, DC/DC converter and DC/AC inverter [45]. Other electronic devices, such as controller, DC disconnect, AC disconnect, grid protection, are excluded from the study, as shown in Fig. 1.

For example, in consideration of the PV system with a centralized configuration illustrated in Fig. 1a, the PV array composed of two strings of two modules each connects a series–parallel configuration. Subsequently, the DC voltage level is combined together in a DC combiner, converted from DC to DC in a DC/DC converter and from DC to AC in a DC/AC inverter, and is finally fed into the electricity grid system. A centralized configuration is mainly used in PV plants, which have a nominal power

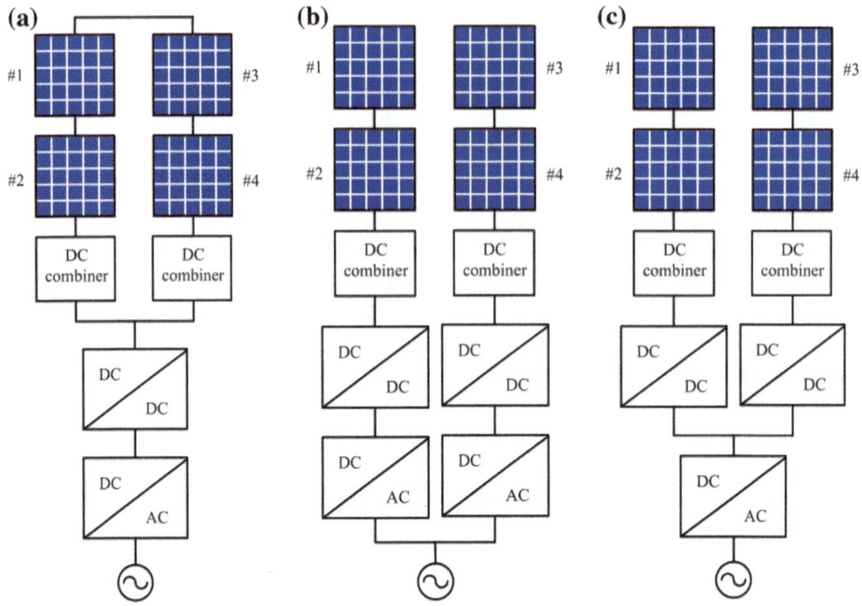

Fig. 1 Three simple PV system configurations: **a** centralized, **b** string, and **c** multistring configurations

higher than 10 kW, a high power conversion efficiency, and low cost. However, the maximum power point tracking (MPPT) efficiency of this central structure sharply decreases in a partial shading condition because it can hardly individually draw the maximum power from each module, thereby decreasing total efficiency [38].

PV module is the packaged, connected assembly of PV cells and is considered the most reliable component in PV systems [6]. DC combiner is used to combine multiple source circuits into a single source, which consists of various electronic devices, e.g., PV fuse, string sensor, DC surge protector, signal surge protector, etc. DC/DC converter and DC/AC inverter are among the vulnerable components in PV systems because they contain semiconductor modules. These components connect the switching components and capacitors. All these components, i.e., PV module, DC combiner, DC/DC converter, and DC/AC inverters, suffer permanent and intermittent faults, which seriously affect the reliability of PV system.

String configuration can connect different PV modules of the same type in every string. If the string voltage does not have the appropriate value, then a boost DC/DC converter or a step-up transformer (usually placed on the AC side) is required [35]. Figure 1b provides a simple PV system with string configuration. The distinguishing feature of this system is that each string has its own DC/DC converter to convert the voltage level and DC/AC inverter to convert DC electricity into an AC output. If a centralized system has the same total capacity as an n-string PV system, then the capacity of each string converter and inverter is only one-nth of that of the

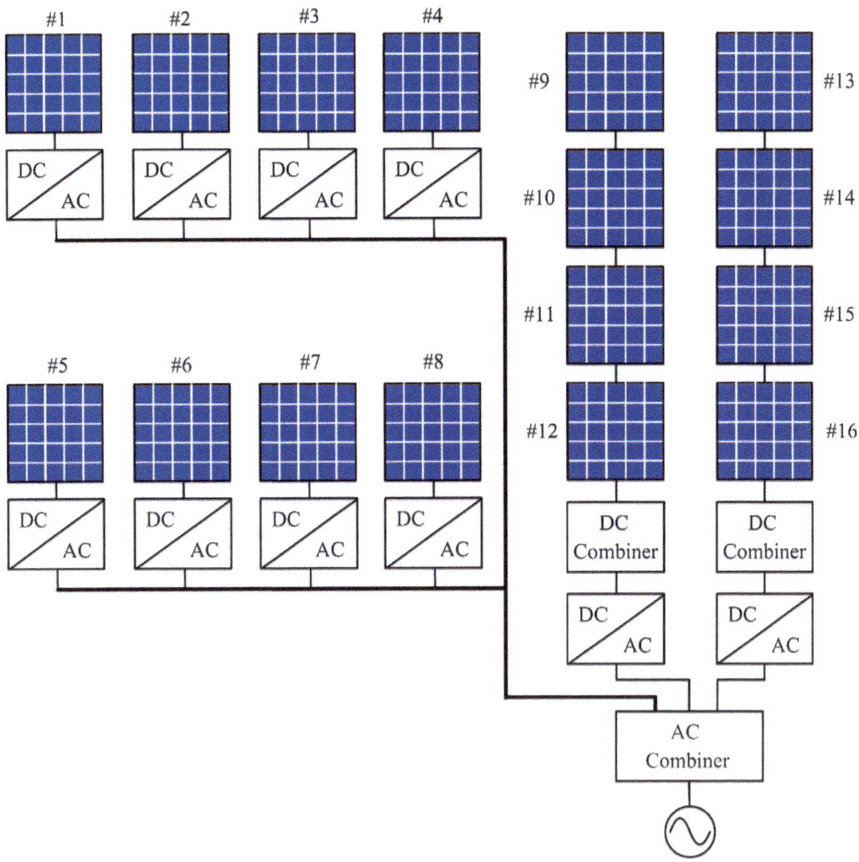

Fig. 2 A complex PV system, including a micro-inverter PV system and two strings

centralized converter and inverter. This event leads to failure rate that significantly varies for different converter and inverter. Figure 1c illustrates a PV system with multistring configuration, wherein the string has its own DC/DC converter to convert the voltage level, but only has one DC/AC converter to convert the DC electricity into an AC output. The string and multistring structures have been used in low power ranges because of their enhanced MPPT efficiency. However, in these configurations, the electrical characteristic difference resulting from PV module's tolerance, partial shading, and reflection problems still hinders the maximum power generation of each module [38].

In this study, a complex system is used to demonstrate the proposed methodology. As shown in Fig. 2, the system comprises a micro-inverter PV system and two strings. The micro-inverter and PV module are integrated as one electrical device, which is directly connected to a distribution grid through an AC bus. The micro-inverter system is adopted to achieve high modularity, easy installation, and enhanced safety. The two PV strings are connected to an AC combiner.

3 DBN Modeling for Reliability Evaluation

3.1 DBN Structure Modeling

A BN is generally constructed through two major procedures, namely the construction of structure models and the definition of parameter models [46]. In the first step, a set of relevant variables and their possible values should be decided. A network structure can then be set up by connecting these variables into a directed acyclic graph. In the second step, the conditional probability table for each network variable should be defined.

The DBN structure models for the PV systems with centralized, string, and multistring configurations in the presence of intermittent faults are constructed (Fig. 3) according to the PV system configurations given in Fig. 1. Figure 3a demonstrates that the failure of any PV component in a PV system with centralized configuration will cause the failure of the entire PV system. This case signifies that the PV components, including four PV modules #1, #2, #3, and #4 (i.e., PV1, PV2, PV3, and PV4), two DC combiners (Comb1 and Comb2), a DC/DC converter (Conv), and a DC/AC inverter (Inve), are considered a series. Therefore, the network structure is built with two layers using the Netica software tool. The first layer consists of eight nodes representing the status of eight PV components. Each node has three states, i.e., the fault not existing state (NF), intermittent faulty state (IF), and permanent faulty state (PF). The second layer includes one node that depicts the status of PV system. This node has two states, i.e., work and fail, which indicate whether the whole PV system is working or not.

DBNs are an extension of the general BNs that allow the explicit modeling of changes over time. In this process, each time step is called a time slice. Figure 3a indicates that the DBNs of the PV system with centralized configuration consist of two time slices, that is, from $t = 0$ to $t = \Delta t$. The nodes PV1, PV2, PV3, PV4, Comb1, Comb2, Conv, and Inve at $t = 0$ are extended to PV5, PV6, PV 7, PV8, Comb3, Comb4, Conv1, and Inve1 at $t = \Delta t$, respectively. The number of time slice and the value of Δt are determined by the purpose of research and the time the Netica runs. A great number of time slices correspond to a smaller value of Δt, and, hence, a longer time at which Netica runs. The DBN structure models for the PV systems with string and multistring configurations are similar to that for the PV system with centralized configuration and are produced based on the series and parallel relationship of the PV components, as shown in Fig. 3b, c. The DBN structure model of the complex PV system is given in Fig. 4. The series and parallel relationship among the PV components establishes the conditional probability tables of nodes, which are described in the subsequent section.

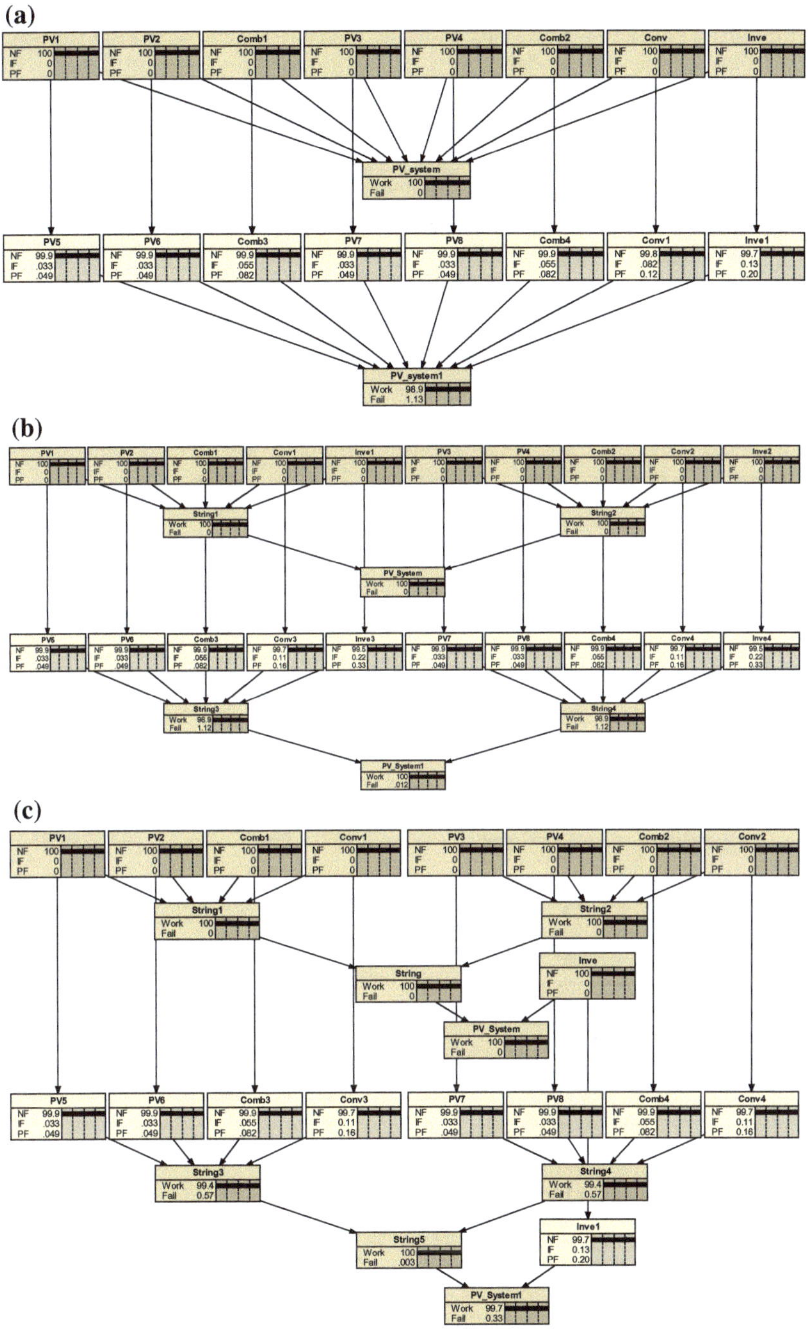

Fig. 3 DBNs of PV systems with **a** centralized, **b** string, and **c** multistring configurations the presence of intermittent faults

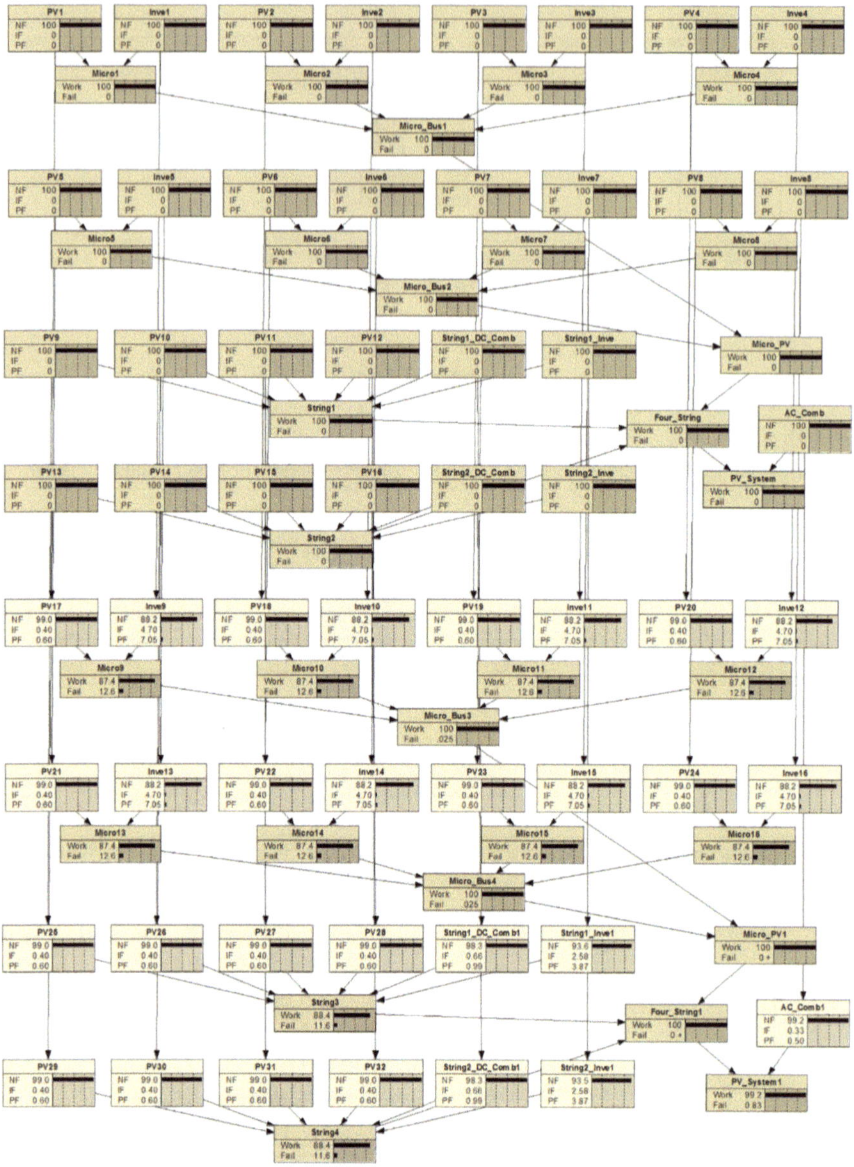

Fig. 4 DBNs of the complex PV system in the presence of intermittent faults

3.2 *Intermittent Fault Modeling*

Intermittent faults can hardly be modeled using a directed DBN structural modeling directed. Therefore, this study proposes a method that fuses the Markov model into a DBN model. The developed method has four basic assumptions specified as follows [47–50]:

(1) The PV systems begin with a perfect operation, in which all PV components are functioning correctly.
(2) The transition rates of the PV components, including the failure and repair rates are different, but constant. The lifetimes of these components are exponentially distributed because they are mainly electronic products.
(3) The states of all components are statistically independent.
(4) The PV systems are considered "as good as new" after repairs.

The idea of intermittent and permanent faults can be incorporated in terms of the three-state Markov models as shown in Fig. 5 [25, 26]. The model stipulates that the NF state can be converted into a PF and IF states with a failure rate λ_1 and λ_2, respectively. An intermittent fault can lead the components into PF and NF states. Therefore, the IF state can become a PF state with a failure rate of λ_3 and to an NF state with a repair rate of μ_1 (autorecovery), as shown in Fig. 5a. If a failed

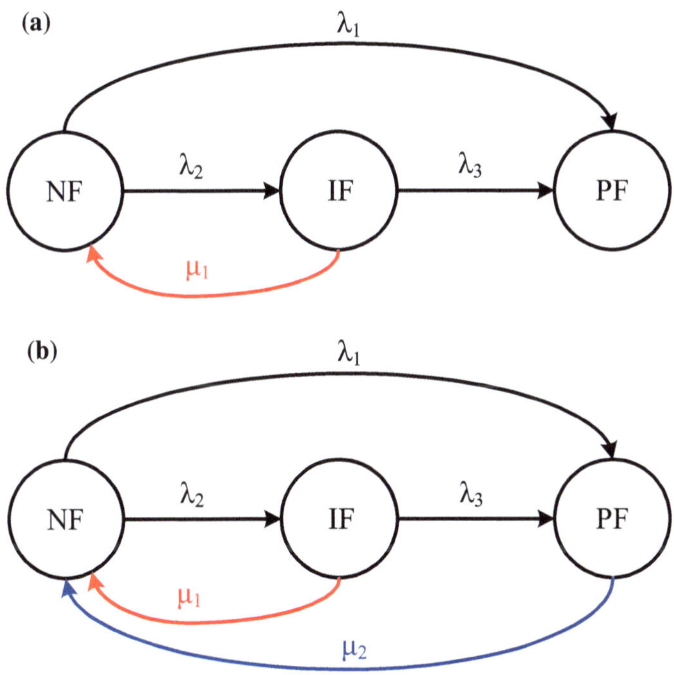

Fig. 5 State transition diagram in the presence of intermittent faults **a** without and **b** with repair

Table 1 Transition relations between the consecutive nodes without repair

t	$t + \Delta t$		
	NF	IF	PF
NF	$e^{-(\lambda_1+\lambda_2)\Delta t}$	$\frac{\left[1-e^{-(\lambda_1+\lambda_2)\Delta t}\right]\lambda_2}{\lambda_2}$	$\frac{\left[1-e^{-(\lambda_1+\lambda_2)\Delta t}\right]\lambda_1}{\lambda_1+\lambda_2}$
IF	$\frac{\left[1-e^{-(\mu_1+\lambda_3)\Delta t}\right]\mu_1}{\mu_1+\lambda_3}$	$e^{-(\mu_1+\lambda_3)\Delta t}$	$\frac{\left[1-e^{-(\mu_1+\lambda_3)\Delta t}\right]\lambda_3}{\mu_1+\lambda_3}$
PF	0	0	1

Table 2 Transition relations between the consecutive nodes with repair

t	$t + \Delta t$		
	NF	IF	PF
NF	$e^{-(\lambda_1+\lambda_2)\Delta t}$	$\frac{\left[1-e^{-(\lambda_1+\lambda_2)}\right]\lambda_2}{\lambda_1+\lambda_2}$	$\frac{\left[1-e^{-(\lambda_1+\lambda_2)\Delta t}\right]\lambda_1}{\lambda_1+\lambda_2}$
IF	$\frac{\left[1-e^{-(\mu_1+\lambda_3)\Delta t}\right]\mu_1}{\mu_1+\lambda_3}$	$e^{-(\mu_1+\lambda_3)\Delta t}$	$\frac{\left[1-e^{-(\mu_1+\lambda_3)\Delta t}\right]\lambda_3}{\mu_1+\lambda_3}$
PF	$1 - e^{-\mu_2\Delta t}$	0	$e^{-\mu_2\Delta t}$

component is repaired once permanent fault occurs, then a repair arc should be added to the state transition diagram. In this case, the PF state can become an NF state with a repair rate of μ_2 (manual repair), as shown in Fig. 5b. When the repair action is not considered, the reliability of the PV system can be calculated. When the repair action is considered, the availability of the PV system can be calculated using the proposed DBN model.

Tables 1 and 2 present the transition relations among the consecutive nodes in the presence of intermittent faults without and with repair, as given in Tables 1 and 2, respectively, given the current time t and the succeeding time $t + \Delta t$ [51].

The total failure rate λ of a single component is constant, and it consists of a permanent failure rate λ_1 from NF to PF and an intermittent failure rate λ_2 from NF to IF. Transition rate λ_3 from IF to PF is part of the intermittent failure rate λ_2 with coefficient y. Repair rate μ_1 from IF to NF is also a component of intermittent failure rate λ_2 with coefficient z. Therefore, failure rates λ_1, λ_2, and λ_3, and repair rate μ_1 are calculated using the following assumptive equations:

$$\lambda = \lambda_1 + \lambda_2 \tag{1}$$

$$\lambda_2 = x \cdot \lambda \tag{2}$$

$$\lambda_3 = y \cdot \lambda_2 \tag{3}$$

$$\mu_1 = z \cdot \lambda_2 \tag{4}$$

Table 3 Failure and repair rates of the PV components

Component	Centralized		String		Multistring	
	Failure rate (10^{-6}/h)	Repair rate	Failure rate (10^{-6}/h)	Repair rate	Failure rate (10^{-6}/h)	Repair rate
PV module	3.2232	0.0667	3.2232	0.0667	3.2232	0.0667
DC combiner	5.3720	0.1667	5.3720	0.1667	5.3720	0.1667
DC/DC converter	8.0580	0.1250	10.7440	0.1250	10.7440	0.1250
DC/AC inverter	12.8928	0.0833	21.4880	0.0833	12.8928	0.0833

Table 4 Failure and repair rates of the PV components for the complex PV system

Component	Failure rate (10^{-6}/h)	Repair rate
PV module	3.2232	0.0667
DC combiner	5.3720	0.1667
DC/AC microconverter	40.2901	0.1000
DC/AC converter	21.4880	0.0833
AC combiner	2.6860	0.0556

where x, y, and z are the intermittent fault, permanent fault, and intermittent repair coefficients, respectively. The sum of y and z should not be larger than 1. The effects of these coefficients on the reliability and availability of the PV system will be identified in this study and are investigated in the following sections.

Total failure rate λ and repair rate μ_2 are obtained from the experiences and judgments of experts, and are demonstrated in Tables 3 and 4. The same component for varied configurations may have different failure rates. If several more accurate failure and repair rate data are available to the readers, they can calculate the reliability and availability of their PV systems using our proposed methodology.

3.3 DBN Parameter Modeling

DBN parameter modeling includes the definition of the prior probabilities of root nodes and the conditional probabilities between the root and leaf nodes. This modeling assumes that the PV systems begin with a perfect operation, in which all components are properly working. Therefore, all root nodes in this modeling technique have a 100% probability of fault not existing. The conditional probabilities are defined based on the series and parallel relationship between the root and leaf nodes.

Table 5 Conditional probability table of node String1 for the PV system with string configuration

No.	PV1	PV2	Comb1	Conv1	Inve1	String1
1	NF	NF	NF	NF	NF	Work
2	NF	NF	NF	NF	IF	Fail
3	NF	NF	NF	NF	PF	Fail
4	NF	NF	NF	IF	NF	Fail
5	NF	NF	NF	IF	IF	Fail
6	NF	NF	NF	IF	PF	Fail
...	...	...	...	...	...	...
238	PF	PF	PF	IF	NF	Fail
239	PF	PF	PF	IF	IF	Fail
240	PF	PF	PF	IF	PF	Fail
241	PF	PF	PF	PF	NF	Fail
242	PF	PF	PF	PF	IF	Fail
243	PF	PF	PF	PF	PF	Fail

Table 6 Conditional probability table of node PV_System for the PV system with string configuration

No.	String1	String2	PV_System
1	Work	Work	Work
2	Work	Fail	Work
3	Fail	Work	Work
4	Fail	Fail	Fail

The PV system with a string configuration in Fig. 3b is used as example. PV modules #1 and #2, the DC combiner, DC/DC converter, and DC/AC inverter in the left string are considered a series. Similarly, PV modules #3 and #4, the DC combiner, DC/DC converter, and DC/AC inverter in the right string are regarded as a series. The left and right strings provide backup for each other. Thus, they are considered parallel. The relationship between nodes PV1, PV2, Comb1, Conv1, Inve1, and node String1 is AND. When the states of all parent nodes are NF, the state of String1 is "work"; otherwise, the state of String1 is "fail", as shown in Table 5 (for simplicity, we omit some terms with the fail state). The relationship between nodes String1, String2, and node PV_System is OR. Only when the states of all parent nodes are "fail", the state of PV_System is denoted as "fail"; otherwise, the state of PV_System is "work", as presented in Table 6. The conditional probability table for the PV systems with centralized and multistring configurations can be obtained using the series and parallel relationship among nodes.

3.4 *Reliability and Availability Evaluation*

The model in the proposed methodology should be validated because it will render a reasonable degree of confidence to the results of the model. In the current study, we adopt a three-axiom-based validation method to validate the DBN model. Correspondingly, three axioms should be satisfied [52].

(1) A slight increase/decrease in the prior subjective probability of each parent node should certainly result in the effect of a relative increase/decrease of the posterior probabilities of the child nodes.
(2) Given the variation in subjective probability distributions of each parent node, its influence magnitude to the child node values should be kept consistent.
(3) The total influence magnitudes of the combination of the probability variations from x attributes on the values should always be greater than that from the set of x-y ($y \in x$) attributes.

3.5 *Reliability and Availability Evaluation*

Using the proposed evaluation framework, this study quantitatively evaluates the reliability and availability of the PV systems with centralized, string, and multi-string configurations in the presence of intermittent faults through the Netica application. Mutual information is investigated to evaluate the sequence of the degree of importance of PV components. Moreover, the effect of intermittent fault parameters including intermittent fault, permanent fault, and intermittent repair coefficients on the reliability and availability of three PV systems is examined. The results of this study can provide important insights into the design of PV systems with improved reliability.

4 Results and Discussions

4.1 *Results of Validation*

Validation is the task of confirming whether a model is a reasonable representation of an actual system. In this study, the proposed model should satisfy the axioms described in Sect. 3.4. The DBNs of the PV system with centralized configuration are taken as example. When the probabilities of the NF and IF states for PV1 node are set to 50%, the failure probability of "PV_system1" increases from 1.13 to 50.6%. When the two probabilities for PV3 node are also set to 50%, the failure probability increases to 75.3%; when the two probabilities for Comb1 node are also set to 50%, the failure probability increases to 87.6%; when the two probabilities for Conv node are also set to 50%, the failure probability increases to 93.8%; finally, when two probabilities for the last parent node Inve are set to 50%, the failure probability increases to 96.9%. Decreasing each influencing node satisfies the axioms, which validates the proposed model.

4.2 *Reliability and Availability*

As shown in Fig. 6, the reliability and availability of the PV systems with centralized, string, and multistring configurations in the presence of intermittent faults are calculated and plotted. Intermittent fault coefficient x, permanent fault coefficient y, and intermittent repair coefficient z are set to 40, 20, and 50%, respectively.

The reliabilities of the three PV systems decrease with the increase of time (Fig. 6a, b). In particular, the reliability of the PV system with centralized configuration rapidly decreases, compared with the other systems. The PV system with string configuration has high reliability in the first five years, whereas the PV system with multistring configuration has a high reliability after five years. The main reason behind this case is the fact that the redundant DC/AC inverters first lead to a high system reliability of the

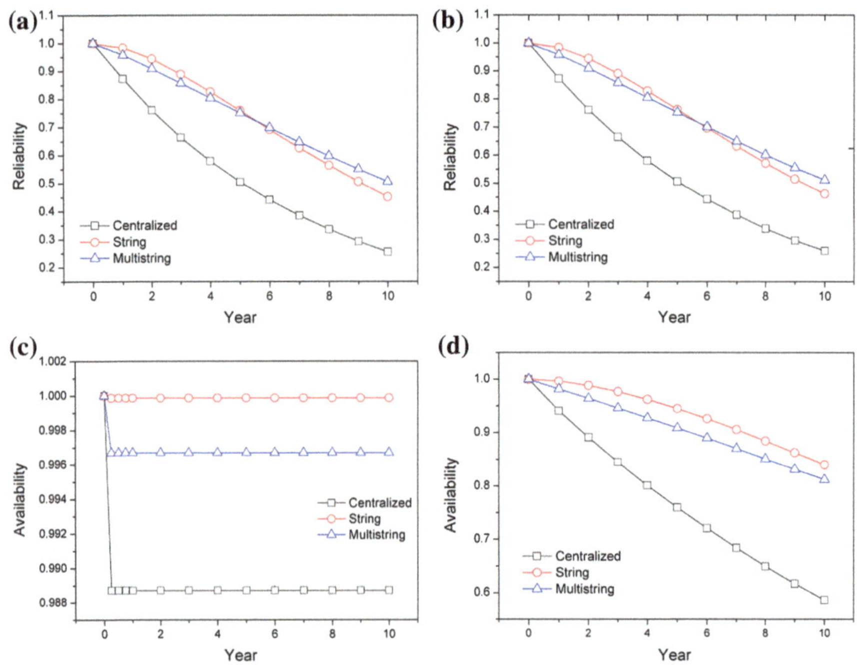

Fig. 6 Reliability **a** without and **b** with intermittent faults, and availability **c** without and **d** with intermittent faults of three PV systems within ten years

PV system with string configuration first. A low failure rate of the DC/AC inverter subsequently leads to a high system reliability of the PV system with multistring configuration Therefore, in terms of reliability, the multistring configuration is the best choice in designing PV systems, whereas the centralized configuration is the worst choice.

The comparison of the reliability in Fig. 6a, b indicates that the intermittent faults just slightly affect the reliability values in the first ten years. The three PV systems with intermittent faults have a slightly higher reliability than those without intermittent faults. The average reliability increments of ten years for centralized, string, and multistring configurations are 0.50, 0.78, and 0.40%, respectively. This is because that the intermittent faults can transform to no faults, which is autorecovery. This finding is attributed to the fact that the intermittent faults can be transformed into "no faults," which is autorecovery.

As shown in Fig. 6c, d, the availabilities of the three PV systems without intermittent faults rapidly decrease at first, and then stabilize at certain levels, whereas those of the three with intermittent faults continuously decrease. The reason for this event is that for PV systems without intermittent faults, all permanent faults are repaired manually once they occur. However, for PV systems with intermittent faults, only the permanent faults are repaired once they occur, but the intermittent faults cannot be repaired manually. Instead, these faults either maintain their current state or can be transformed into a no fault state or permanent fault state. Among all systems, the availability of the PV system with centralized configuration decreases the fastest, followed by that of the PV system with string configuration. The availability of the PV system with multistring configuration decreases with the phase between that of the centralized and string configurations. Therefore, in terms of availability, the string configuration is the best choice in designing PV systems, whereas the centralized configuration is the worst choice.

In summary, PV systems with string and multistring configurations have high reliability and availability and can be used in high-performance applications.

The reliability and availability of the complex PV system with intermittent faults are calculated and plotted in Fig. 7. The reliability and availability of the system continuously decrease with the increase of time, but the latter decreases slower than the former because of the repair actions conducted whenever necessary.

4.3 Mutual Information Investigation

Mutual information measures the information shared by two variables and determines the degree of uncertainty of reduction of one variable by knowing one of the other variables [34]. This information can be used to identify the degree of importance of each PV component to the entire PV system. In this study, the degree of importance in three moments, i.e., first, fifth, and tenth years, is investigated, as shown in Fig. 8. The degree of importance of the components of the three PV systems is the same, which is in the order of DC/AC inverter, DC/DC converter, DC combiner, and PV

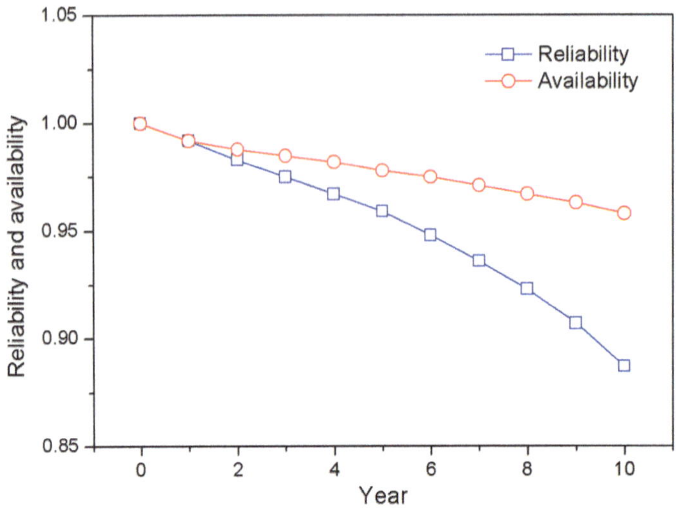

Fig. 7 Reliability and availability of the complex PV system with intermittent faults within ten years

module arranged from largest to smallest. This degree increases with the increase of time. The DC/AC inverter is determined to affect the reliability of the PV system significantly with multistring configuration, whereas the other components only exert a few contributions. Therefore, the DC/AC inverter should be given considerable attention to improve the reliability and availability of PV systems and to prevent their possible failures. The failure rates of the components of a PV system with a specified configuration should be reduced, but their repair rates should be increased to improve the reliability and availability of such system. Hence, the DC/AC inverter with low failure rates should be used in design and manufacturing stages of PV systems. Moreover, the repair rate of this component should be increased in the usage stage.

4.4 Effect of Model Parameters

The effects of three important model parameters, i.e., intermittent fault coefficient x, permanent fault coefficient y, and intermittent repair coefficient z, on the reliability and availability of PV systems in the fifth year are examined. Figure 9a, b, c indicate that the intermittent fault coefficient largely affect the reliability of the systems, followed by the intermittent repair coefficient. By contrast, the permanent fault coefficient almost has no effects on the reliability of the systems. However, the effects of these coefficients on reliability are extremely small. Reliability improves with the increase of the intermittent fault coefficient because the total failure rate of

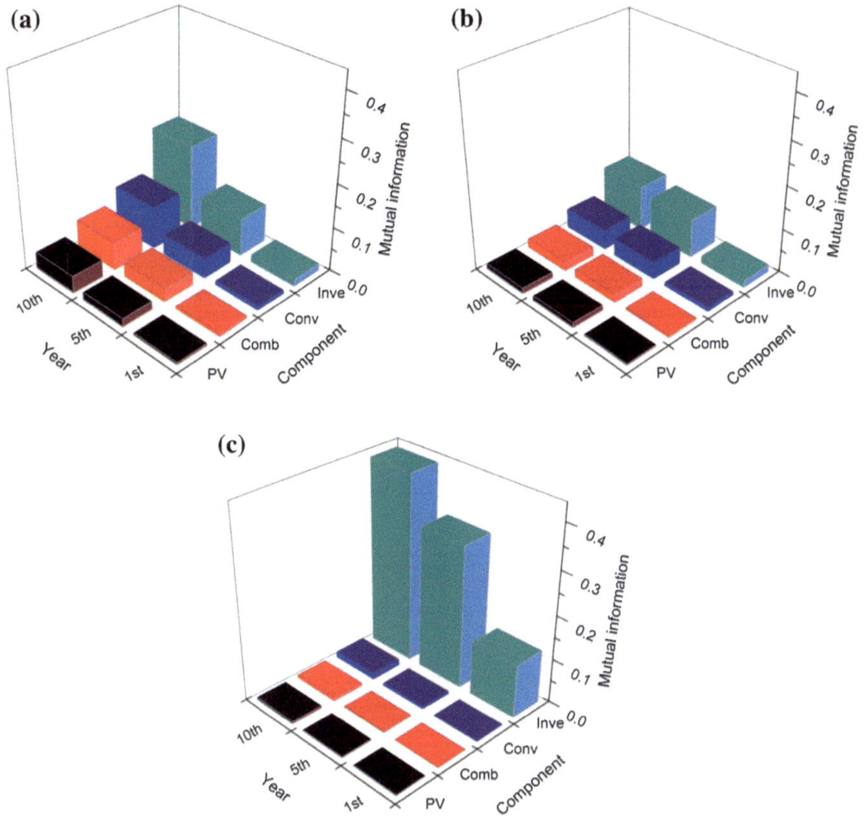

Fig. 8 Mutual information of PV components and PV system **a** centralized, **b** string, and **c** multi-string configurations in the presence of intermittent faults in the first, fifth, and tenth years

a single component is constant. In this case, the intermittent fault coefficient rises, which increases the intermittent failure rate, but reduces the permanent failure rate. Nonetheless, the intermittent fault can transform from the IF state into the NF state because of autorecovery. Thus, the reliability of the system is improved. However, the reliability values are not significantly important in reality, because repair actions are performed once faults occur. Corresponding, availability value is more important than the reliability value. Figure 9d illustrates that the intermittent fault coefficient significantly affects the availability of PV system, particularly those with centralized configuration. Availability decreases from 92.5 to 66.8% when the intermittent fault coefficient increases from 10 to 60%. By contrast, the permanent fault and intermittent repair coefficient only incur minimal effects on the availability of PV system, even for those with a centralized configuration, the availabilities of the system only increase from 75.87 to 76.09% and 75.82 to 76.05%, respectively. In summary, the effects of intermittent fault, permanent fault, and intermittent repair coefficients on the reliability and availability of PV systems are arranged from largest to smallest.

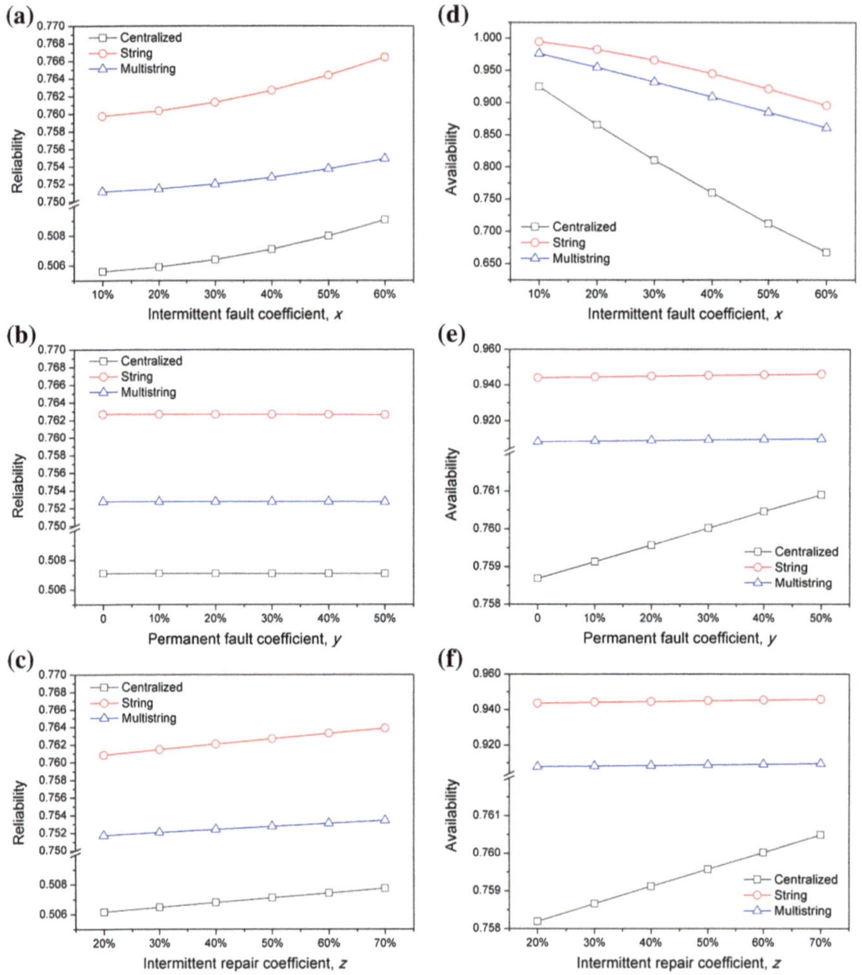

Fig. 9 Effect of model parameters **a** x, **b** y, **c** z on the reliability, **d** x, **e** y, and **f** z on the availability of PV systems in fifth year

4.5 Discussions

Previous studies have analyzed the reliability evaluation of grid-connected PV systems and components using various methods. However, this line of research failed to consider a significant feature of electronic products, that is, intermittent faults. Therefore, the current study proposes a novel DBN-based reliability evaluation methodology with intermittent faults to predict the reliability and availability of PV systems based on the collected failure data of components. The research results indicate that intermittent faults only slightly affect the reliability value of three simple PV systems, and systems with intermittent faults have a slightly higher reliability than those without intermittent faults because of autorecovery. Nevertheless, the reliability values are not remarkably important because repair actions are performed once faults occur in reality. Thus, the availability values are more important than the reliability values. The availabilities of the three PV systems continuously decrease, but they do not stabilize at certain levels. This is because intermittent faults cannot be repaired manually. Therefore, these faults should be diagnosed and repaired timely. The degree of importance of PV components is another important issue. Unlike other reliability evaluation methods, DBN is a powerful tool for examining mutual information and for identifying the degree of importance. The component with high degree of important should be given considerable attention to improve the reliability and availability of PV systems and to prevent their possible failures.

Future research can look into the optimization of reliability and cost issues. Reliability evaluation intends to identify the reliability and availability of an entire system. The life-cycle cost analysis aims to select a cost-effective approach to achieve the least cost of ownership. From the customer's perspective, reliability and cost are proportional. That is if we need high reliability, the configuration of PV components should be redundant, and therefore the cost will be high. There is a balance between reliability and cost, which is an optimization problem. Correspondingly, future works can combine the DBN-based reliability evaluation model with the life-cycle cost analysis model to optimize PV systems from the perspective of reliability and cost.

5 Conclusion

To integrate the significant feature of electronic products, i.e., intermittent faults, in the reliability research of grid-connected PV systems, this study presented a DBN-based reliability evaluation methodology to predict the reliability and availability of PV systems based on the collected failure data of components. The DBN structure and parameter models were constructed, and the intermittent faults were modeled by fusing the Markov model into the DBNs model. The reliability and availability of PV systems with centralized, string, and multistring configurations were analyzed using the proposed method.

The results show that the reliability and availability of the PV system with centralized configuration rapidly decrease compared with those with string and multistring configurations. Therefore, the string and multistring configurations are a good choice in designing PV system in terms of system reliability and availability. The degree of importance of PV components to the PV systems is arranged from largest to smallest, that is, DC/AC inverter, DC/DC converter, DC combiner, and PV module. This case indicates that the DC/AC inverter should be given considerable attention to improve the reliability and availability of PV systems and to prevent their possible failures. The failure rates of the components of a PV system with a specified configuration should be reduced, and their repair rates should be increased to improve the reliability and availability of such system. Hence, the DC/AC inverter with low failure rate should be selected in the design and manufacturing stages of PV systems. Moreover, the repair rate of DC/AC inverter should be increased in the usage stage. The degree of effect of model parameters on the reliability and availability of PV systems is arranged from largest to smallest, that is, intermittent fault, permanent fault, and intermittent repair coefficients.

The proposed DBN-based reliability evaluation methodology has an advantage in handling the intermittent faults of electronic products. However, a large amount of failure data of the PV components should be collected to obtain failure rate, repair rate, intermittent fault coefficient, permanent fault coefficient, and intermittent repair coefficient. These data greatly affect the accuracy of reliability prediction. If several accurate failure and repair rate data are available to the readers, then the system reliability and availability can be easily updated using the proposed methodology.

References

1. G.G. Pillai, G.A. Putrus, T. Georgitsioti, N.M. Pearsall, Near-term economic benefits from grid-connected residential PV (photovoltaic) systems. Energy **68**, 832–843 (2014)
2. K.Y. Lau, C.W. Tan, A.H.M. Yatim, Photovoltaic systems for Malaysian islands, Effects of interest rates, diesel prices and load sizes. Energy **83**, 204–216 (2015)
3. R.A. Mastromauro, M. Liserre, A. Dell'Aquila, Control issues in single-stage photovoltaic systems: MPPT, current and voltage control. IEEE Trans. Ind. Inf. **8**(2), 241–254 (2012)
4. E. Koutroulis, F. Blaabjerg, Design optimization of transformerless grid-connected PV inverters including reliability. IEEE Trans. Power Electron. **28**(1), 325–335 (2013)
5. G. Petrone, G. Spagnuolo, R. Teodorescu, M. Veerachary, M. Vitelli, Reliability issues in photovoltaic power processing systems. IEEE Trans. Ind. Electron. **55**(7), 2569–2580 (2008)
6. P. Zhang, W. Li, S. Li, Y. Wang, W. Xiao, Reliability assessment of photovoltaic power systems: Review of current status and future perspectives. Appl. Energy **104**, 822–833 (2013)
7. V. Sharma, S.S. Chandel, Performance and degradation analysis for long term reliability of solar photovoltaic systems: A review. Renew. Sustain. Energy Rev. **27**, 753–767 (2013)
8. N.K. Gautam, N.D. Kaushika, Reliability evaluation of solar photovoltaic arrays. Sol. Energy **72**(2), 129–141 (2002)
9. A. Urbina, T.L. Paez, C. O'Gorman, P. Barney, R.G. Jungst, D. Ingersoll, Reliability of rechargeable batteries in a photovoltaic power supply system. J. Power Sources **80**(1–2), 30–38 (1999)
10. W.M. Rohouma, I.M. Molokhia, A.H. Esuri, Comparative study of different PV modules configuration reliability. Desalination **209**(1–3), 122–128 (2007)
11. P.S. Shenoy, K.A. Kim, B.B. Johnson, P.T. Krein, Differential power processing for increased energy production and reliability of photovoltaic systems. IEEE Trans. Power Electron. **28**(6), 2968–2979 (2013)
12. F. Chan, H. Calleja, Design strategy to optimize the reliability of grid-connected PV systems. IEEE Trans. Ind. Electron. **56**(11), 4465–4472 (2009)
13. R.M. Moharil, P.S. Kulkarni, Reliability analysis of solar photovoltaic system using hourly mean solar radiation data. Sol. Energy **84**(4), 691–702 (2010)
14. Q. Zhao, P. Wang, L. Goel, Y. Ding, Evaluation of nodal reliability risk in a deregulated power system with photovoltaic power penetration. IET Gener. Transm. Distrib. **8**(3), 421–430 (2014)
15. Z. Qin, W. Li, X. Xiong, Incorporating multiple correlations among wind speeds, photovoltaic powers and bus loads in composite system reliability evaluation. Appl. Energy **110**, 285–294 (2013)

16. S. Harb, R.S. Balog, Reliability of candidate photovoltaic module-integrated-inverter (PV-MII) topologies—A usage model approach. IEEE Trans. Power Electron. **28**(6), 3091–3097 (2013)
17. P. Zhang, Y. Wang, W. Xiao, W. Li, Reliability evaluation of grid-connected photovoltaic power systems. IEEE Trans. Sustain. Energy **3**(3), 379–389 (2012)
18. Y.A. Katsigiannis, P.S. Georgilakis, G.J. Tsinarakis, A novel colored fluid stochastic Petri net simulation model for reliability evaluation of wind PV diesel small isolated power systems. IEEE Trans. Syst., Man, Cybern. A, Syst., Humans **40**(6), 1296–1309 (2010)
19. G. Zini, C. Mangeant, J. Merten, Reliability of large-scale grid-connected photovoltaic systems. Renew. Energy **36**(9), 2334–2340 (2011)
20. A. Ahadi, N. Ghadimi, D. Mirabbasi, Reliability assessment for components of large scale photovoltaic systems. J. Power Sources **264**, 211–219 (2014)
21. S.V. Dhople, A.D. Domínguez-García, Estimation of photovoltaic system reliability and performance metrics. IEEE Trans. Power Syst. **27**(1), 554–563 (2012)
22. M. Theristis, I.A. Papazoglou, Markovian reliability analysis of standalone photovoltaic systems incorporating repairs. IEEE J. Photovoltaics **4**(1), 414–422 (2014)
23. S.V. Dhople, A. Davoudi, A.D. Dominguez-Garcia, P.L. Chapman, A unified approach to reliability assessment of multiphase DC–DC converters in photovoltaic energy conversion systems. IEEE Trans. Power Electron. **27**(2), 739–751 (2012)
24. H. Qi, S. Ganesan, M. Pecht, No-fault-found and intermittent failures in electronic products. Microelectron. Reliab. **48**(5), 663–674 (2008)
25. V.B. Prasad, Computer networks reliability evaluations and intermittent faults, in *Proceedings of the 33rd Midwest Symposium on Circuits and Systems*, vol. 1, pp. 327–330, Calgary, Alta, Aug. 12–14, 1990
26. V.B. Prasad, Markovian model for the evaluation of reliability of computer networks with intermittent faults, in IEEE International Symposium on Circuits and Systems, vol. 4, pp. 2084–2087, Jun. 11–14, 1991
27. C.S. Cheng, Y.T. Hsu, C.C. Wu, An improved neural network realization for reliability analysis. Microelectron. Reliab. **38**(3), 345–352 (1998)
28. A. Habib, R. Alsieidi, G. Youssef, Reliability analysis of a consecutive r-out-of-n: F system based on neural networks. Chaos, Solitons Fractals **39**(2), 610–624 (2009)
29. C. Constantinescu, Dependability evaluation of a fault-tolerant processor by GSPN modeling. IEEE Trans. Reliab. **53**(3), 468–474 (2005)
30. P. Morris, D. Vine, L. Buys, Application of a Bayesian Network complex system model to a successful community electricity demand reduction program. Energy **84**, 63–74 (2015)
31. J. Pearl, *Probabilistic reasoning in intelligent systems: networks of plausible inference* (Morgan Kaufmann Publishers Inc., San Francisco, CA, 1988)
32. B. Cai, Y. Liu, Y. Ma, Z. Liu, Y. Zhou, J. Sun, Real-time reliability evaluation methodology based on dynamic Bayesian networks, a case study of a subsea pipe ram BOP system. ISA Trans. http://dx.doi.org/10.1016/j.isatra.2015.06.011
33. B. Cai, Y. Liu, Y. Zhang, Q. Fan, S. Yu, Dynamic Bayesian networks based performance evaluation of subsea blowout preventers in presence of imperfect repair. Expert Sys. Appl. **40**(18), 7544–7554 (2013)
34. B. Cai, Y. Liu n, Q. Fan, Y. Zhang, S. Yu,Z. Liu, and X. Dong, Performance evaluation of subsea BOP control systems using dynamic Bayesian networks with imperfect repair and preventive maintenance. Eng. Appl. Artif. Intell. **26**(10), 2661–2672 (2013)
35. E. Romoro-Cadaval, G. Spagnuolo, L.G. Franquelo, C. Ramos-Paja, T. Suntio, W. Xiao, Grid-connected photovoltaic generation plants, components and operation. IEEE Ind. Electron. Mag. **7**(7), 6–20 (2013)
36. A. Arabkoohsar, L. Machado, M. Farzaneh-Gord, R.N.N. Koury, The first and second law analysis of a grid connected photovoltaic plant equipped with a compressed air energy storage unit. Energy **87**, 520–539 (2015)
37. E.M. Saber, S.E. Lee, S. Manthapuri, W. Yi, C. Deb, PV (photovoltaics) performance evaluation and simulation-based energy yield prediction for tropical buildings. Energy **71**, 588–595 (2015)

38. G.S. Seo, J.W. Shin, B.H. Cho, K.C. Lee, Digitally controlled current sensorless photovoltaic micro-converter for DC distribution. IEEE Trans. Ind. Inf. **10**(1), 117–126 (2014)
39. B. Liu., S. Duan, T. Cai, Photovoltaic DC-building-module-based BIPV system—concept and design considerations. IEEE Trans. Power Electron. **26**(5), 1418–1429 (2011)
40. A.M. Pavan, A. Mellit, D.D. Pieri, S.A. Kalogirou, A comparison between BNN and regression polynomial methods for the evaluation of the effect of soiling in large scale photovoltaic plants. Appl. Energy **108**, 392–401 (2013)
41. M. Tanrioven, Reliability and cost-benefits of adding alternate power sources to an independent micro-grid community. J. Power Sources **150**, 136–149 (2005)
42. E. Román, R. Alonso, P. Ibañez, S. Elorduizapatarietxe, D. Goitia, Intelligent PV module for grid-connected PV systems. IEEE Trans. Ind. Electron. **53**(4), 1066–1073 (2006)
43. C. Paravalos, E. Koutroulis, V. Samoladas, T. Kerekes, D. Sera, R. Teodorescu, Optimal design of photovoltaic systems using high time-resolution meteorological data. IEEE Trans. Ind. Inf. https://doi.org/10.1109/tii.2014.2322814
44. S.B. Kjaer, J.K. Pedersen, F. Blaabjerg, A review of single-phase grid-connected inverters for photovoltaic modules. IEEE Trans. Ind. Appl. **41**(5), 1292–1306 (2005)
45. H. Wang, F. Blaabjerg, Reliability of capacitors for DC-link applications in power electronic converters—an overview. IEEE Trans. Ind. Appl. **50**(5), 3569–3578 (2014)
46. Z. Liu, Y. Liu, D. Zhang, B. Cai, C. Zheng, Fault diagnosis for a solar assisted heat pump system under incomplete data and expert knowledge. Energy **87**, 41–48 (2015)
47. B.P. Cai, Y.H. Liu, Z.K. Liu, X.J. Tian, Y.Z. Zhang, J. Liu, Performance evaluation of subsea blowout preventer systems with common-cause failures. J. Pet. Sci. Eng. **90–91**, 18–25 (2012)
48. B.P. Cai, Y.H. Liu, Z.K. Liu, X.J. Tian, H. Li, C.K. Ren, Reliability analysis of subsea blowout preventer control systems subjected to multiple error shocks. J. Loss Prev. Process Ind. **25**(6), 1044–1054 (2012)
49. Y.L. Liu, M. Rausand, Reliability assessment of safety instrumented systems subject to different demand modes. J. Loss Prev. Process Ind. **24**(1), 49–56 (2011)
50. A. Khatab, N. Nahas, M. Nourelfath, Availability of K-out-of-N: G systems with non-identical components subject to repair priorities. Reliab. Eng. Syst. Saf. **94**(2), 142–151 (2009)
51. T. Kohda, W. Cui, Risk-based reconfiguration of safety monitoring system using dynamic Bayesian network. Reliab. Eng. Syst. Saf. **92**, 1716–1723 (2007)
52. B. Jones, I. Jenkinson, Z. Yang, J. Wang, The use of Bayesian network modelling for maintenance planning in a manufacturing industry. Reliab. Eng. Syst. Saf. **95**(3), 267–277 (2010)

Reliability Evaluation of Auxiliary Feedwater System by Mapping GO-FLOW Models into Bayesian Networks

Abstract Bayesian network (BN) is a widely used formalism for representing uncertainty in probabilistic systems, and it has become a popular tool in reliability engineering. The GO-FLOW method is a success-oriented system analysis technique and capable of evaluating system reliability and risk. To overcome the limitations of GO-FLOW method and add new method for BN model development, this paper presents a novel approach on constructing a BN from GO-FLOW model. GO-FLOW model involves with several discrete time points, and some signals change at different time points. But it is a static system at one time point, which can be described with BN. Therefore, the developed BN with the proposed method in this paper is equivalent to GO-FLOW model at one time point. The equivalent BNs of the fourteen basic operators in the GO-FLOW methodology are developed. Then, the existing GO-FLOW models can be mapped into equivalent BNs on basis of the developed BNs of operators. A case of auxiliary feedwater system of a pressurized water reactor is used to illustrate the method. The results demonstrate that the GO-FLOW chart can be successfully mapped into equivalent BNs.

Keywords Reliability · GO-FLOW methodology · Bayesian network · Auxiliary feedwater system

1 Introduction

A Bayesian network (BN) is an acyclic directed graph, consisting of nodes and arcs between the nodes. In the networks, random variables are assigned to each node, together with the conditional dependence on the parent nodes. The probabilistic dependences are quantified by a conditional probability table for each node. Each conditional probability table contains the probability of a node, given any possible combination of its parent nodes. Root nodes are nodes without parents, and prior probabilities are assigned to them. Given the values of the observed variables as evidences, the posterior probabilities of the unobserved variables could be obtained by inferences.

B. Cai et al., *Bayesian Networks for Reliability Engineering*,
https://doi.org/10.1007/978-981-13-6516-4_3

BN is a widely used formalism for representing uncertainty in probabilistic systems. Recently, BNs have been popular for reliability evaluation as a robust and viable alternative to most traditional methods such as fault tree, reliability block diagrams, and so on [1]. The properties of the modeling framework that make BNs particularly well suited for reliability applications are discussed [2]. Honari et al. [3] developed a new approach using BN to evaluate the reliability of an (r, s)-out-of-(m, n): F system. Norrington et al. [4] used a BN method to model the reliability of search and rescue operations within UK Coastguard Coordination Centers. Cai et al. [5] proposed a novel real-time reliability assessment method by combing root cause diagnosis phase based on BNs and reliability evaluation phase based on dynamic BN. Musharraf et al. [6] proposed a data collection method using a virtual environment for a simplified BN model of offshore emergency reliability evaluation to handle the data scarcity problem.

BNs are graphical and qualitative illustration of relationships among different nodes using directed arcs. A BN can be obtained by machine learning using data sets or deducing from expert knowledge [7]. These two methods can be used individually or jointly. In order to make use of the powerful representation in uncertainty, BNs can be developed by converting the other reliability models. Bobbio et al. [8] presented an approach to convert fault tree into a BN. Dynamic fault tree model has also been mapped into BNs [9]. Montani et al. [10] developed a software tool (RADYBAN), which can automatically translate a dynamic fault tree into the corresponding dynamic BN. Beardield and Marsh [11] proposed the method for mapping event tree into BN. Lo et al. [12] presented a novel approach for constructing BN based on a bond graph model. Kim [13] provided a method of mapping a reliability block diagram with general gates model into an equivalent BN model without losing the one-to-one matching characteristic. Khakzad et al. [14] presented a method of mapping bow-tie into BN. If a BN is involved with temporal factors, it is a dynamic network. Static BN can be extended into dynamic Bayesian networks (DBNs) by introducing relevant temporal dependencies between representations of the static network at different times. Several DBN models have been proposed for assessing reliability of technical systems. Portinale et al. [15] presented an approach to reliability modeling and analysis based on the automatic conversion of the dynamic fault tree into DBN. Cai et al. [16, 17] developed DBNs for performance evaluation of subsea blowout preventer (BOP) and its control system in presence of imperfect repairs.

The GO-FLOW method is a success-oriented system analysis technique and capable of evaluating system reliability and risk [18]. It is applicable to systems with complex sequences of system operation or changes of system state over time. Therefore, GO-FLOW is suitable for the analysis of a phased mission problem or a time-dependent system [19, 20]. Besides, it can perform common cause failure analysis and sensitivity analysis of a system. Reliability analysis by GO-FLOW includes two steps: first constructing a GO-FLOW chart for the target system and then computing the system reliability quantitatively [21]. The GO-FLOW model can be easily developed from operators, which denote particular functional logical gates. Muhammad et al. [22] performed reliability evaluation with GO-FLOW method to determinate the influence of automatic depressurization system in dynamic reliability of passive core cooling system of AP1000. Common cause failure issues in the reliability analysis and the procedure of analysis have been treated by GO-FLOW method [23, 24]. The GO-FLOW method is a success-oriented system analysis technique and capable of evaluating system reliability and risk [18]. Reliability analysis by GO-FLOW mainly includes two steps, namely construction of a GO-FLOW chart for the system and calculation of the reliability quantitatively. The GO-FLOW method has the following features: (a) The GO-FLOW chart corresponds to the physical layout of a system and is easy to construct and validate, (b) alterations and updates to a GO-FLOW chart are readily finished, and (c) GO-FLOW includes all possible system operational states. Like the other reliability evaluation methods, GO-FLOW has its drawbacks. It has too many different operators, which will be difficult to select appropriate ones to describe the components. Besides, GO-FLOW is not able to construct hierarchical charts. Further, identifying the scenarios is not an intuitive process [25].

In this paper, a novel method on constructing BNs from GO-FLOW models is presented. On the one hand, it can overcome the mentioned limitations of GO-FLOW model. On the other hand, a novel method to develop BN can be proposed. There is no specific semantic to guide the model development, which becomes one weak point of BN [26]. For solving this issue, one research direction concerns the translation of the classical dependability model into a BN model. Mapping GO-FLOW model into BN will have some improvements. Firstly, beyond the usual measures available in GO-FLOW analysis, the BN method is able to perform backward analysis that concerns the computation of the posterior probability of any given set of variables given some observation. Besides, the modeling flexibility of BN can accommodate different kinds of statistical dependencies that cannot be included in the GO-FLOW method. The corresponding BNs of classical operators in GO-FLOW methodology are developed firstly. Then a case study of subsea BOP control system is given to illustrate the proposed method. The reminder of this paper is organized as follows. Section 2 presents background concerning GO-FLOW methodology and BNs. In Sect. 3, equivalent BNs of operators in GO-FLOW method are developed. Section 4 illustrates the method by a case study. Section 5 summarizes the paper.

2 Background

2.1 Bayesian Networks

A BN is a directed acyclic graph composed of nodes and arcs [27]. Nodes represent random variables, and directed arcs between pairs of nodes denote dependencies between the variables. Conditional probability distribution is specified at each node that has parents, while prior probability is specified at node that has no parents. The nodes ($X_1, \ldots, X_N$) in the network are labeled by related random variables. Assuming Pa(X_i) is the parent node of X_i in the model, the conditional probability distribution of X_i is denoted by $P(X_i|\mathrm{Pa}(X_i))$. The joint probability distribution $P(X_1, \ldots, X_N)$ can be written as Eq. (1).

$$P(X_1, \ldots, X_N) = \prod_{Xi \in \{X_1,\ldots,X_N\}} P(X_i|\mathrm{Pa}(X_i)) \quad (1)$$

BN allows both forward (or predictive) analysis and backward (diagnostic) analysis [28], where the posterior probability of any set of variables can be calculated.

Inference in BNs is to calculate the probability of each node when other variables are known. Bayesian theorem is used to compute conditional probabilities. Given the variable Y, the conditional probability of X is given by

$$P(X|Y) = \frac{P(X)P(Y|X)}{P(Y)} \quad (2)$$

2.2 GO-FLOW Methodology

The GO-FLOW method is a success-oriented system analysis technique and capable of evaluating system reliability. It uses a set of standardized operators to describe the logic operation, interaction, and combination of physical equipment. The operators are used to model function or failure of physical equipments, logical gates and a signal generator. As shown in Fig. 1, fourteen different types of operators are defined.

A system is modeled by selecting operators and interrelating their inputs and outputs. Consisting of operators and signal lines, GO-FLOW chart represents the engineering function of the system. The signal lines identify the inputs and outputs to the operators, with which the operators are connected. Signals denote some physical quantity or information, for example, the electric current in the conductor, a command signal, etc. A quantity defined as "intensity" is associated with a signal line. Generally, the intensity represents the probability of signal existence. When a signal is a subinput signal to the type-35, -37, or -38 operator, the intensity is used for denoting a time interval.

A finite number of discrete time values are required for expressing the system operation sequence. The values do not necessarily denote real time but correspond to it and represent an ordering. The total number of time points is a user-defined parameter.

The GO-FLOW method has the following features: (a) The GO-FLOW chart corresponds to the physical layout of a system and is easy to construct and validate, (b) alterations and updates to a GO-FLOW chart are readily finished, and (c) GO-FLOW includes all possible system operational states.

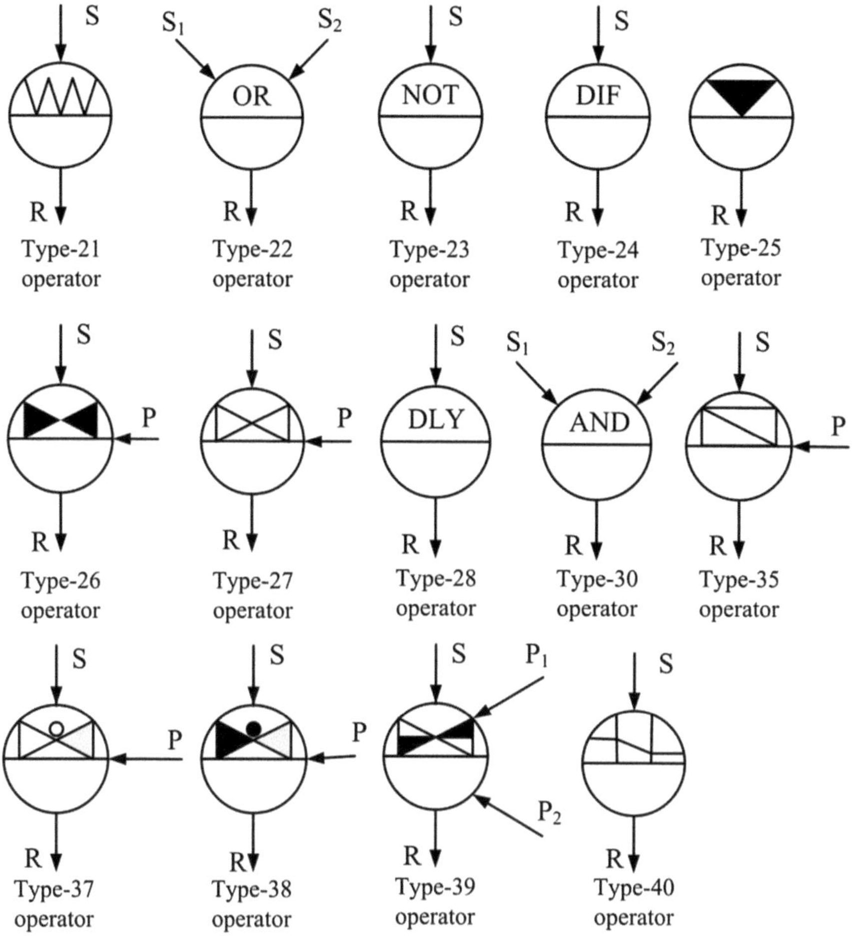

Fig. 1 Symbols of operators used in GO-FLOW methodology

2.3 *Reliability Analysis Process*

GO-FLOW model of a system is assumed as follows: (1) It is a coherent system, which means all the components are relevant to the system. (2) The components are mutually independent. (3) When a failed component is repaired, it will be as good as new.

The procedures for reliability analysis with GO-FLOW method are included as follows.

(1) Define the system. In this step, the scope of the system and reliability index is defined. Besides, its function and structure including the components are determined.

(2) Determine the boundary conditions. The input, output, and interface of the system are defined.
(3) Establish success criteria. It means clearly defining the success state and failure state of the system. Different success criteria will lead to different evaluation results.
(4) Develop the GO-FLOW chart. The following steps are listed. First, select proper types of operators to represent the components. Second, connect the operators with signal lines from input to output based on the logical relations among them. Third, number the operators and signals, respectively. Fourth, determine the state probabilities of all the operators.
(5) Input the data according to the operator numbers.
(6) Calculate the probabilities based on the operational rules of the operators.
(7) Evaluate the system. The calculation results and success criteria are used to obtain reliability of the system. Based on its function and requirements, the system is evaluated.

3 Mapping the Operators in GO-FLOW Method into BNs

Generally, three types of signals are used to connect to an operator: the main input signal $S(t)$, subinput signal $P(t)$, and an output signal $R(t)$. Different operators have different logics for combining the inputs and producing the outputs. The equivalent BNs of the fourteen operators are proposed in this paper. As the equations of output signal intensity are complicated and time variant, it is hard to develop a complete equivalent DBN or BN for a GO-FLOW model. A GO-FLOW model involves with several discrete time points and some signals change at different time points. But the GO-FLOW model at one time point is a static system, it is able to be described with BN. Therefore, the developed BN with the proposed method in this paper is equivalent to GO-FLOW model at one time point. Defining the parameters of BN based on the analysis of GO-FLOW model, an equivalent BN can be developed at all time points. In the following, each of those fourteen different types of operators is introduced and its equivalent BNs is presented.

Type-21 operator describes a "good/bad" component, which has one input and one output signal line. The output signal will be present when the input signal is present and the operator is in good state. P_g is the probability for the good state of a component that the operator describes. So, $1 - P_g$ is the probability for the bad state of a component. Hence, the intensity of the output signal is obtained by

$$R(t) = P_g \cdot S(t) \tag{3}$$

Its equivalent BN is shown in Fig. 2. Node S denotes the input signal S, and node R represents the output signal. For node S, "Yes" and "No" denote the presence and absence of input signal, respectively. For node R, "Yes" and "No" denote the

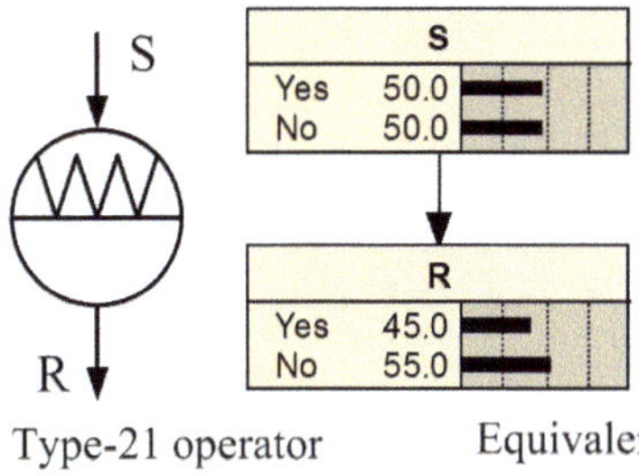

S	R: Yes	R: No
Yes	P_g	$1-P_g$
No	0	1

Fig. 2 Equivalent BN of type-21 operator

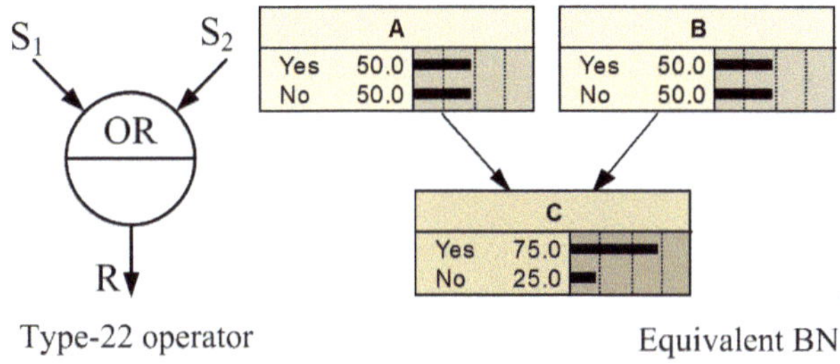

S1	S2	R: Yes	R: No
Yes	Yes	1	0
Yes	No	1	0
No	Yes	1	0
No	No	0	1

Fig. 3 Equivalent BN of type-22 operator

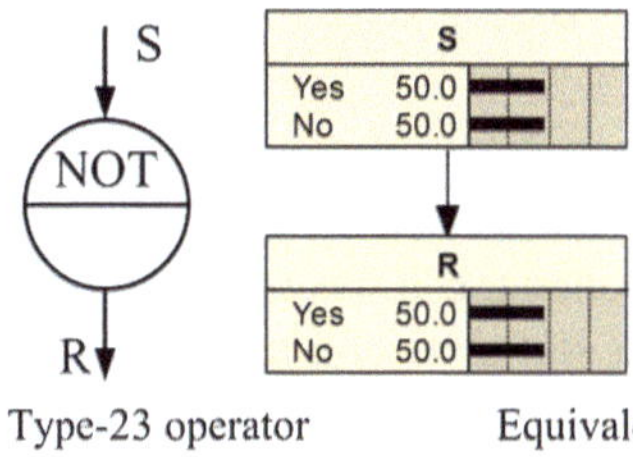

S	P(R|S)
Yes	No
No	Yes

Equivalent BN

Fig. 4 Equivalent BN of type-23 operator

presence and absence of output signal, respectively. Conditional probability of node R is given in Fig. 2. According to the Bayesian inference,

$$P(R = \text{Yes}) = \sum P(S) \cdot P(R = \text{Yes}|S) = S(t) \cdot P_g \tag{4}$$

which is the same as the value obtained by GO-FLOW method. For example, at time t, if Pg = 0.9, $S(t) = 0.5$, then "Yes = 0.45" and "No = 0.55" in R as shown in Fig. 2.

Type-22 operator models an OR gate, which has more than one input signal line. The intensity of the output signal is the probability that at least one input signal S_i exists. Assuming there are two input signals, its equivalent BN is shown in Fig. 3.

Type-23 operator models a NOT gate, which implements logical negation. It has an input signal and an output signal. Its equivalent BN is shown in Fig. 4.

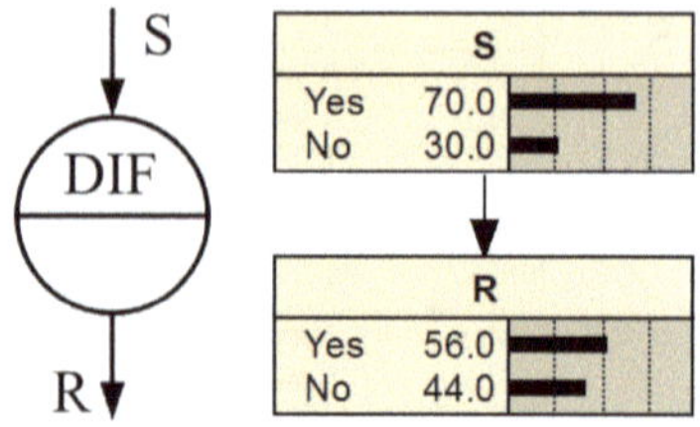

S	R: Yes	R: No
Yes	m	1-m
No	0	1

Type-24 operator Equivalent BN

Fig. 5 Equivalent BN of type-24 operator

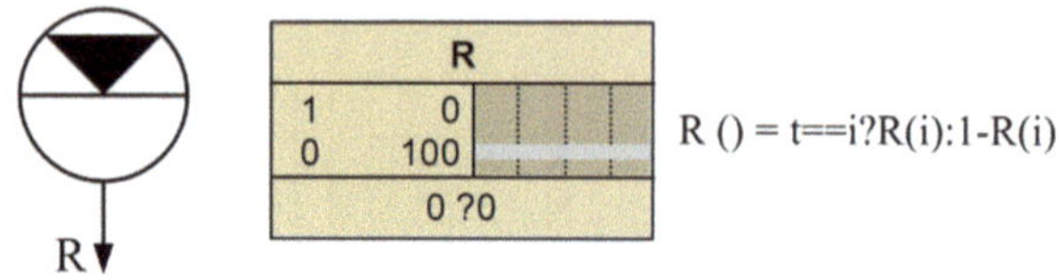

Type-25 operator Equivalent BN

Fig. 6 Equivalent BN of type-25 operator

Type-24 operator models difference operation between output and input signals. It has one input and one output signal. Hence, the intensity of the output signal is obtained by

$$R(t) = mS(t) \tag{5}$$

where m is a constant. Its equivalent BN is shown in Fig. 5. If $m = 0.8$, $S(t) = 0.7$, then $R(t) = 0.56$.

Type-25 operator is a signal generator. It has only one output signal line, which is commonly used to generate a signal at one time point. This operator is used to control the timing of component action, and the intensity of the output signal represents the probability that generate a signal at one time point. The signal line can also be connected to a type-35, -37, or -38 operator through a subinput signal P. In this case, the intensity of the output signal denotes a period. Its equivalent BN is shown in Fig. 6. In Netica software, the conditional probability between a node and its parent nodes can be entered using an equation if desired. This is possible whether the nodes are continuous or discrete, and whether the relationship is probabilistic or deterministic. Netica will convert all equations into tables (CPT or function table) before compiling a net, doing net transforms, or expanding a DBN. The tables are then used in the same way as if the user had entered them by hand. The probability is defined by the equation, where $t{=}{=}i?$ is a conditional statement. It means that if the time is i, then the intensity of the output signal is $R(i)$.

Type-26 operator models a normally closed valve. It has one main input, one subinput, and one output signal line. Probabilities for successful operation (P_g) and premature operation (P_p) are required. The output intensity can be obtained by Eqs. (6) and (7).

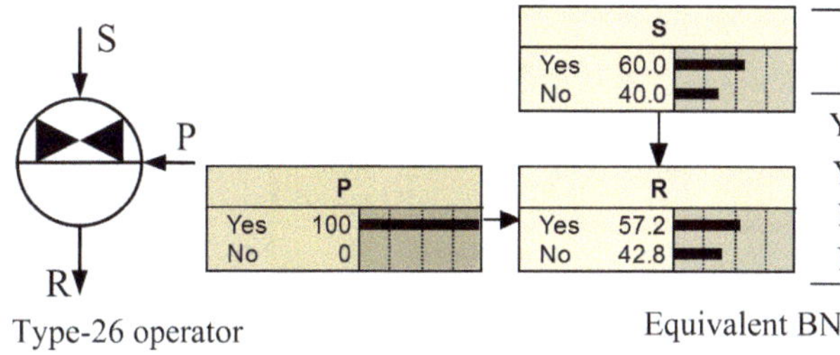

S	P	R: Yes	R: No
Yes	Yes	$O(t)$	$1-O(t)$
Yes	No	$O(t)$	$1-O(t)$
No	Yes	0	1
No	No	0	1

Fig. 7 Equivalent BN of type-26 operator

$$R(t) = S(t) \cdot O(t) \tag{6}$$

$$O(t) = O(t') + [1.0 - O(t')] \cdot P(t) \cdot P_g \tag{7}$$

where $O(t')$ is the probability for a valve in the open state at a time point immediately before the time point t. $O(t_1)$is equal to P_p, where t_1 is the initial time point. Its equivalent BN is shown in Fig. 7. Conditional probability is shown in the figure. According to the Bayesian inference,

$$P(R = \text{Yes}) = \sum P(S, P) \cdot P(R = \text{Yes}|S, P) = S(t) \cdot O(t) \tag{8}$$

which is the same as *R(t)* in Eq. (6). $O(t)$ and $O(t')$are obtained from the analysis of GO-FLOW model. An example is made for the illustration. If $P_p = 0.08$, $P_g = 0.95$, $S(t) = 0.6$, and $P(t) = 1$ at time point $t = 2$, then $O(t_1) = 0.08$, $O(t_2) = 0.08 + (1 - 0.08) * 1 * 0.95 = 0.954$, $R(t) = 0.6 * 0.965 = 0.5724$. So, the parameters of the equivalent BN can be defined and the results are shown in Fig. 7.

Type-27 operator models a normally open valve. It has one main input, one subinput, and one output signal line. Like type-26 operator, P_g and P_p are required. The output intensity can be calculated by Eqs. (9) and (10).

$$R(t) = S(t) \cdot O(t) \tag{9}$$

$$O(t) = O(t')[1.0 - P(t) \cdot P_g] \tag{10}$$

where $O(t')$ has the same meaning as in Eq. (7). $O(t_1)$ is equal to $1 - P_p$. Its equivalent BN is shown in Fig. 8. If $P_p = 0.08$, $P_g = 0.95$, $S(t) = 0.6$ and $P(t) = 1$ at time point $t = 2$, then $O(t_1) = 0.92$, $O(t_2) = 0.92 * [1.0 - 1.0 * 0.95] = 0.046$, $R(t) = 0.6 * 0.046 = 0.0276$.

Type-28 operator models delay operation. It has one input signal and one output signal, representing the delay relationship between them. Its equivalent BN is shown in Fig. 9. Conditional probability of node R is defined by the equation, which means that the delay time is i. It means that state of node R is the same as node S after delay time i. Thus, the delay operation is realized.

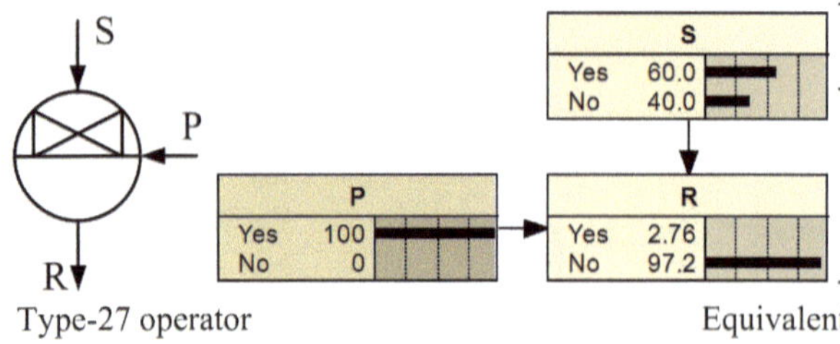

S	P	R: Yes	R: No
Yes	Yes	$O(t)$	$1-O(t)$
Yes	No	$O(t)$	$1-O(t)$
No	Yes	0	1
No	No	0	1

Fig. 8 Equivalent BN of type-27 operator

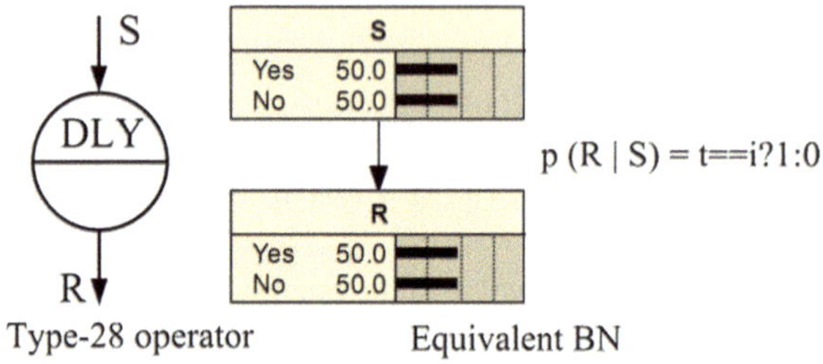

Fig. 9 Equivalent BN of type-28 operator

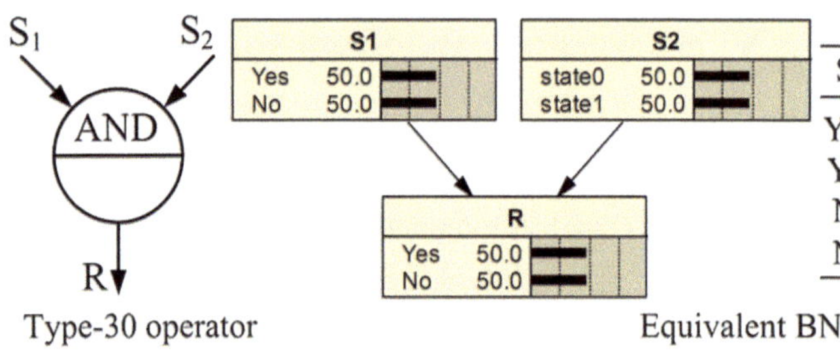

S1	S2	P(R|S)
Yes	Yes	Yes
Yes	No	No
No	Yes	No
No	No	No

Fig. 10 Equivalent BN of type-30 operator

Type-30 operator describes an AND gate. It has more than one main input signal lines but only one output signal line. The output signal intensity is the probability that all the inputs S_i exist. Assuming there are two input signals, its equivalent BN is shown in Fig. 10.

Type-35 operator modes a component with an increasing failure rate over time, such as a luminous light bulb. It has one main input, several subinputs, and one output signal line. A component failure rate λ is required, which is assumed to be constant. The output intensity is

$$R(t) = S(t)\exp\{-\lambda \sum_{i} \sum_{t_k \le t} P_i(t_k) \cdot \min[1.0, S(t_k)/S(t)]\} \quad (11)$$

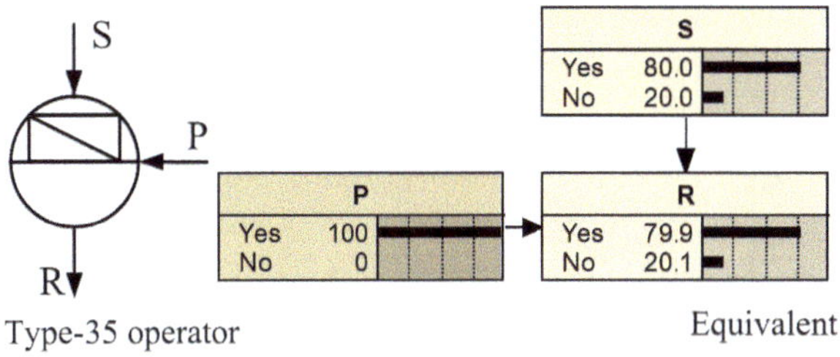

S	P	R: Yes	R: No
Yes	Yes	$F(t)$	$1-F(t)$
Yes	No	$F(t)$	$1-F(t)$
No	Yes	0	1
No	No	0	1

Fig. 11 Equivalent BN of type-35 operator

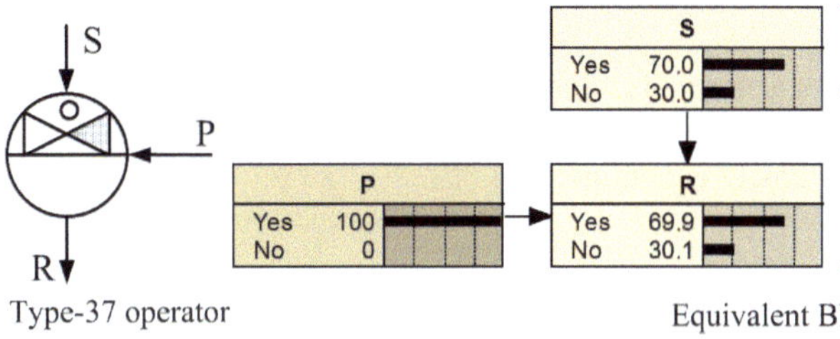

S	P	R: Yes	R: No
Yes	Yes	$F_1(t)$	$1-F_1(t)$
Yes	No	$F_1(t)$	$1-F_1(t)$
No	Yes	0	1
No	No	0	1

Fig. 12 Equivalent BN of type-37 operator

The subinput signal denotes a time duration, which has the same unit time like the failure rate. Define $F(t)$ as $\exp\{-\lambda \sum_i \sum_{t_k \le t} P_i(t_k) \cdot \min[1.0, S(t_k)/S(t)]\}$. Its equivalent BN is shown in Fig. 11. Similarly, the information of $S(t_k)$and $P_i(t_k)$ come from the analysis of GO-FLOW model and then the parameters of BN are defined based on these values. Suppose there is only one subinput signal, and it means $i = 1$. If $P(1) = P(2) = 0$, $P(3) = 1$, $\lambda = 0.001/h$, $S(1) = S(2) = 0.5$ and $S(3) = 0.8$, then $R(1) = S(1) = 0.5$, $R(2) = S(2) = 0.5$, $R(3) = 0.8 * \exp(-0.001 * 1.0 * \min[1.0, 0.8/0.8]) = 0.7992$). Therefore, the parameters of BN at time point 3 can be defined.

Type-37 operator models normally open valve with a time-dependent failure rate. It has one main input, several subinputs, and one output signal line. Like type-35 operator, its failure rate is constant. The output intensity can be calculated by

$$R(t) = S(t) \exp(-\lambda \sum_i \sum_{t_k \le t} P_i(t_k)) \tag{12}$$

The subinput signal denotes a time duration, which has the same unit time like the failure rate. Define $F_1(t)$ as $\exp\{-\lambda \sum_i \sum_{t_k \le t} P_i(t_k) \cdot\}$. Its equivalent BN is shown in Fig. 12. Similarly, the information of $P_i(t_k)$ comes from the analysis of GO-FLOW model and then the parameters of BN are defined based on these values. Suppose there is only one subinput signal, and it means $i = 1$. If $P(1) = 0$, $P(2) = 1$, $\lambda = 0.001/h$, $S(1) = S(2) = 0.7$, then $R(1) = S(1) = 0.7$, $F_1(2) = \exp(-0.001 * (0 + 1)) = 0.999$, $R(2) = 0.7 * 0.999 = 0.6993$. Therefore, the parameters of BN at time point 2 can be defined.

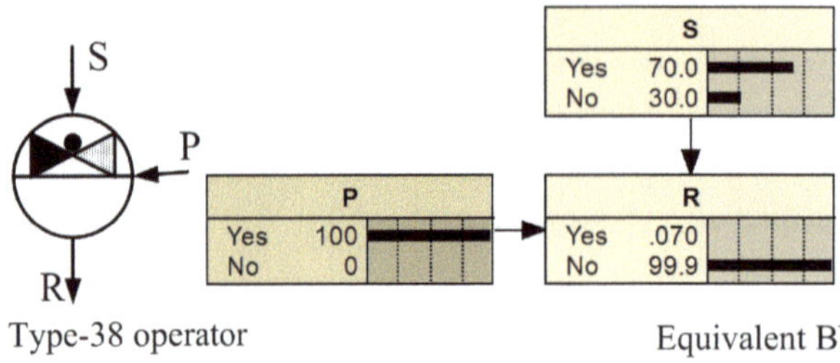

S	P	R: Yes	R: No
Yes	Yes	$F_2(t)$	$1-F_2(t)$
Yes	No	$F_2(t)$	$1-F_2(t)$
No	Yes	0	1
No	No	0	1

Fig. 13 Equivalent BN of type-38 operator

Type-38 operator models a normally closed valve with time-dependent failure rate. It has one main input, several subinputs, and one output signal line. Failure rate λ is required. The output intensity is obtained by

$$R(t) = S(t)(1 - \exp(-\lambda \sum_{i} \sum_{t_k \le t} P_i(t_k))) \tag{13}$$

Its equivalent BN is shown in Fig. 13. Define $F_2(t)$ as $(1 - \exp(-\lambda \sum_i \sum_{t_k \le t} P_i(t_k)))$. Suppose there is only one subinput signal, and it means $i = 1$. If $P(1) = 0$, $P(2) = 1$, $\lambda = 0.001/h$, $S(1) = S(2) = 0.7$, then $R(1) = 0.7 * (1.0 - 1.0) = 0$, $F_2(2) = 1.0 - \exp(-0.001 * (0 + 1)) = 0.001$, $R(2) = 0.7 * 0.001 = 0.0007$. Therefore, the parameters of BN at time point 2 can be defined.

Type-39 operator models the opening and closing actions of a valve. It has one main input, two subinputs, and one output signal line. Subinput P_1 is responsible for opening action and subinput P_2 control closing function. The probabilities of the valve successfully opened and closed are defined as P_o and P_c, respectively. The output intensity is

$$R(t) = S(t) \cdot O(t) \tag{14}$$

If P_1 arrives, $O(t)$ is equal to $O(t') + (1 - O(t')) \cdot P_1(t) \cdot P_o$. When P_2 arrives, $O(t)$ is equal to $O(t') \cdot (1 - P_2(t) \cdot P_c)$. Equivalent BN of type-39 operator is shown in Fig. 14. According to the Bayesian inference,

$$P(R = \text{Yes}) = \sum P(S, P) \cdot P(R = \text{Yes}|S, P) = S(t) \cdot O(t) \tag{15}$$

which is the same as $R(t)$ in Eq. (14). Suppose subinput P_1 is present at the second time point, so $P_1(1) = 0$, $P_1(2) = 1.0$. If $S(1) = 7$, $S(2) = 0.6$, $P_O = 0.9$, $P_C = 0.88$, then $O(1) = 0.9$, $O(2) = 0.9 + (1 - 0.9) * 1.0 * 0.9 = 0.99$, $R(2) = 0.6 * 0.99 = 0.594$. Therefore, the parameters of BN at time point 2 can be defined.

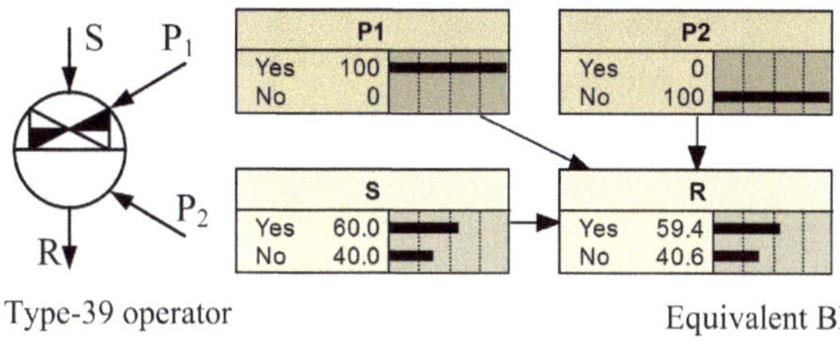

S	P1	P2	R: Yes	R: No
Yes	Yes	Yes	$O(t)$	$1-O(t)$
No	Yes	Yes	0	1
Yes	No	Yes	$O(t)$	$1-O(t)$
No	No	Yes	0	1
Yes	Yes	No	$O(t)$	$1-O(t)$
No	Yes	No	0	1
Yes	No	No	$O(t)$	$1-O(t)$
No	No	No	0	1

Fig. 14 Equivalent BN of type-39 operator

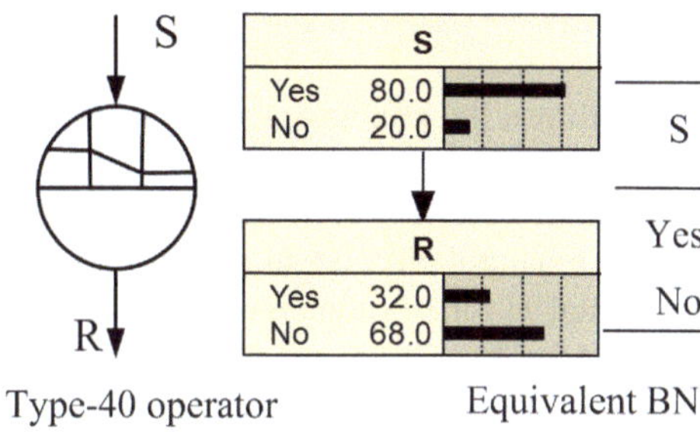

S	R: Yes	R: No
Yes	0.4	0.6
No	0	1

Fig. 15 Equivalent BN of type-40 operator

Type-40 operator has one input and one output signal. It is used to simulate that input signal is converted to output signal according to different specific relations. As there is not specific relationship between the input signal and output signal, the intensity of phased output signal at different time points can be defined by equations. Equivalent BN of type-40 operator is shown in Fig. 15. Suppose there is a linear relation between output signal and input signal, then $R(t) = k(t)S(t)$. If $k(1) = 0.2$, $k(2) = 0.4$, $S(1) = 0$, $S(2) = 0.8$, then $R(1) = 0$, $R(2) = 0.32$. Therefore, the parameters of BN at time point 2 can be defined. It shows that the probability of node R in BN is the same as R(2).

4 Case Study

4.1 Auxiliary Feedwater System and Its GO-FLOW Model

To illustrate the mapping method based on the developed BNs of the operators, a case of auxiliary feedwater system of a pressurized water reactor has been selected.

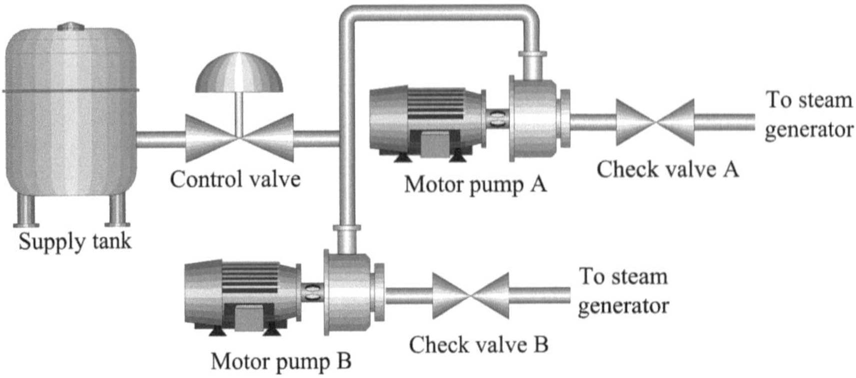

Fig. 16 Simplified scheme of auxiliary feedwater system

As one of the engineered safety facilities, the auxiliary feedwater system acts as a heat sink to remove the decay heat from the reactor core during accident scenarios [29]. It cools down the steam generator secondary side and eventually removes the decay heat from the reactor core by a natural circulation mechanism, for example, condensing steam in nearly horizontal U-tubes submerged inside a pool [30]. Its simplified scheme is shown in Fig. 16. The system consists of two motor pumps, which take suction from a common condensate storage tank and provide auxiliary feedwater flow to steam generators. A normally closed valve is employed between the supply tank and pumps. Two identical motor pumps are used for redundancy. The outlet of each pump is connected with a normally closed check valve. Each motor pump can provide flow through successful valve opening to the steam generator. Therefore, two pumps compose a parallel system. For the successful operation of the system, it needs to supply flow to at least one steam generator.

A total of five time points are defined in the case. Time point 1 is an initial time, and no actions are taken. At time point 2, the system starts to work and motor pump A is activated. Time point 3 is 3 h later than time point 2, and time point 4 immediately succeeds time point 4. At time point 4, motor pump B is also activated. Time point 5 is 12 h later than time point 4.

According to the function and structure of the system, GO-FLOW model of the case is shown in Fig. 17. A total of 15 operators in total are used, and the sequence number is labeled in each operator. The model initiates from operator 1, and the output line of the final operator 15 corresponds to the probability of the successful operation of the system. As shown in Fig. 17, type-25 operators are used to generate input signals at different time points. The supply tank is denoted by type-21 operator. Control valve and two check valves are denoted by type-21 operators. For the motor pump, Type-26 operator is used to describe the motor, while type-35 operator is used to model the pump. Parameters of the all the operators are given in Table 1. Based on the operational rules of operators, their probabilities at different time points are

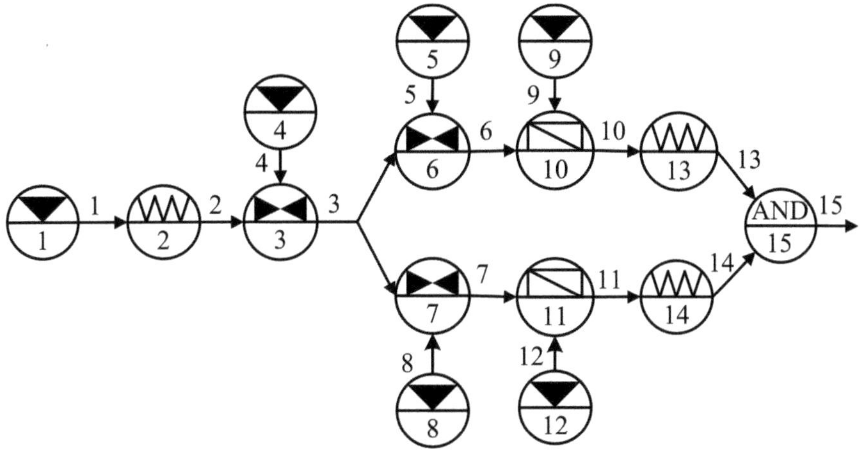

Fig. 17 GO-FLOW chart of the case

Table 1 Parameters of the operators in GO-FLOW model

No.	Type	Parameters
1	25	$R(1) = 0, R(t) = 1.0(t \neq 1)$
2	21	$P_g = 0.99995$
3	26	$P_p = 0.00376, P_g = 0.999893$
4	25	$R(1) = 0, R(t) = 1.0(t \neq 1)$
5	25	$R(2) = 1, R(t) = 0\ (t \neq 2)$
6, 7	26	$P_p = 0.00332, P_g = 0.999756$
8	25	$R(4) = 1, R(t) = 0\ (t \neq 4)$
9	25	$R(3) = R(5) = 10$ h, $R(t) = 0\ (t \neq 3)$
10, 11	35	$\lambda = 1.5e{-}5$
12	25	$R(5) = 10$ h, $R(t) = 0\ (t \neq 5)$
13, 14	21	$P_g = 0.999846$
15	30	–

calculated. The details about the steps of the analysis and results are shown in Table 2. The intensity of 15 output signals at time points 1 through 12 is listed in Table 2.

4.2 *The Equivalent BN*

Based on the developed BNs of the operators in the GO-FLOW method, the equivalent BN of the case at time point 2 is presented in Fig. 18. In the BN, node t is defined as a constant node with five states denoting the five time points. Based on the equations

Table 2 Intensity of output signals at different time points

Output	Time points				
No.	1	2	3	4	5
1	0	1	1	1	1
2	0	0.99995	0.99995	0.99995	0.99995
3	0	0.999843408	0.999843408	0.999843408	0.999843408
4	0	1	0	0	0
5	0	1	0	0	0
6	0	0.999600256	0.999600256	0.999600256	0.999600256
7	0	0.00331948	0.00331948	0.999600256	0.999600256
8	0	0	0	1	0
9	0	0	10	0	10
10	0	0.999600256	0.999450327	0.999450327	0.999300421
11	0	0.00331948	0.00331948	0.999600256	0.999450327
12	0	0	0	0	10
13	0	0.999446317	0.999296412	0.999296412	0.999146528
14	0	0.003319467	0.003319467	0.999596257	0.999446329
15	0	0.999447636	0.999298228	0.999843272	0.999843131

defined in some nodes, Netica software will convert all equations related to t into tables. Therefore, the CPT related to time points will be established once the time point t is chosen. That is to say, the structure of the BN for the system is fixed, but the CPTs of the nodes are varying at different time points. In Fig. 18, probabilities of nodes in the BN are displayed in the percentage format. It can be seen that the probability is the same as the corresponding value in Table 1. For example, output intensity of signal 15 at time point 2 in Table 1 is identical to the probability of S15 = Yes shown in Fig. 18. For both models, the probability of the successful operation is 99.945%. It is worth to note that only three digits of the probability are shown in the node, but the value with high precision can be checked by clicking "Calibration" option of the node in the model. The successful probability of pump A is 99.96%, which is also the same as the value in Table 2.

By setting the value of node t as 5, probabilities of nodes at time point 5 in BN are shown in Fig. 19. The probabilities are the same as those corresponding values at time point 5 in Table 2. For example, output intensity of signal 15 at time point 5 in Table 2 is identical to the probability of S15 = Yes shown in Fig. 19. For both models, the probability of the successful operation is 99.984%. The other probabilities are also the same as those in Table 2. The results demonstrate that the GO-FLOW chart has been successfully mapped into equivalent BNs.

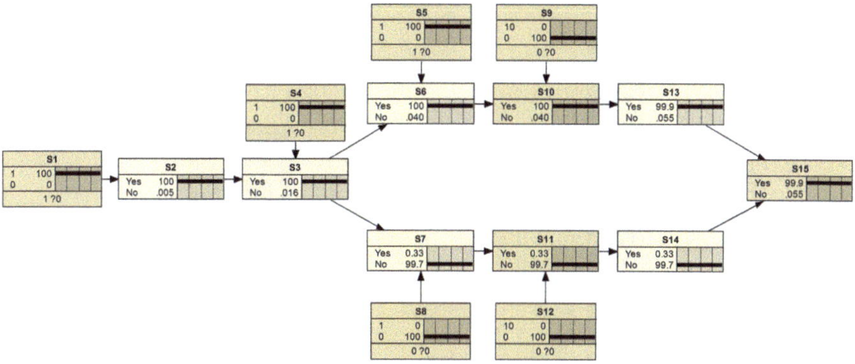

Fig. 18 Equivalent BN of the case at time point 2

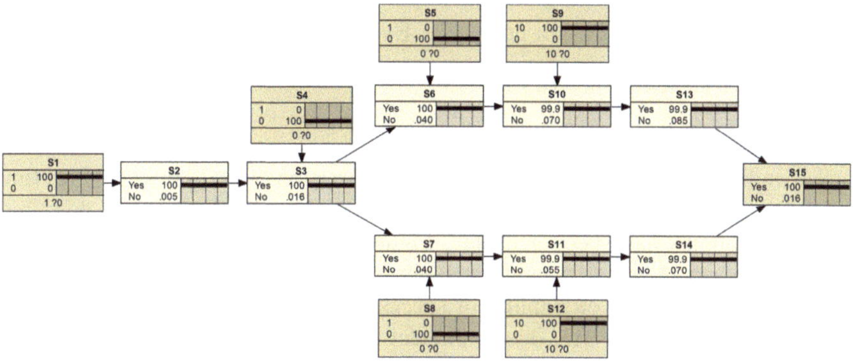

Fig. 19 Equivalent BN of the case at time point 5

4.3 Mutual Information Investigation

The important degree of events to the system successful operation can be assessed by using Shannon's mutual information (entropy reduction), which is one of the most widely used measurement for ranking information sources [31]. The mutual information is the total uncertainty-reducing potential of random variable X, given the original uncertainty in T prior to consulting X. Intuitively, mutual information measures the information that T and X share: It measures how much knowing one of these variables reduces our uncertainty about the other [32]. The mutual information of T and X is given by

$$I(T, X) = -\sum_{x}\sum_{t} P(t, x) \log \frac{P(t, x)}{P(t)P(x)} \tag{16}$$

where $P(t, x)$ is the joint probability distribution function of T and X and $P(t)$ and $P(x)$ are the marginal probability distribution functions of T and X, respectively.

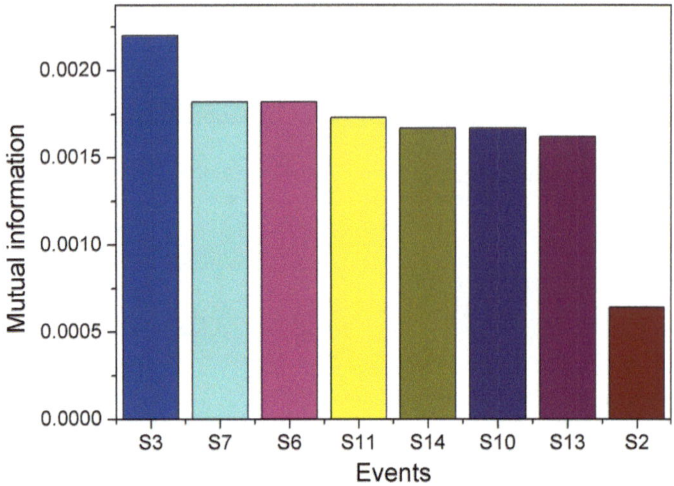

Fig. 20 Mutual information of events and S15

The individual contribution of events to the operation of the system (S15) at time point 5 is described using mutual information, as shown in Fig. 20. At this time, two motor pumps have been activated. Figure 20 shows that S3 has the greatest contribution to S15, which means that control valve should be given more attention in order to improve the probability of successful operation of the system. S2 has the least contribution to S15, due to the fact that supply tank has very low failure probability.

5 Conclusions

Beyond the usual measures available in GO-FLOW analysis, the BN method is able to perform backward analysis that concerns the computation of the posterior probability of any given set of variables given some observation. Besides, the modeling flexibility of BN can accommodate different kinds of statistical dependencies that cannot be included in the GO-FLOW method. In this paper, a method of mapping GO-FLOW model into equivalent BNs is proposed. Firstly, the equivalent BNs of fourteen operators in the GO-FLOW methodology are developed. Then a case of auxiliary feedwater system of a pressurized water reactor is used to illustrate the mapping process. With the equivalent BNs of basic operators, BN of the case is presented. Based on BN inference, probabilities of all nodes are calculated. The results demonstrate that a BN has been successfully constructed based on the GO-FLOW model. Mutual information investigation is performed to assess the important degree of events to the system successful operation.

References

1. N. Khakzad, F. Khan, P. Amyotte, Risk-based design of process systems using discrete-time Bayesian networks. Reliab. Eng. Sys. Saf. **109**, 5–17 (2013)
2. H. Langseth, L. Portinale, Bayesian networks in reliability. Reliab. Eng. Sys. Saf. **92**, 92–108 (2007)
3. B. Honari, J. Donovan, E. Murphy, Using Bayesian networks in reliability evaluation for an (r, s)-out-of-(m, n): F distributed communication system. J. Stat. Plan. Inference **139**, 1756–1765 (2009)
4. L. Norrington, J. Quigley, A. Russell, R. Van der Meer, Modelling the reliability of search and rescue operations with Bayesian Belief Networks. Reliab. Eng. Sys. Saf. **93**, 940–949 (2008)
5. B. Cai, Y. Liu, Y. Ma, Z. Liu, Y. Zhou, J. Sun, Real-time reliability evaluation methodology based on dynamic Bayesian networks: A case study of a subsea pipe ram BOP system. ISA T **58**, 595–604 (2015)
6. M. Musharraf, D. Bradbury-Squires, F. Khan, B. Veith, S. MacKinnon, S. Imtiaz, A virtual experimental technique for data collection for a Bayesian network approach to human reliability analysis. Reliab. Eng. Sys. Saf. **132**, 1–8 (2014)
7. Y. Zhao, F. Xiao, S. Wang, An intelligent chiller fault detection and diagnosis methodology using Bayesian belief network. Energy Build **57**, 278–288 (2013)
8. A. Bobbio, L. Portinale, M. Minichino, E. Ciancamerla, Improving the analysis of dependable systems by mapping fault trees into Bayesian networks. Reliab. Eng. Sys. Saf. **71**, 249–260 (2001)
9. H. Boudali, J.B. Dugan, A discrete-time Bayesian network reliability modeling and analysis framework. Reliab. Eng. Sys. Saf. **87**, 337–349 (2005)
10. S. Montani, L. Portinale, A. Bobbio, D. Codetta-Raiteri, RADYBAN: A tool for reliability analysis of dynamic fault trees through conversion into dynamic Bayesian networks. Reliab. Eng. Sys. Saf. **93**, 922–932 (2008)
11. G. Bearfield, W. Marsh, Generalising event trees using Bayesian networks with a case study of train derailment. Lect. Notes Comput. Sci. **3688**, 52–66 (2005)
12. C.H. Lo, Y.K. Wong, A.B. Rad, Bond graph based Bayesian network for fault diagnosis. Appl. Soft Comput. **11**, 1208–1212 (2011)
13. M.C. Kim, Reliability block diagram with general gates and its application to system reliability analysis. Ann. Nucl. Energy. **38**, 2456–2461 (2011)
14. N. Khakzad, F. Khan, P. Amyotte, Dynamic safety analysis of process systems by mapping bow-tie into Byesian network. Process Saf. Environ. Prot. **91**, 46–53 (2013)
15. L. Portinale, D.C. Taiteri, S. Montani, Supporting reliability engineers in exploiting the power of Dynamic Bayesian Networks. Int. J. Approximate Reasoning **51**, 179–195 (2010)
16. B. Cai, Y. Liu, Y. Zhang, Q. Fan, S. Yu, Dynamic Bayesian networks based performance evaluation of subsea blowout preventers in presence of imperfect repair. Expert. Sys. Appl. **40**, 7544–7554 (2013)
17. B. Cai, Y. Liu, Q. Fan, Y. Zhang, S. Yu, Z. Liu, X. Dong, Performance evaluation of subsea BOP control systems using dynamic Bayesian networks with imperfect repair and preventive maintenance. Eng. Appl. Artif. Intell. **26**, 2661–2672 (2013)
18. T. Matsuoka, M. Kobayashi, GO-FLOW: A new reliability analysis methodology. Nucl. Sci. Eng. **98**, 64–78 (1988)
19. T. Matsuoka, M. Kobayashi. A phased mission analysis by the GO-FLOW methodology, in *Proceedings of the International ANS/ENS Tropical Meeting Probability, Reliability and Safety Assessment*, 1989, Pittsburgh, USA
20. T. Matsuoka, M. Kobayashi, GO-FLOW methodology: A reliability analysis of the emergency core cooling system of a marine reactor under accident conditions. Nucl. Technol. **84**, 285–295 (1989)
21. J. Yang, M. Yang, Y. Hidekazu, F.Q. Yang, Development of a risk monitoring system for nuclear power plants based on GO-FLOW methodology. Nucl. Eng. Des. **278**, 255–267 (2014)

22. M. Hashim, Y. Hidekazu, M. Takeshi, Y. Ming, Application case study of AP1000 automatic depressurization system(ADS) for reliability evaluation by GO-FLOW methodology. Nucl. Eng. Des. **278**, 209–221 (2014)
23. H. Muhammad, Y. Hidekazu, M. Takeshi, Y. Ming, Common cause failure analysis of PWR containment spray system by GO-FLOW methodology. Nucl. Eng. Des. **262**, 350–357 (2013)
24. T. Matsuoka, M. Kobayashi, The GO-FLOW reliability analysis methodology -analysis of common cause failures with uncertainty. Nucl. Eng. Des. **175**, 205–214 (1997)
25. P.E. Labeau, C. Smidts, S. Swaminathan, Dynamic reliability: Towards an integrated platform for probabilistic risk assessment. Reliab. Eng. Sys. Saf. **68**, 219–254 (2000)
26. P. Weber, G. Medina-Oliva, C. Simon, B. Iung, Overview on Bayesian networks applications for dependability, risk analysis and maintenance areas. Reliab. Eng. Sys. Saf. **25**, 671–682 (2012)
27. P.R. Kannan, Bayesian networks: Application in safety instrumentation and risk reduction. ISA T **46**, 255–259 (2007)
28. R. Gonzalez, B. Huang, E. Lau, Process monitoring using kernel density estimation and Bayesian networking with an industrial case study. ISA T **58**, 330–347 (2015)
29. L.L. Tong, G. Shao, M.F. Wang, X.W. Cao, Safety evaluation of the modified auxiliary feedwater system for the Chinese improved PWR. Ann. Nucl. Energy **70**, 169–174 (2014)
30. S. Kim, B.U. Bae, Y.J. Cho, Y.S. Park, K.H. Kang, B.J. Yun, An experimental study on the validation of cooling capability for the Passive Auxiliary Feedwater System (PAFS) condensation heat exchanger. Nucl. Eng. Des. **260**, 54–63 (2013)
31. J. Pearl, *Probabilistic reasoning in intelligent systems: Networks of plausible inference* (Morgan Kaufmann, San Mateo, 1988)
32. Y.F. Wang, S.F. Roohi, X.M. Hu, M. Xie, Investigations of human and organizational factors in hazardous vapor accidents. J. Hazard. Mater. **191**, 69–82 (2011)

Dynamic Bayesian Network Modeling of Reliability of Subsea Blowout Preventer Stack in the Presence of Common Cause Failures

Abstract A subsea blowout preventer (BOP) stack is used to seal, control, and monitor oil and gas wells. It can be regarded as a series–parallel system consisting of several subsystems. This paper develops the dynamic Bayesian network (DBN) of a parallel system with n components, taking account of common cause failures and imperfect coverage. Multiple-error shock model is used to model common cause failures. Based on the proposed generic model, DBNs of the two commonly used stack types, namely the conventional BOP and modern BOP, are developed. In order to evaluate the effects of the failure rates and coverage factor on the reliability and availability of the stacks, sensitivity analysis is performed.

Keywords Dynamic Bayesian network · Reliability · Subsea blowout preventer · Common cause failures · Imperfect coverage

1 Introduction

A blowout preventer (BOP) is a specialized mechanical device, usually installed redundantly in stacks, used to seal, control, and monitor oil and gas wells. It is developed to deal with extreme erratic pressures and uncontrolled flow gushing from a well reservoir during drilling. BOPs play an important role in the safety of crew, rig, and environment, and they are critical to the monitoring and maintenance of well integrity. Once the BOP fails, kicks or blowout in the process of drilling will lead to serious consequences. For example, the semisubmersible drilling platform Deepwater Horizon in the Gulf of Mexico exploded and sank on April 20, 2010. This tragedy not only caused huge property losses and casualties, but also brought irreparable disaster to the ecological environment of the Gulf of Mexico [1]. One important cause of this accident is that the subsea BOP fails to function. Hence, reliability research of subsea BOP is of significance and it attracts more and more attentions recently.

Several methods have been proposed for reliability analysis of subsea BOP system. Fowler and Roche [2] use failure mode and effects analysis (FMEA) and fault tree analysis (FTA) techniques for reliability analysis of a BOP and a hydraulic control

B. Cai et al., *Bayesian Networks for Reliability Engineering*,
https://doi.org/10.1007/978-981-13-6516-4_4

system. Historical data about subsea BOP failures and malfunctions are collected and estimated by using the FTA method [3, 4]. But, the two methods are only suitable for non-repair systems and lack of time element is their limitation [5]. Besides, FMEA technique cannot differentiate situation of common failures or severe failure caused by compound failures [6]. Owing to their flexibility, Markov methods for performance evaluation of subsea BOP stack configuration and mounting types for control pods [7]. Due to the exponential growth of the state space with the number of components, Markov method is faced with state explosion problem [8]. With Bayesian network (BN), there is no longer such a constraint since the number of parameters within the conditional probabilities table is considerable lower compared to a Markov model [9].

Recently, BNs have been popular for reliability and risk evaluation as a robust and viable alternative to most traditional methods such as fault tree, reliability block diagrams, and so on [10]. Martins and Maturana [11] present a method based on BN for analyzing human reliability and apply this methodology to the operation of an oil tanker, focusing on the risk of collision accidents. Li et al. [12] develop a fuzzy BN approach to improve the quantification of organizational influences in human reliability analysis frameworks. Morales-Napoles and Steenbergen [13] have presented the potential of hybrid BNs for modeling complex data such as the one generated by the weigh-in-motion system in the Netherlands. Doguc and Ramirez-Marquez [14] develop a new method for estimating grid service reliability, which does not need prior knowledge about the grid system structure unlike the previous studies. Daemi et al. [15] use BN for reliability assessment of composite power systems with emphasis on the importance of system components.

If a BN is involved with temporal factors, it is a dynamic network. Static BN can be extended into dynamic Bayesian networks (DBN) by introducing relevant temporal dependencies between representations of the static network at different times, which allows modeling the dynamic behavior of the systems [16]. Thus, DBN is more appropriate for monitoring and predicting values of random variables and capable of representing the system state at any time with respect to BN [17]. Several DBN models have been proposed for assessing reliability of technical systems. Portinale et al. [18] present an approach to reliability modeling and analysis based on the automatic conversion of the dynamic fault tree into DBN, which is implemented in a software tool called RADYBAN [19]. Cai et al. [20] present a quantitative reliability and availability evaluation method for subsea BOP system by translating fault tree into DBN directly, taking account of imperfect repair. Boudali and Dugan [8] propose a new reliability and analysis framework based on the BN formalism, and the method is to investigate timed BNs and to find a suitable reliability framework for dynamic systems.

Subsea BOP stack is composed of annular BOPs, ram BOPs, LMRP connector, and wellhead connector. In order to improve the reliability of the BOP sack, several annular and ram BOPs are used for redundancy. Therefore, it can be regarded as a series–parallel system. This paper presents a method to develop the DBNs for reliability analysis of the subsea BOP stack. With common cause failures and imperfect coverage taken into account, two commonly used types, conventional BOP stack and modern BOP stack, are discussed. Sensitivity analysis is performed to research the influences of failure rates and imperfect coverage on system reliability and availability. The paper is structured as follows: Section 2 describes the subsea BOP stack in detail. Section 3 presents the method to develop DBNs of subsea BOP stacks. Section 4 covers the analytical results and discussions. Section 5 summarizes the paper.

2 Description of Subsea BOP Stack

BOPs come in two basic types, ram and annular. Both are often used together in drilling rig BOP stacks, typically with at least one annular BOP along with several ram BOPs. An annular-type BOP can close around the drill string or casing. Drill pipe including the larger-diameter tool joints (threaded connectors) can be "stripped" (i.e., moved vertically while pressure is contained below) through an annular preventer by careful control of the hydraulic closing pressure. Annular BOPs are also effective at maintaining a seal around the drill pipe even as it rotates during drilling. Regulations typically require that an annular preventer be able to completely close a wellbore, but annular preventers are generally not as effective as ram preventers in maintaining a seal on an open hole. Annular BOPs are typically located at the top of a BOP stack, with one or two annular preventers positioned above a series of several ram preventers. A ram-type BOP is similar in operation to a gate valve, but uses a pair of opposing steel plungers, rams. The rams extend toward the center of the wellbore to restrict flow or retract open in order to permit flow. The inner and top faces of the rams are fitted with packers that press against each other, against the wellbore, and around tubing running through the wellbore. Outlets at the sides of the BOP housing are used for connection to choke and kill lines or valves. Rams are of four common types: pipe, blind, shear, and blind shear. Pipe rams close around a drill pipe, restricting flow in the annulus (ring-shaped space between concentric objects) between the outside of the drill pipe and the wellbore, but do not obstruct flow within the drill pipe. Variable-bore pipe rams can accommodate tubing in a wider range of outside diameters than standard pipe rams, but typically with some loss of pressure capacity and longevity. Blind rams (also known as sealing rams), which have no openings for tubing, can close off the well when the well does not contain a drill string or other tubing, and seal it. Shear rams cut through the drill string or casing with hardened steel shears. Blind shear rams (also known as shear seal rams or sealing shear rams) are intended to seal a wellbore, even when the bore is occupied by a drill string, by cutting through the drill string as the rams close off the well. The upper portion of the severed drill string is freed from the ram, while the lower portion may be crimped and the "fish tail" captured to hang the drill string off the BOP. Two hydraulic connectors are used to connect the BOP stack with the lower marine riser package (LMRP) and wellhead. LMRP connector joins the LMRP to the top of the lower BOP stack, while wellhead connector joins the stack to the subsea wellhead.

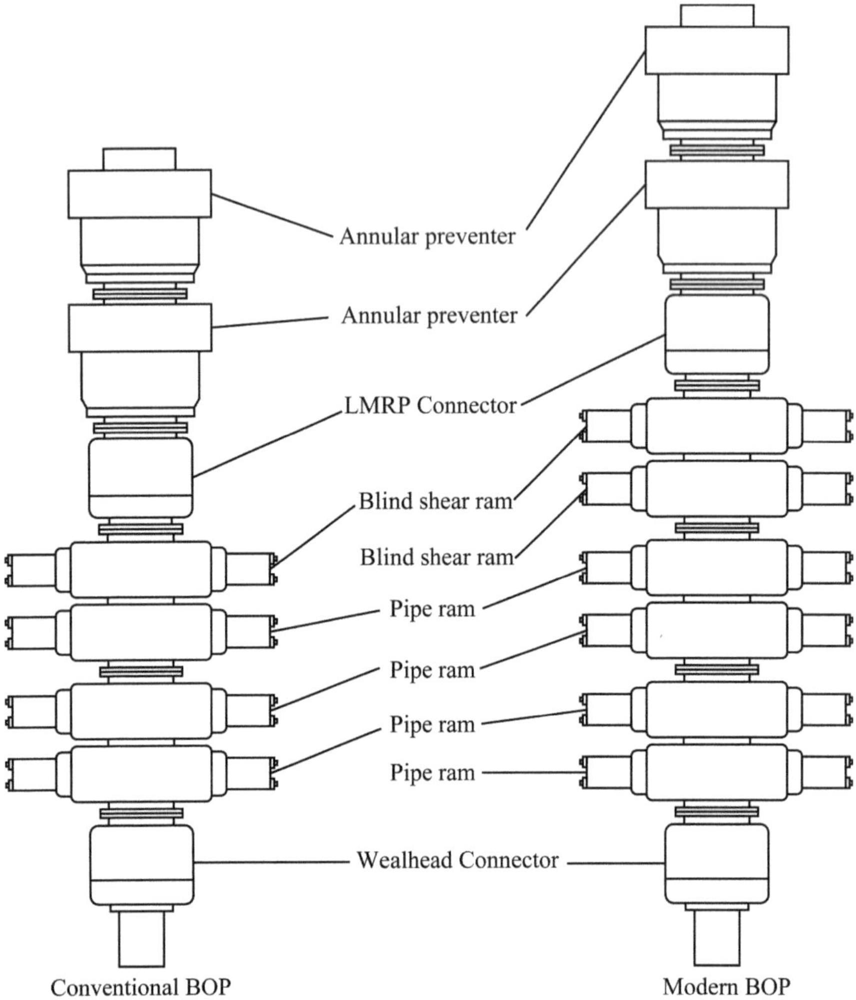

Fig. 1 Typical BOP configuration

Figure 1 demonstrates typical BOP configurations for a conventional and a modern BOP, respectively. However, these are representative sketches of BOPs as configuration may vary from rig to rig. A modern subsea BOP typically has six ram preventers, while a conventional subsea BOP has four ram preventers. As shown in Fig. 1, a conventional BOP configuration has two annular preventers, three pipe ram preventers, one blind shear ram preventer, one LMRP connector, and wellhead connector. A modern BOP configuration has two annular preventers, four pipe ram preventers, two blind shear ram preventers, one LMRP connector, and wellhead connector. Compared with the conventional configuration, the modern BOP has one more pipe ram preventer and one more blind shear ram preventer.

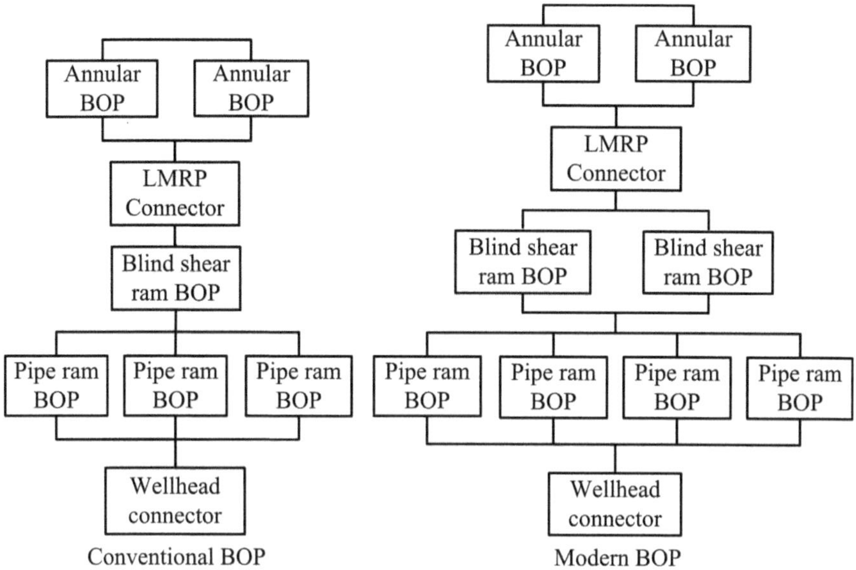

Fig. 2 System structure of the BOP configurations

According to the configuration, a BOP stack can be regarded as a series–parallel system composed of five subsystems, which is shown in Fig. 2. For conventional BOP, annular BOP subsystem is a parallel subsystem with two components and pipe ram BOP subsystem is a parallel subsystem with three components. For modern BOP, annular BOP subsystem and blind shear ram BOP subsystem are parallel subsystems with two components and pipe ram BOP subsystem is a parallel subsystem with four components.

3 Dynamic Bayesian Network Modeling

3.1 Dynamic Bayesian Network

BNs are probabilistic models based on directed acyclic graphs which are used for representing and reasoning with uncertain knowledge [21]. A BN is made up of a set of nodes representing the system variables and directed arcs representing the dependencies or influence among the variables. In BNs, a variable is defined over several mutually exclusive states and a probability is associated to each state. The probabilistic dependences are quantified by a conditional probability table for each node [22]. Each conditional probability table contains the probability of a node, given any possible combination of its parent nodes. Without parent nodes, root nodes only have priori probabilities. The nodes (X_1, …, X_N) in the network are labeled by related random variables. Assuming $\mathrm{Pa}(X_i)$ is the parent node of X_i in the model, the conditional probability distribution of X_i is denoted by $P(X_i|\mathrm{Pa}(X_i))$. The joint probability distribution $P(X_1, \ldots, X_N)$ can be written as Eq. (1).

$$P(X_1, \ldots, X_N) = \prod_{Xi \in \{X_1, \ldots, X_N\}} P(X_i|\mathrm{Pa}(X_i)) \quad (1)$$

BN allows both forward (or predictive) analysis and backward (diagnostic) analysis, where the posterior probability of any set of variables can be calculated.

A DBN extends the static BN by introducing temporal dependencies that describe the dynamic behaviors. DBNs comprise a sequence of time slices, and each slice consists of a static BN describing the system in the corresponding time step. Temporal links between variables in different time slices denote a temporal probabilistic dependence between the variables. A DBN is defined as a pair $(B_1, B_{\rightarrow})$, where B_1 is a BN that defines the prior $P(X_t)$, and $B_{\rightarrow}$ represents the transition probability $P(X_t|X_{t-1})$ as a two-slice temporal BN, which is shown in Eq. (2).

$$P(X_t|X_{t-1}) = \prod_{i=1}^{N} P\left(X_t^i|\mathrm{Pa}\left(X_t^i\right)\right) \quad (2)$$

Herein, X_t^i denotes the ith node at time t ($i = 1,2, \ldots, N$), and $\mathrm{Pa}(X_t^i)$ denotes the parents of X_t^i in the networks. Generally, DBNs compromise two assumptions, namely (i) the system is first-order Markov, which means that the edges between nodes is located in the same time slice or two neighboring time slices; (ii) time-homogeneous, which means that the parameters of the conditional probability distribution are time-invariant. Unrolling the two-slice temporal BN until T time slices, the resulting joint distribution probability can be obtained by Eq. (3).

$$P(X_{1:T}) = \prod_{t=1}^{T} \prod_{i=1}^{N} \left(P(X_t^i|Pa\left(X_t^i\right)\right) \quad (3)$$

3.2 DBN Modeling for Parallel Systems

The whole DBN of a series–parallel system can be developed by integrating the models of subsystems. Hence, DBN of the subsystems is developed firstly. It is easy to develop the model of the subsystem with one component. Besides, it can be regarded as a special case of parallel system, which includes only one component. For parallel system, it will not work until all the components fail. Besides, an important consideration for reliability of the parallel system is the possible occurrence of common cause failures (CCF), where multiple components fail simultaneously for the same underlying cause [23]. Causes of potential CCF might be introduced in design as well as in the operational phase [24]. Due to inadequate understanding of failure mechanisms and responses, improper selections of components, CCF may occur in the design phase. In the operational phase, CCF may be caused by improper testing, human errors, and environmental stresses beyond the design envelope. Several methods have been used to model CCF, such as the β factor model and alpha-factor model [25]. In this paper, multiple-error shock (MESH) model is used. Compared with beta factor model, it can distinguish between failures of two, three, or more units [6].

The objective of the MESH model is to calculate rates of one, two, three, or more failures per stress event. These failure rates $\lambda(n)$ are defined as the failure rate where n units fail per stress event. P_n is the probability that n units will fail per stress event. The stress event rate is denoted by ν. The relationship between stress event rate and individual failure rate is given by:

$$\nu = n\lambda/M \tag{4}$$

where M is the average number of units failed per stress event. It can be calculated by the following equation:

$$M = \sum_{i=1}^{n} nP_n \tag{5}$$

Once M is obtained, the stress event rate can be calculated by:

$$\lambda(n) = \nu P_n \tag{6}$$

For a parallel subsystem consisting of n components, there are n types of failures, namely from one unit to n units fail per stress event. The number of cases N_i of each type is given by:

$$N_i = \frac{n(n-1)\cdots(n-i+1)}{1 \times 2 \times \cdots \times i}, \quad i = 1, 2, \ldots, n. \tag{7}$$

Define the cases of each failure type as the basic events in the DBN model. Then, the total number of basic events associated with the modeling is:

$$N_{\text{Total}} = \sum_{i=1}^{n} N_i \tag{8}$$

In this paper, Netica software is used to develop the DBN models. Netica designed by NORSYS Software Corporation is a powerful, easy-to-use, complete program for working with Bayesian belief networks and influence diagrams. It has an intuitive

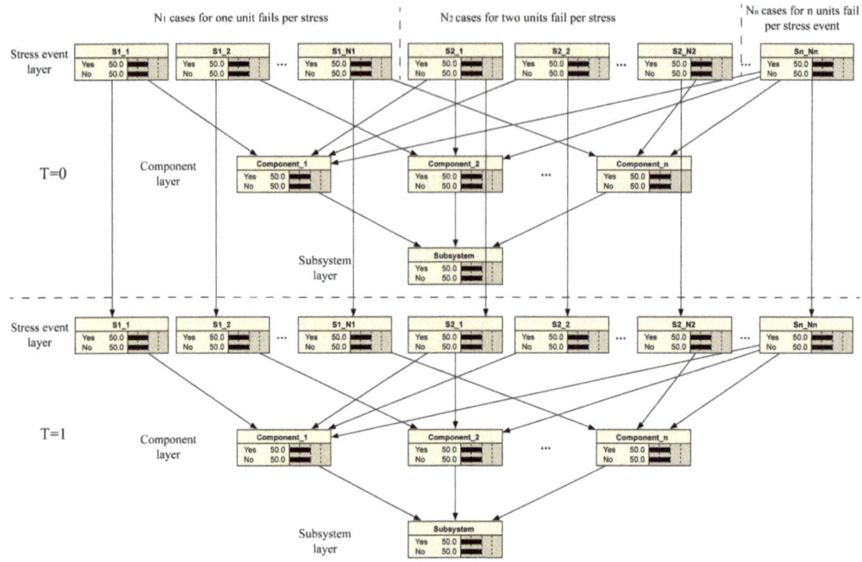

Fig. 3 DBN of a parallel subsystem with n components

and smooth user interface for drawing the networks, and the relationships between variables may be entered as individual probabilities, in the form of equations, or learned from data files. With CCF, the DBN of a subsystem with n components is developed in Fig. 3. It shows that the model consists of three layers, namely subsystem layer, component layer, and stress event layer. Each node in the layers has two states: Yes and No. "Yes" denotes the normal state, and "No" denotes the failure state. In the stress event layer, nodes represent cases of different types of failures. For example, node S1_1 denotes the case that Component1 fails caused by the stress event. The number of cases for one unit fails per stress event is N1, which can be calculated by Eq. (7). In the component layer, nodes denote the states of components, which are dependent on the parent nodes in the stress event layer. Node "subsystem" in the subsystem layer fails when all the component nodes fail in the component layer. Figure 3 shows two time slices and the inter-slice arcs between variables in different time slices denote a temporal probabilistic dependence between the variables.

After the structure is developed, parameters of the DBN will be established. Firstly, conditional probability of the nodes in stress event layer is needed. As mentioned above, the system is first-order Markov. Therefore, the state transitions of the nodes associated with temporal links follow Markov process. Failure rate and repair rate of the component are denoted by λ and μ, respectively. Assume the current time is t and the time interval between two time slices is Δt. Then, the transition probability of the nodes with temporal links between two time slices is given by [26]:

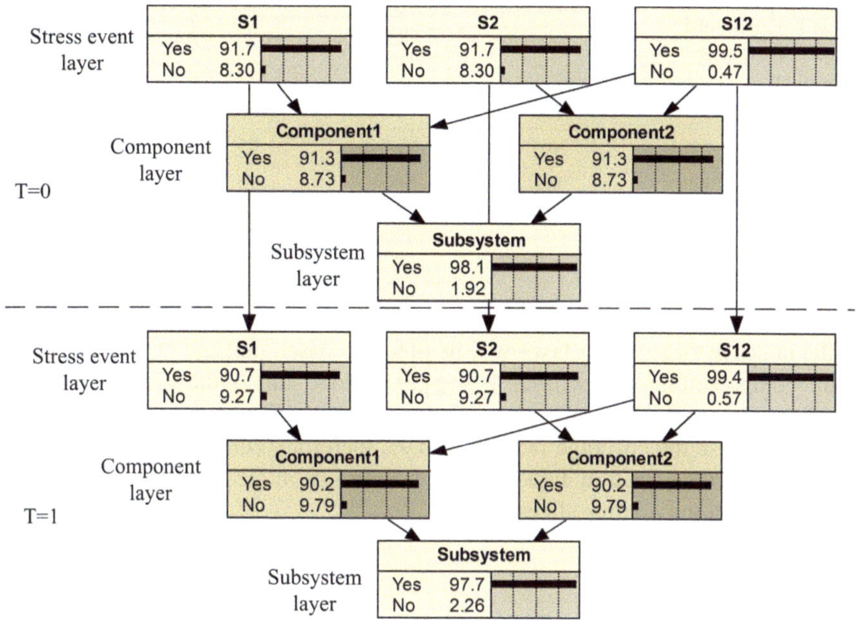

Fig. 4 DBN of a simple case

$$\begin{cases} P(X_i(t+\Delta t) = \text{Yes}|X_i(t) = \text{Yes}) = \mathrm{e}^{-\lambda\Delta t} \\ P(X_i(t+\Delta t) = \text{No}|X_i(t) = \text{Yes}) = 1 - \mathrm{e}^{-\lambda\Delta t} \\ P(X_i(t+\Delta t) = \text{Yes}|X_i(t) = \text{No}) = 1 - \mathrm{e}^{-\mu\Delta t} \\ P(X_i(t+\Delta t) = \text{No}|X_i(t) = \text{No}) = \mathrm{e}^{-\mu\Delta t} \end{cases} \tag{9}$$

Secondly, conditional probability of the nodes in the component layer is needed. Any failure state of the parent nodes in the stress event layer will lead to the failure of child nodes in the component layer. For example, componet_1 will be in the state "Yes", when all its parent nodes are in state "Yes". Lastly, conditional probability of the node subsystem is established. As the components are connected in parallel, the subsystem will fail when all the components fail. The imperfect coverage and CCF are two important factors in reliability issues [27]. Due to the uncertainty, sometimes the system might not be able to recover from the occurrence of a fault. The fault coverage value crucially affects the dependability of a system. Thus, fault coverage is one of the most critical factors in reliability evaluation and probabilistic safety assessment [28]. A coverage factor c is defined, which can be expressed as $c =$ P (system recovers | fault occurs) [29].

A case of parallel system with two components is given to illustrate the DBN modeling process. Hence, there are two types of failures in total. Based on Eq. (7), $N_1 = 2$ and $N_2 = 1$, so the number of stress events is determined. DBN of the case is developed in Fig. 4. λ(1) and λ(2) can be calculated according to Eqs. (4)–(6),

Table 1 Conditional probability of node Subsystem

Component1	Component2	Subsystem	
		Yes (%)	No (%)
Yes	Yes	100	0
No	Yes	95	5
Yes	No	95	5
No	No	0	100

where $\lambda = 1.0e{-}4$, $P_1 = 0.95$, $P_2 = 0.05$. If $\Delta t = 1000\ h$, conditional probability of the nodes in stress event layer will be obtained based on Eq. (9). For each node in the component layer, it will fail if any parent node fails in the stress event layer. Conditional probability of the nodes Component1 and Component2 can be determined. Suppose the coverage factor $c = 0.95$, therefore, conditional probability of subsystem node is listed in Table 1. As shown in Fig. 4, all the probabilities of the DBN are listed in two time slices.

3.3 DBN Modeling of Subsea BOP Stacks

As the BOP stack is a series–parallel system, its DBN model can be developed by integrating the DBNs of the subsystems, which are established using the modeling method described in the above section. DBNs of the conventional and modern BOP stacks are developed in Figs. 4 and 5, respectively. As shown in Fig. 5, nodes Shear_BOP, LMRP_Connector, Wellhead_Connector, PipeRam_BOP, and Annular_BOP denote the states of blind shear ram BOP, LMRP connector, wellhead connector, pipe ram BOP subsystem, and annular BOP subsystem. Any failure of the five nodes will lead to the failure of the stack. For annular BOP subsystem, node Annular_S1 represents the event that leads to the failure of node Annular1 and Annular_S12 denotes the event that leads to the failure of nodes Annular1 and Annular2 at the same time. Similar definitions are applied to Fig. 6. As shown in Fig. 5, DBN of the pipe ram subsystem has seven basic events, which is the same as the value calculated by Eq. (8).

Failure rates and repair rates of the stack are obtained listed in Table 2. These values are obtained based on the collected reliability data and reference review [4]. Conditional probability of the nodes associated with temporal links is established according to Eq. (9). For the parallel subsystems, imperfect coverage is considered and the coverage factor is set to 0.95 in this paper. It means that the probability of the system recovering from the occurrence of a fault is 95%. But if all the components fail, the system cannot recover.

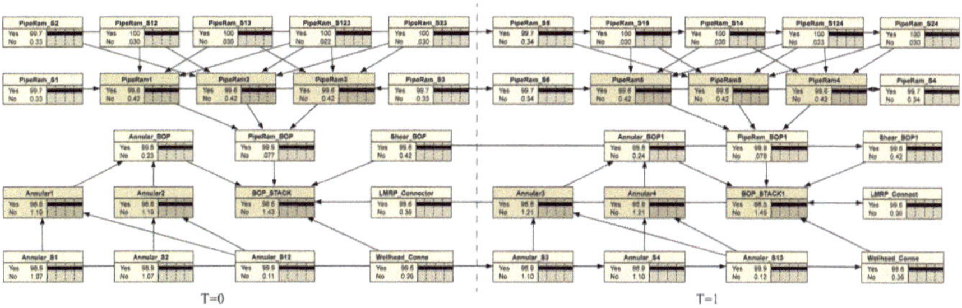

Fig. 5 Extended DBN of the conventional BOP stack form 0th to 1st

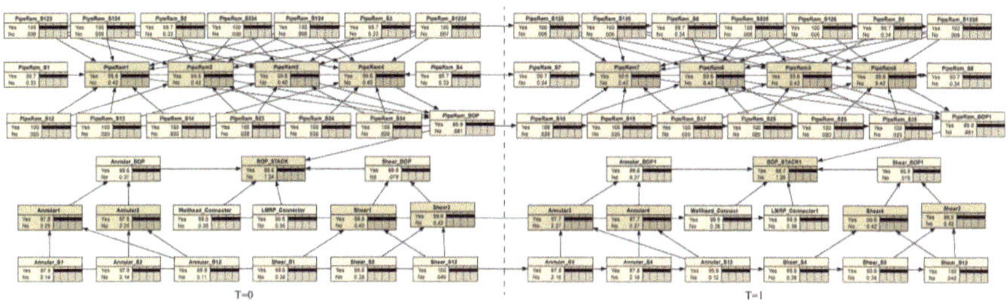

Fig. 6 Extended DBN of the modern BOP stack form 0th to 1st

Table 2 Failure rates and repair rates of the components

Component	Failure rate (/h)	Repair time (h)
Annular BOP	3.552e−5	98
Pipe ram BOP	1.240e−5	77
Shear ram BOP	1.240e−5	77
LMRP connector	1.070e−5	80
Wellhead connector	1.070e−5	80

4 Results and Discussions

4.1 Reliability and Availability

Figure 7 shows reliability of the two types of stacks. As indicated, as time progresses, the reliability decreases. The reliability drops to 0.7910 and 0.82943 for conventional BOP and modern BOP in the 30th week, respectively. Through calculation, the two types of BOP have similar availability.

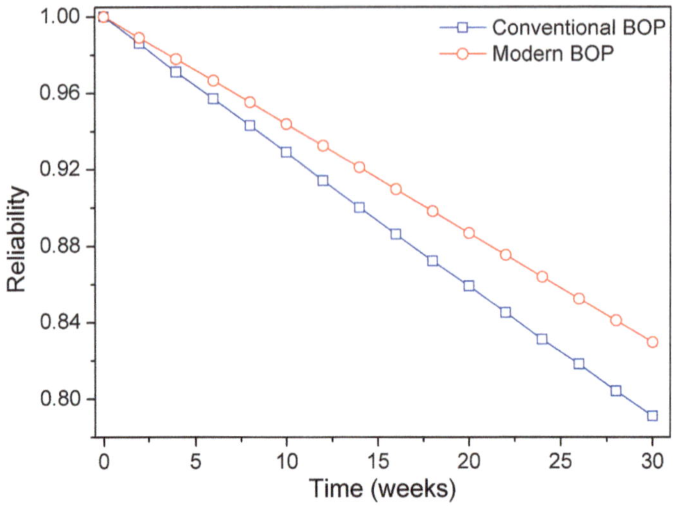

Fig. 7 Reliability of the conventional BOP and modern BOP

4.2 *Sensitivity Analysis*

Because the parameters of the DBN are calculated based on the failure rates, repair rates, and coverage factor, sensitivity analysis is performed. Sensitivity analysis assumes that input parameters, such as failure rates, are not accurate and show the designer the variation of the input parameters. Reliability and availability of the two types of stacks at the given time of 10th week are computed, assuming that the each failure rate is subject to an uncertainty of $\pm 20\%$. The upper and lower bounds of the reliability and availability of the conventional BOP are plotted in Figs. 8 and 9, respectively. It can be seen that failure rates of the components have similar influences on the reliability and availability of the system. Failure rate of pipe ram BOP has the slightest influence on reliability and availability, because pipe ram BOP subsystem is a parallel system with three components. Failure rate of shear ram BOP has the greatest influence, because it is not a parallel system. Failure rates in order of influence degree to the reliability and availability are: Shear ram BOP > LMRP connector = wellhead connector > annular BOP > pipe ram BOP.

Effects of failure rates on the reliability and availability of the modern BOP are shown in Figs. 10 and 11, respectively. Because one more blind shear ram BOP is used in the modern BOP, its influence on reliability and availability has decreased, compared with conventional BOP. Failure rates of LMRP connector and wellhead connector have the greatest influence on reliability and availability. As parallel system with two components, annular BOP has greater influence than blind shear ram BOP because its failure rate is higher.

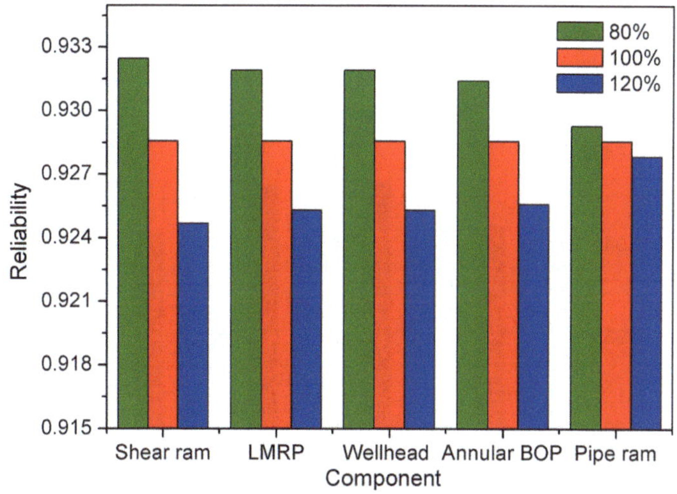

Fig. 8 Effects of failure rates on reliability of conventional BOP

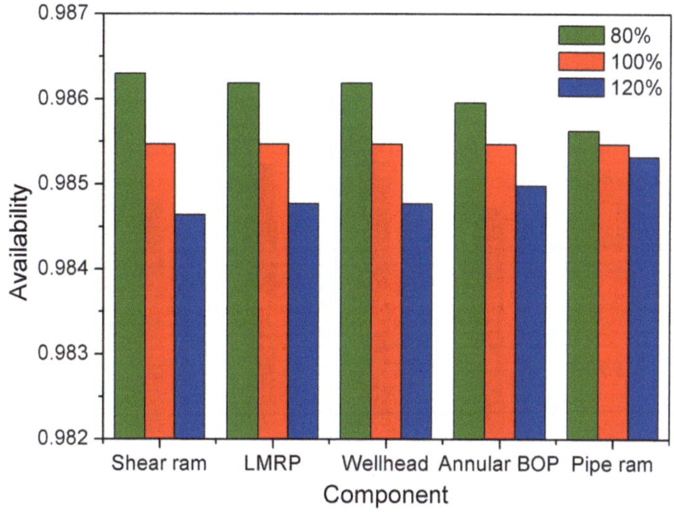

Fig. 9 Effects of failure rates on availability of conventional BOP

Finally, the effects of coverage factor are shown in Fig. 12. As coverage factor increases, reliability and availability of the system increase. It shows that the coverage factor has greater influence on reliability than on availability. Compared with conventional BOP, modern BOP is more easily influenced by the coverage factor.

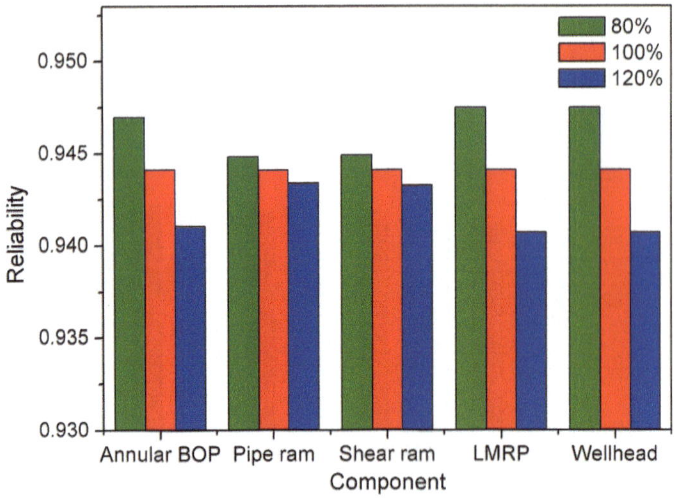

Fig. 10 Effects of failure rates on reliability of modern BOP

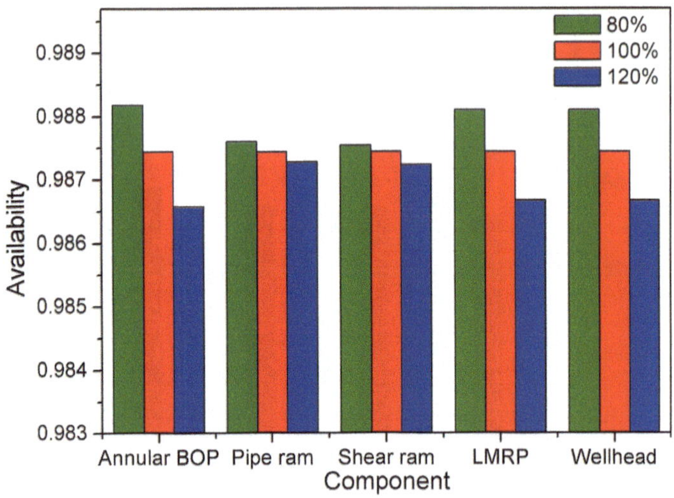

Fig. 11 Effects of failure rates on availability of modern BOP

5 Conclusions

In this paper, a generic DBN of a parallel system with n components is proposed, taking into account CCF and imperfect coverage. MESH model is employed to model the CCF. DBNs of two commonly used BOP types, namely conventional BOP and modern BOP, are developed. The proposed model can also be applied to develop the DBNs of other series–parallel systems composed of several subsystems.

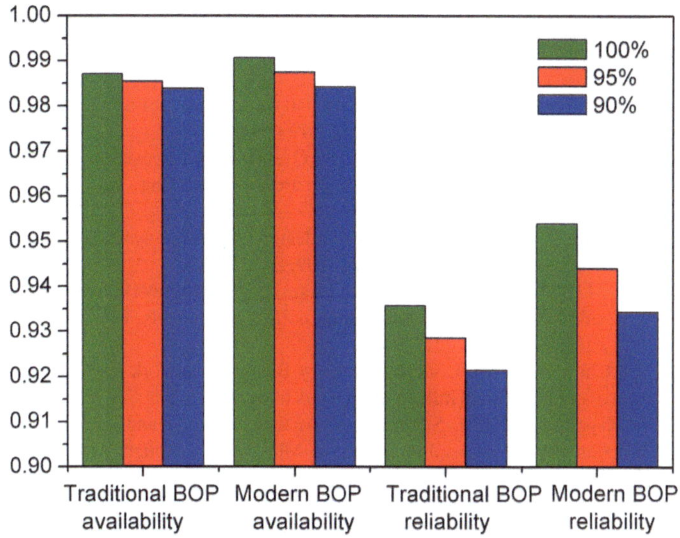

Fig. 12 Effects of coverage factor on availability and reliability

(1) Based on the developed DBNs, availability and reliability of the conventional BOP and modern BOP are obtained. Compared with conventional BOP, modern BOP has higher reliability and similar availability.
(2) Sensitivity analysis is performed to evaluate the effects of the failure rates and coverage factor. For the conventional BOP, failure rate of blind shear ram BOP has the greatest influence on the reliability and availability. As one more shear ram BOP is used in modern BOP, failure rate of connectors has the greatest influence on system performance.
(3) Coverage factor has greater influence on reliability than on availability. Compared with the conventional BOP, the modern BOP is more easily influenced by the coverage factor.

References

1. J.E. Skogdalen, I.B. Utne, J.E. Vinnem, Developing safety indicators for preventing offshore oil and gas deepwater drilling blowouts. Saf. Sci. **49**, 1187–1199 (2011)
2. J.H. Fowler, J.R. Roche, System safety analysis of well-control equipment. SPE Drill Complet **3**, 193–198 (1994)
3. P. Holand, M. Rausand, Reliability of subsea BOP systems. Reliab. Eng. **19**, 263–275 (1987)
4. P. Holand, H. Awan, Reliability of deepwater subsea BOP systems and well kicks. ExproSoft Report ES, 52/02. (Unrestricted version) (2012)
5. N. Sadou, H. Demmou, Reliability analysis of discrete event dynamic systems with Petrinets. Reliab. Eng. Syst. Safety **94**, 1848–1861 (2009)
6. W.M. Globe, *Control systems safety evaluation and reliability*, 3rd edn. (ISA, North Carolina, 2010)

7. B.P. Cai, Y.H. Liu, Z.K. Liu, X.J. Tian, Y.Z. Zhang, J. Liu, Performance evaluation of subsea blowout preventer systems with common-cause failures. J. Petrol. Sci. Eng. **90–91**, 18–25 (2012)
8. H. Boudali, J.B. Dugan, A discrete-time Bayesian network reliability modeling and analysis framework. Reliab. Eng. Syst. Safety **87**, 337–349 (2005)
9. P. Weber, G. Medina-Oliva, C. Simon, B. Iung, Overview on Bayesian networks applications for dependability, risk analysis and maintenance areas. Reliab. Eng. Syst. Safety **25**, 671–682 (2012)
10. N. Khakzad, F. Khan, P. Amyotte, Risk-based design of process systems using discrete-time Bayesian networks. Reliab. Eng. Syst. Safety **109**, 5–17 (2013)
11. M.R. Martins, M.C. Maturana, Application of Bayesian Belief networks to the human reliability analysis of an oil tanker operation focusing on collision accidents. Reliab. Eng. Syst. Safety **110**, 89–109 (2013)
12. P.C. Li, G.H. Chen, L.C. Dai, L. Zhang, A fuzzy Bayesian network approach to improve the quantification of organizational influences in HRA frameworks. Saf. Sci. **50**, 1569–1583 (2012)
13. O. Morales-Napoles, R.D.J.M. Steenbergen, Analysis of axle and vehicle load properties through Bayesian networks based on Weigh-in-Motion data. Reliab. Eng. Syst. Safety **125**, 153–164 (2014)
14. O. Doguc, J.E. Ramirez-Marquez, An automated method for estimating reliability of grid systems using Bayesian networks. Reliab. Eng. Syst. Safety **104**, 96–105 (2012)
15. T. Daemi, A. Ebrahimi, M. Fotuhi-Firuzabad, Constructing the Bayesian network for components reliability importance ranking in composite power systems. Electr. Power Energy Syst. **43**, 474–480 (2012)
16. S.F. Galán, F. Aguadob, F.J. Díeza, J. Miraa, NasoNet, modeling the spread of nasopharyngeal cancer with networks of probabilistic events in discrete time. Artif. Intell. Med. **5**, 247–264 (2002)
17. P. Weber, L. Jouffe, Reliability modelling with dynamic bayesian networks. In: Fifth IFAC symposium on fault detection, supervision and safety of technical processes (SAFEPROCESS'03), pp 57–62, Washington, D.C., USA (2003)
18. L. Portinale, D.C. Ratiteri, S. Montani, Supporting reliability engineers in exploiting the power of dynamic Bayesian Networks. Int. J. Approximate Reasoning **51**, 179–195 (2010)
19. S. Montani, L. Portinale, A. Bobbio, D. Codetta-Raiteri, RADYBAN: A tool for reliability analysis of dynamic fault trees through conversion into dynamic Bayesian networks. Reliab. Eng. Syst. Safety **93**, 922–932 (2008)
20. B. Cai, Y.H. Liu, Y.W. Zhang, Q. Fan, S.L. Yu, Dynamic Bayesian networks based performance evaluation of subsea blowout preventers in presence of imperfect repair. Expert Syst. Appl. **40**, 7544–7554 (2013)
21. P.A.P. Ramírez, I.B. Utne, Using dynamic Bayesian networks for life extension assessment of aging systems. Reliab. Eng. Syst. Safety **133**, 119–136 (2015)
22. O. Arsene, I. Dumitrache, I. Mihu, Medicine expert system dynamic Bayesian network and ontology based. Expert Syst. Appl. **38**, 15253–15261 (2011)
23. F.P.A. Coolen, T. Coolen-Maturi, Predictive inference for system reliability after common-cause component failures. Reliab. Eng. Syst. Safety **135**, 27–33 (2015)
24. M.A. Lundteigen, M. Rausand, Common cause failures in safety instrumented systems on oil and gas installations: implementing defense measures through functions testing. J. Loss Prev. Process Ind. **20**, 218–229 (2007)
25. H. Jin, M. Rausand, Reliability of safety-instrumented systems subjected to partial testing and common-cause failures. Reliab. Eng. Syst. Safety **121**, 146–151 (2014)
26. T. Kohda, W. Cui, Risk-based reconfiguration of safety monitoring system using dynamic Bayesian network. Reliab. Eng. Syst. Safety **92**, 1716–1723 (2007)
27. S.J. Kim, P.H. Seong, J.S. Lee, M.C. Kim, H.G. Kang, S.C. Jang, A method for evaluating fault coverage using simulated fault injection for digitalized systems in nuclear power plants. Reliab. Eng. Syst. Safety **91**, 614–623 (2006)

28. B.G. Kim, H.G. Kang, H.E. Kim, S.J. Lee, P.H. Seong, Reliability modeling of digital component in plant protection system with various fault-tolerant techniques. Nucl. Eng. Des. **265**, 1005–1015 (2013)
29. J.B. Duagn, K. Trivedi, Coverage modeling for dependability analysis of fault-tolerant systems. IEEE Trans. Comput. **38**, 775–787 (1989)

Reliability Evaluation Methodology of Complex Systems Based on Dynamic Object-Oriented Bayesian Networks

Abstract A novel reliability evaluation methodology of complex systems is proposed using dynamic object-oriented Bayesian networks (DOOBNs). This modeling methodology consists of two main phases, namely construction phases for object-oriented Bayesian networks (OOBNs) and for DOOBNs. In the first phase, the network fragments with similar structures and parameters are divided into classes; these classes are then encapsulated. The construction of OOBNs is completed according to the relationship among the encapsulated classes. In the second phase, every fragment of the dynamic Bayesian networks (DBNs) that was constructed by the first phase is encapsulated as a class called DOOBN. The construction of DOOBNs is completed according to the relationship among the time fragments. The accuracy of this methodology is validated using the all-series, all-voting, voting-after-series, series-after-voting, parallel-after-series, and series-after-parallel systems. This methodology is further illustrated and verified by using deepwater blowout preventer system and can model the system from global to local levels, thereby effectively reducing modeling difficulty and adopting efficient arithmetic reasoning algorithms.

Keywords Complex system · DOOBNs · Reliability evaluation · Availability

1 Introduction

Modern industry systems are becoming increasingly complex and large. Thus, an increase in faults significantly affects the reliability and availability of these systems. Complex industry systems include copies of similar or even nearly identical subsystems. For example, a subsea production system consists of numerous similar subsea control systems, and each subsea control system includes many similar feedback control loops [1]. Therefore, a reliability and availability evaluation method for this type of complex industry system should be developed [2].

Bayesian network (BN) is an important probabilistic graphical model that can effectively handle various uncertainty problems based on probabilistic information representation and inference [3]. BN is relatively attractive as a representational tool for three reasons. First, BN is consistent and completely represents and defines a

B. Cai et al., *Bayesian Networks for Reliability Engineering*,
https://doi.org/10.1007/978-981-13-6516-4_5

unique probability distribution over network variables. Second, the network is modular, hence its consistency and completeness are ensured by using localized tests, which are only applicable to variables and their direct causes. Third, BN is a compact representation because it allows specifying exponentially sized probability distribution using a polynomial of probabilities [4]. Thus, this tool is intensely researched and extensively used in the domains of reliability, risk, and fault diagnoses [5–7]. For example, Cai et al. [8] used BN models to perform the reliability evaluation of subsea blowout preventer (BOP) control systems. Tien et al. [9] developed novel algorithms for the modeling and reliability assessment of infrastructure systems using BNs. Barua et al. [10] presented a BN-based operational risk assessment methodology for dynamic systems. Daemi [11] proposed a BN-based reliability assessment methodology of composite power systems by emphasizing the importance of system components. Moreover, Wu et al. [12] proposed a probabilistically analysis method for natural gas pipeline network (NGPN) accidents by using BNs. The combination of BNs and Dempster–Shafer evidence theory was alternatively used to evaluate NGPN accidents. Bhandari et al. [13] investigated the applicability of BNs in conducting a dynamic safety analysis of deepwater managed pressure drilling operations and underbalanced drilling. Important factors were addressed to distinguish the importance of identifying each cause of potential accidents.

BNs are appropriate because evaluating the reliability of objects in real time excludes temporal features. However, if the reliability is predicted in the future, then temporal features are involved, and dynamic BNs (DBNs) are required. For example, Cai et al. [14] used DBNs to analyze the reliability of a series, parallel, and two-out-of-three (2003) voting systems given the common cause failure, imperfect coverage, imperfect repair, and preventive maintenance; furthermore, these researchers presented a DBN-based quantitative risk assessment method that specifically considers human factors on offshore blowouts [15]. Luque and Straub [16] presented a modeling and computational framework to study the reliability of deteriorating structural systems using DBNs. Zhu and Collette [17] proposed a novel scheme based on the DBNs to estimate the low-probability failure event. This new method compensates for the deficiency of traditional methods, which cannot accurately track low-probability failure events when discretizing certain types of continuous random variables is required. Cai et al. [18] introduced a novel safety integrity level determination methodology for safety-instrumented systems by using multiphase DBNs. Khakzad [19] proposed the application of DBN to the risk analysis of domino effects in chemical infrastructures to model the spatial and temporal evolutions of domino effects, which are characterized by low-frequency, high-consequence chain of accidents, and to quantify the most probable sequence of accidents in a potential domino effect. Foulliaron et al. [20] introduced several specific DBN structures to improve degradation modeling and perform reliability analysis.

If an evaluated object is excessively complex for modeling using BNs, especially when the object consists of collections of identical or similar components, then establishing a BN-based reliability evaluation model is difficult and tedious. Alternatively, object-oriented BNs (OOBNs) are suitable tools in evaluating the reliability of objects with large, complex, and hierarchical structures [21]. For example, Cai et al. [22] proposed a real-time fault diagnosis methodology for complex systems with repetitive structures by using OOBNs; this methodology can reduce the overall complexities of BNs for fault diagnosis and report the faults that occur immediately. Weber and Jouffe [23] presented a methodology that can help develop dynamic OOBNs given the increasing requirement for optimizing the diagnosis and maintenance policies. Jensen et al. [24] used BNs to estimate the risk indexes of leg disorders for finisher herd. An object-oriented structure was used to simplify the specification because the information originated from two levels. Weidl et al. [25] developed a methodology that uses generic OOBNs for digesting and screening a pulp to meet the requirement of the diagnosis processes in plant operation and maintenance. Liu et al. [26] used an extended OOBN as the underlying mathematical tool for presenting a modeling risk management process in a large, complex, and dynamic system.

The concepts of "dynamic" and "object-oriented" should be used to solve the problems of degraded components and repetitive systems in reliability evaluation. This work proposes a reliability evaluation methodology of complex systems based on DOOBNs. The remainder of this paper is structured as follows: Sect. 2 presents the DOOBN-based complex system reliability evaluation methodology. Section 3 discusses the reliability of a series, parallel, and 2003 voting system. Section 4 analyzes a case of subsea BOP system to demonstrate the applicability the proposed DOOBN modeling. Section 5 summarizes the paper.

2 Proposed Reliability Evaluation Method for Complex Systems

A. Overview of BNs, OOBNs, and DBNs

BNs, or static BNs, are probabilistic-directed acyclic graphical models that use nodes to represent a random variable, arcs to signify direct dependencies between the linked nodes, and conditional probabilities to quantify dependencies. Static BNs are most extensively used in the field of reliability evaluation because many monographs have introduced BNs in detail. However, this introduction is not repeated in this paper.

Similar network segments can be observed constantly in complex BN systems. In Fig. 1, all the state space of A and B are similar, and C, D, and E demonstrate similar conditional probability distributions, thereby indicating that Inst. 1–4 are similar network segments. The OOBNs use the concept of classes and objects in programming ideas. A class is a generic network fragment, that is, a BN; an object is a fragment generated by instantiating the class. The nodes in a class or object can be classified into three categories, namely input, output, and encapsulated. Input and output nodes are considered object interfaces. OOBNs provide an approach to achieving a hierarchical representation of the model when an object is encapsulated in other objects, and each level corresponds to a particular level of abstraction, thereby revealing the encapsulated nodes for the current layer of the object.

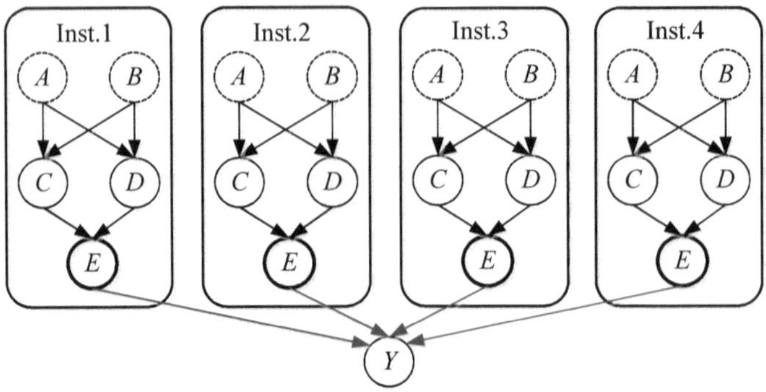

Fig. 1 Typical OOBN

DBNs combine the static BN with time information, thereby forming a random data model. The time step of each model is called a time slice. The structure is easy to find, and the parameters of each time slice in a DBN are exactly the same. The difference is the metabolic change in the time of the node parameters, which is suitable for the application of the OOBN. Figure 2a illustrates a simple BN with two slices. Each time slice is considered a class. Figure 2b displays that nodes A_{pre} and B_{pre} (added to nodes A and B) varied with time. A and B are input nodes, whereas A_{pre} and B_{pre} are output nodes. Input nodes are not real nodes and only act as placeholders. The state space and father nodes of input nodes demonstrate a one-to-one correspondence relationship. Output nodes have the features of ordinary nodes. The network in Fig. 2a with applied OOBN is presented in Fig. 2c. The parameters of A_{pre} and B_{pre} in Inst. 2 are determined by nodes A and B in Inst. 1, respectively. The OOBNs apply the concept of irrelevant information to be simplified in visual perception and modeling operations.

B. DOOBNs for reliability evaluation

Repeated segments for the reliability evaluation of complex dynamic systems for BN models not only exist in static BNs but also in the entire DBNs. That is, many

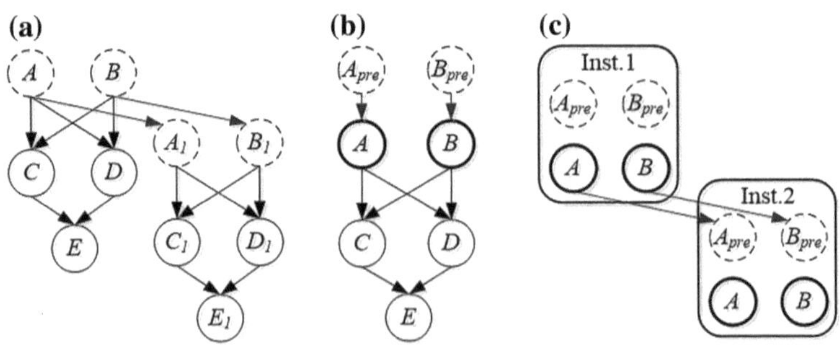

Fig. 2 Concept of an OOBN used in a DBN: **a** DBN, **b** class, and **c** OOBN

network segments are repeated in a single time slice, whereas the entire time slice is considered a repeated network segment for the entire complex system. Therefore, DOOBNs are developed to describe this type of complex dynamic systems.

The modeling process of the methodology has two steps. The first step is finding the same or similar fragments in a static BN. Encapsulated irrelevant information is subsequently defined as a class according to the relationship between a class and other segments of a network to establish the static OOBNs. In Fig. 3, the network segment of C1–C2 is encapsulated as the object, and then the static network is established. The second step is observing at the entire time slice. Time0 is a class, and all of the time slices belong to this class. A secondary node should be added before the nodes change over time because an encapsulated node cannot change. Figure 3 also illustrates that A1I1 and A1O1 are considered the input and output nodes in the DOOBN nodes, respectively. Node A is only a placeholder. Its value is determined by its parent node. Ultimately, the entire OOBN is established according to the transfer conditional probability tables (CPTs) of each time slice.

In the proposed DOOBN-based reliability evaluation model, the root nodes represent the state of components in an evaluated system, and the final leaf node represents the state of the entire system. Many intermediate nodes are set between the root and the final leaf nodes to simplify the networks. The structures and parameters of the networks are established based on the causal relationship of the components and system state, that is, reliability and availability. The failure and repair rates of the components directly affect their reliability and availability. Moreover, these components are modeled using the CPT. The reliability and availability of the system can be calculated by using suitable inference algorithms, including exact and approximate. Inference can be performed using commercial or free software, such as Hugin.

DOOBNs provide the following advantages by comparing general BNs and DBNs. First, the DOOBNs support top-down model construction process. Second, DOOBNs are constructed by integrating minimal and understandable network fragments, thereby benefiting knowledge acquisition and communication between modelers and domain experts. Third, this approach reduces the complexity of building BNs and improves the reusability of models. Finally, DOOBNs exhibit a high average rate of convergence and time efficiency because of encapsulation and hierarchy.

3 Reliability Evaluation of Series, Parallel, and 2003 Voting Systems

Series, parallel, and voting systems are commonly used for demonstrating reliability evaluation methodology. In the current work, the series, parallel, and voting systems with three similar components are combined to form new systems, as depicted in Fig. 4. The BN model of these systems is displayed in Fig. 5, and their reliability and availability are calculated to verify the accuracy of the proposed method. This model can be used to represent the all-series, all-voting, voting-after-series, series-

Fig. 3 Modeling method of a DOOBN

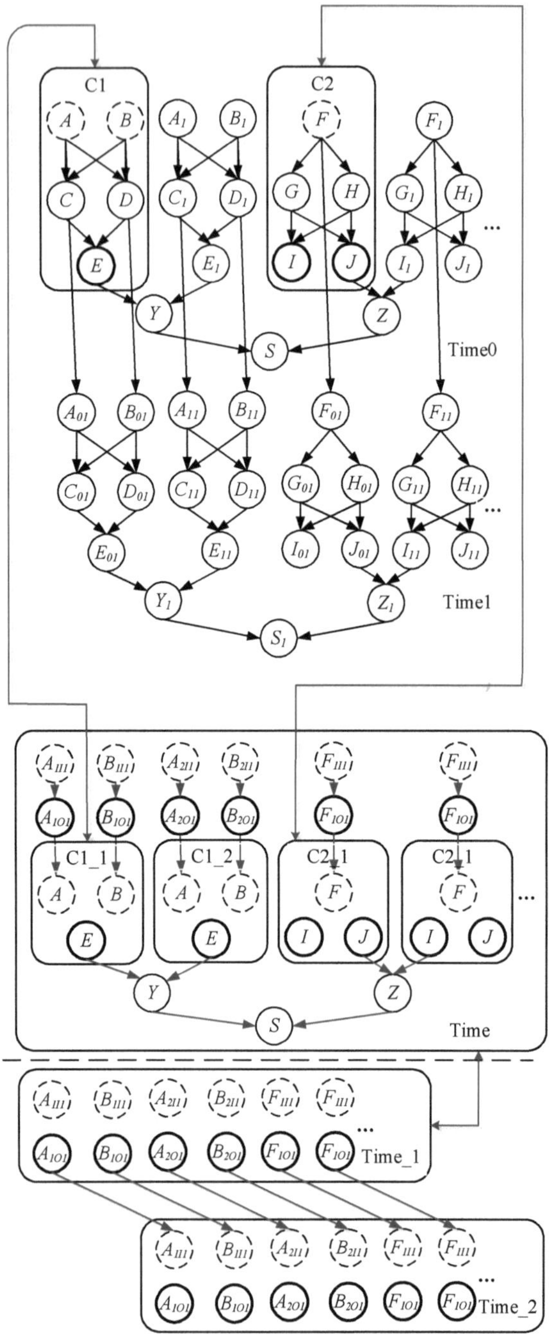

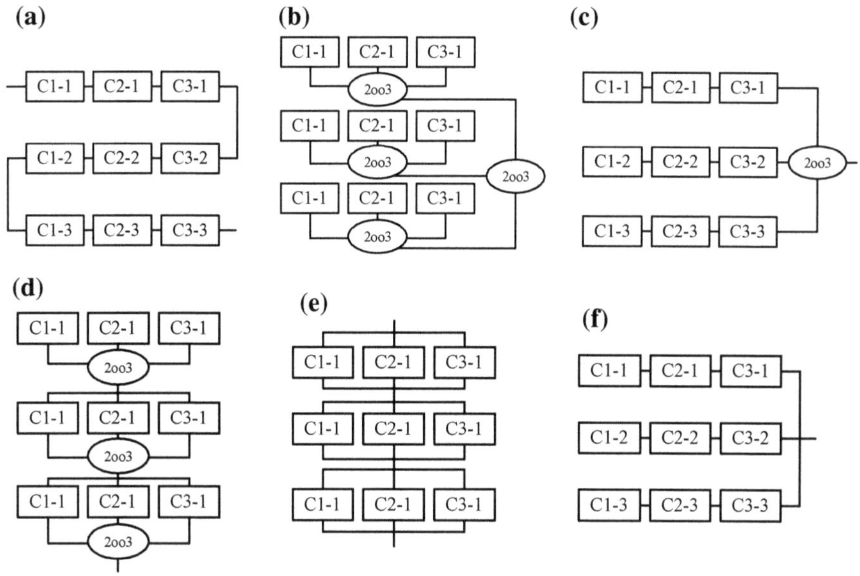

Fig. 4 Reliability block diagram of series, parallel, and voting systems. **a** All-series system, **b** all-voting system, **c** voting-after-series system, **d** series-after-voting system, **e** series-after-parallel system, and **f** parallel-after-series system

after-voting, parallel-after-series, and series-after-parallel systems by changing the relationships of C1_X, C2_X, and C3_X (X represents 1, 2, and 3) and of C4_1, C4_2, and C4_3. The two states of the root and intermediate nodes are "No" and "Yes," while the two states of node "C_System," which represents the entire system, are "Work" and "Fail." Each network is similar in structure but different in the relationship between parent and child nodes, which is represented by CPT.

A. DOOBN-based reliability evaluation for systems with exponential distribution

In Fig. 6, the BN (Time_1) is extended to the DBN (Time_2). The CPT is summarized in Table 1. The root nodes are under exponential distribution; hence, the removal of CPT has no correlation. The root node is determined by Δt.

Reliability and availability are important indicators for evaluating industrial systems. Reliability is for non-repairable systems, and availability is for repairable systems. Figure 7 illustrates the reliability and availability curves of the six systems in 140 weeks. The three figures depict that the availabilities decreased rapidly and reached stable levels at different times in various systems. The reliabilities decreased constantly because no repair actions occur. The reliabilities of the all-series, all-voting, voting-after-series, series-after-voting, parallel-after-series, and series-after-parallel systems gradually decrease, with their values recorded at 87.20, 99.9999021, 99.468, 99.803, 99.99334, and 99.999086% in the 140th week, respectively. The availabilities of the six systems after a certain time are maintained at the stable values of approximately 97.24, 99.99999982, 99.9814, 99.9874, 99.9999716, and

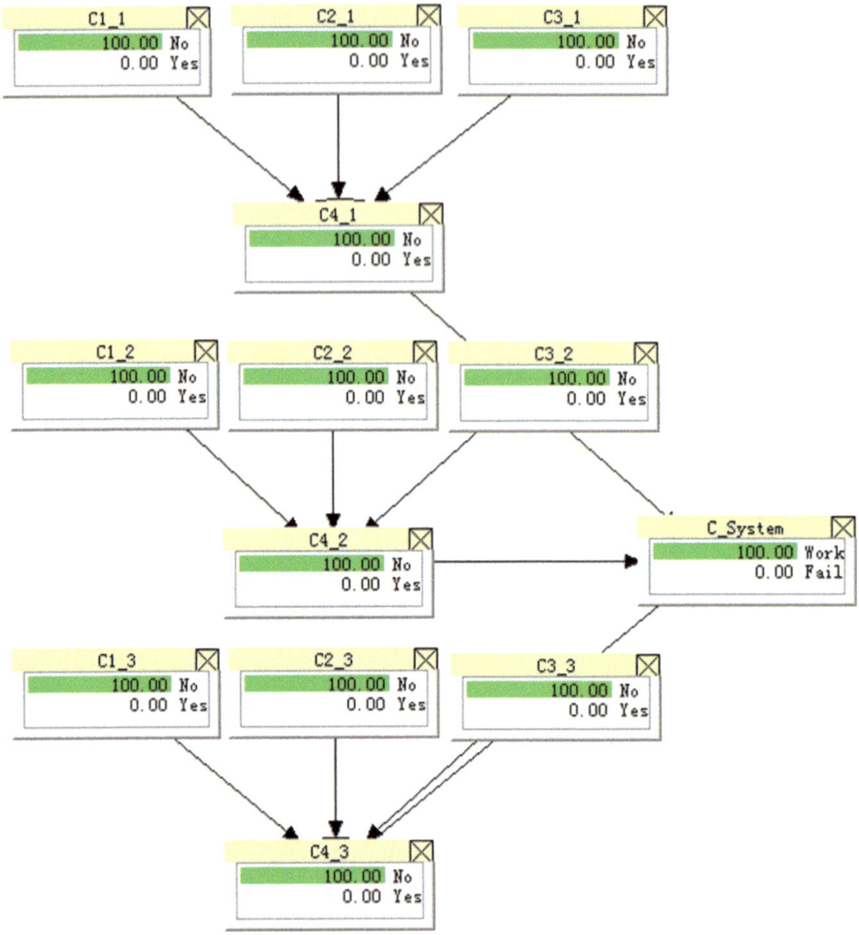

Fig. 5 Structure of BNs of the case in Fig. 4

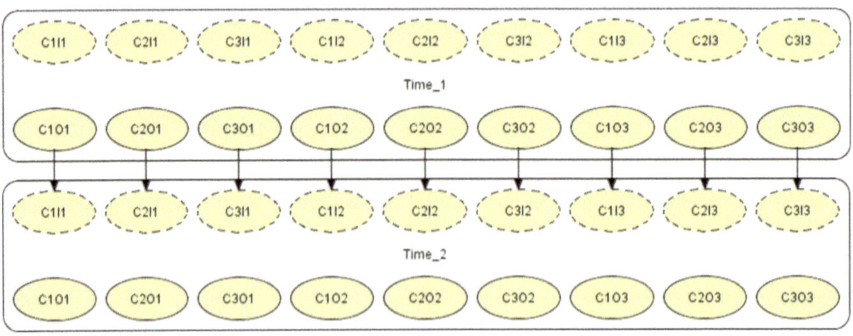

Fig. 6 DOOBN-based model of systems with exponential distribution

Table 1 Transition CPT of the systems of exponential distribution

No maintenance	t_n	t_{n+1}	
		No	Yes
	No	$\exp(-\lambda \Delta t)$	$1 - \exp(-\lambda \Delta t)$
	Yes	0	1
Maintenance	t_n	t_{n+1}	
		No	Yes
	No	$\exp(-\lambda \Delta t)$	$1 - \exp(-\lambda \Delta t)$
	Yes	$1 - \exp(-\mu \Delta t)$	$\exp(-\mu \Delta t)$

99.9999785%. Moreover, complete maintenance can significantly improve the performance of the system. Among the six systems, the reliability and availability of the all-series system are at the minimum, whereas those of the all-voting system are at the maximum. The results are found to be completely consistent by comparing the proposed method with the traditional methods, such as Markov, thus proving the accuracy of the proposed method.

B. DOOBN-based reliability evaluation for systems with Weibull distribution

In the reliability evaluation, the exponential distribution is generally applicable to the electronic control system, whereas the failure law of mechanical components is near the Weibull distribution. In Table2, the transition CPT is related not only to Δt but also to t, which is no longer a constant. Consequently, the removal of time state cannot be accomplished by simply changing the relationship between nodes C1I1 and C1O1. The intermediate node C1M1 should be added between nodes C1I1 and C1O1. The transition CPT is represented by the relationship between nodes C1O1 and C1M1, and nodes C1M1 and C1I1 are entirely consistent, as depicted in Fig. 8.

Figure 9 presents the reliability and availability curves of the six mechanical systems. The three figures show that the availability curves of the six mechanical systems rapidly and more dramatically decreased than the curves of the exponential distribution. The availability curves of the systems with Weibull distribution are not maintained at stable values but decline continuously. The tendency of the availability curves is slower in the systems with Weibull distribution than in the systems without maintenance; this result is consistent with the failure regularity of the mechanical systems. Figure 8 depicts that the reliabilities of the all-series, all-voting, voting-after-series, series-after-voting, parallel-after-series, and series-after-parallel systems decline to 86.87, 99.999895, 99.5, 99.75, 99.99544, and 99.99864% in the 140th week, correspondingly, and the availabilities of these systems decline to 97.24, 99.999999822, 99.9814, 99.9874, 99.9999716, and 99.9999785% in the 140th week, respectively. These results are entirely consistent compared with the conventional methods, thereby validating the accuracy of the proposed method.

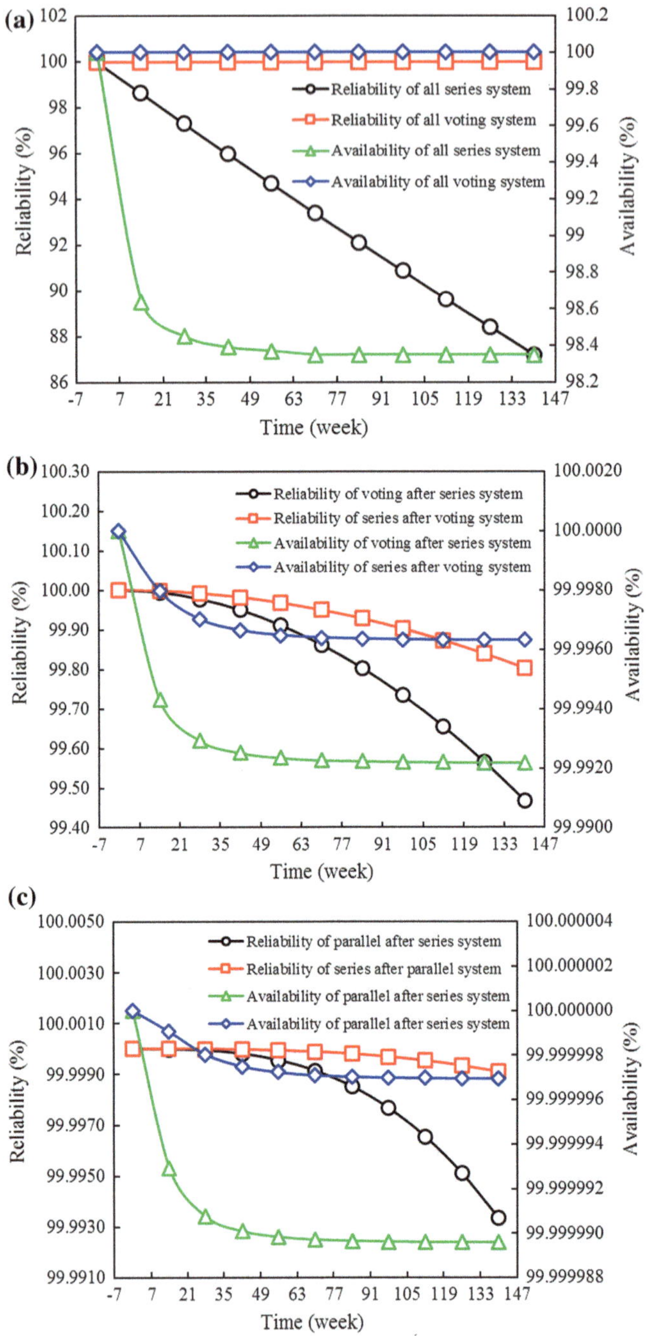

Fig. 7 Reliability and availability curves of the six systems with exponential distribution

Table 2 Transition CPT in the systems of Weibull distribution

t_n	No maintenance/t_{n+1}	
	Work	Fail
Work	exp[−(tn+1/η)β]/exp[−(tn/η)β]	1 − exp[−(tn + 1/η)β]/ exp[−(tn/η)β]
Fail	0	1
t_n	Maintenance/t_{n+1}	
	Work	Fail
Work	exp[−(tn+1/η)β]/exp[−(tn/η)β]	1 − exp[−(tn+1/η)β]/exp[−(tn/η)β]
Fail	$1 - \exp(-\mu\Delta t)$	$\exp(-\mu\Delta t)$

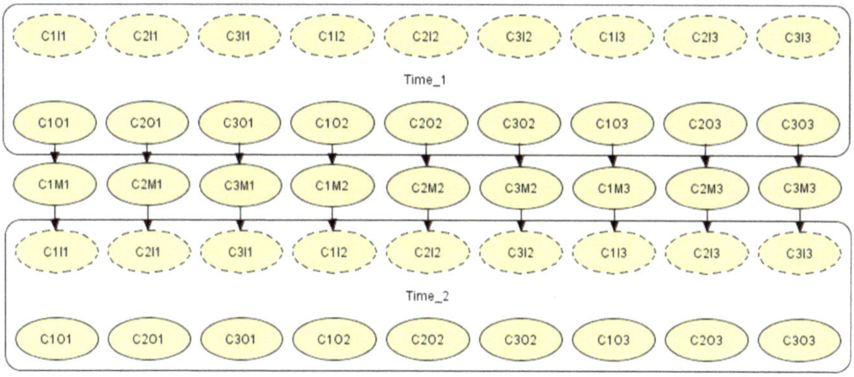

Fig. 8 DOOBN-based model of systems with Weibull distribution

4 Reliability Evaluation of Deepwater BOP System

Introduction of Configuration of deepwater BOP system

A typical configuration of the BOP mechanical system is displayed in Fig. 10. Generally, a traditional BOP mechanical system comprises two annular, one shear-blind ram, and three pipe ram BOPs. Moreover, a modern BOP mechanical system consists of two annular, two shear-blind ram, one testing pipe ram, and three pipe ram BOPs. The plurality of valve groups, elastic joint, riser connector, and wellhead connector is observed in a BOP mechanical system. Several basic assumptions are formulated:

(1) Only the different combinations of subsea BOP are considered, and the influence of the other parts on the system, such as joints and connectors, is ignored; this condition is due to the main faults occur in the annular, shear-blind ram, pipe ram, and testing pipe ram BOPs [8, 14, 27, 28] but not in the various joints and connectors because the current work aims to study only the configurations of BOPs.
(2) Only the failures with a high probability of occurrence or significant influence on the system are considered and investigated, whereas the failure with a low

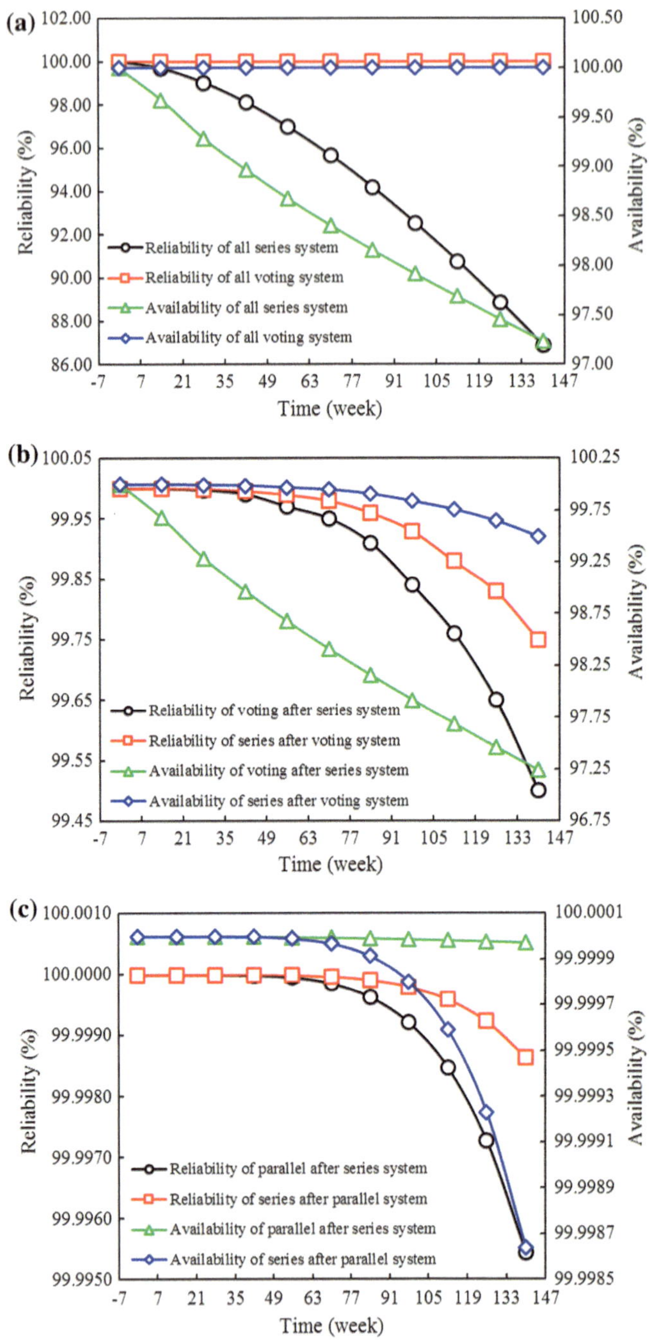

Fig. 9 Reliability and availability curves of the six systems with Weibull distribution

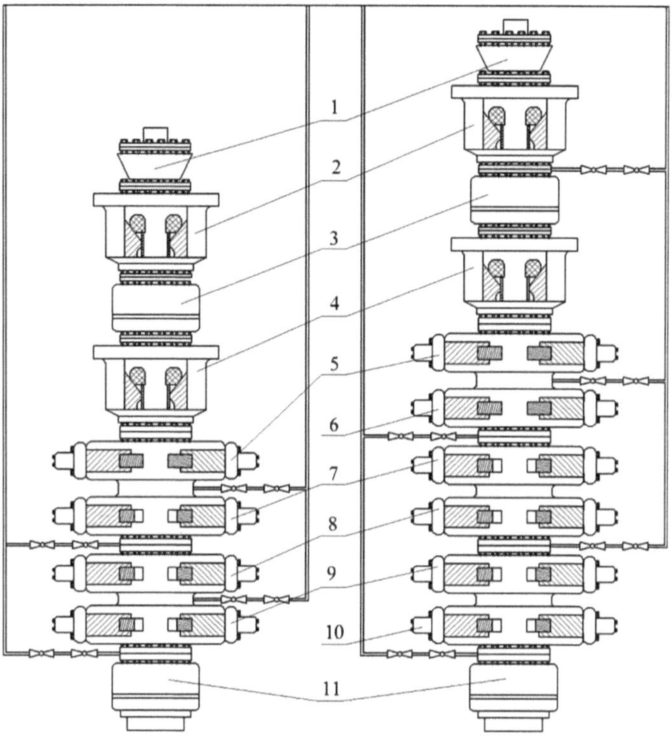

Fig. 10 Typical subsea BOP stack **a** traditional BOP mechanical system and **b** modern BOP mechanical system (1—elastic joint, 2, 4—annular BOP, 3–riser connector, 5, 6—shear-blind ram BOP, 7, 8, 9—pipe ram BOP, 10—testing pipe ram BOP, 11—wellhead connector)

probability of occurrences and slight influences on the system are ignored for simplification.

(3) The distribution of the different failures is presented as follows. The wear-out failure in rubber satisfies the Weibull distribution. Maintenance satisfies the exponential distribution. Random events, such as locking by drill cutting, are calculated according to the probability of occurrence.

B. Reliability model of the deepwater BOP System

An annular BOP, or a universal BOP, is mainly used for sealing different sizes of drill pipes, tubing and casings, and open holes. A common annular BOP has three types, namely spherical, conical, and combined rubber cores. Conical rubber core BOP is widely used because of its advantages, such as extended stroke and simple structure. This type of BOP mainly consists of a shell, rubber core, support cylinder, and hydraulic components. This work investigates the conical rubber core BOP. Table 3 summarizes the common failure modes of the annular BOP, which mainly consists of

Table 3 Common failures and several basic parameters of an annular BOP

Node	Event	η	β	μ	a
Annular 1	Rubber core rubber failure	3500	2.15	5.78E−03	–
Annular 2	Rubber core bracing steel bar fracture failure	4500	2.70	5.78E−03	–
Annular 3	Rubber core failure	–	–	–	–
Annular 4	Drill cutting seizing between rubber core and drill pipe	–	–	–	8.30E−07
Annular 5	Untightened sealing of well	–	–	–	–
Annular 6	Support cylinder failure	–	–	–	8.20E−09
Annular 7	Drill cutting seizure	–	–	–	3.33E−08
Annular 8	Piston seizure	–	–	–	–
Annular 9	Hydraulic system leakage	40,000	2.40	3.33E−03	–
Annular 10	Liquid blockage	20,0000	2.50	4.17E−03	–
Annular 11	Hydraulic component failure	–	–	–	–
Annular 12	Switch failure	–	–	–	–

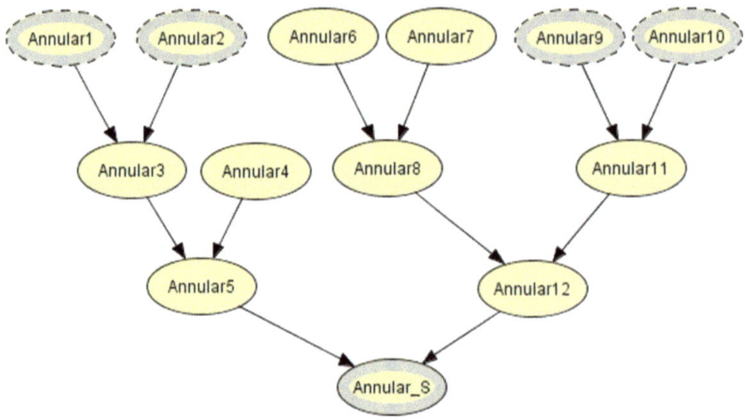

Fig. 11 Static BNs of an annular BOP

the rubber core rubber failure, rubber core bracing steel bar fracture failure, support cylinder failure, and blockage of the liquid channel.

Figure 11 displays the static BNs of the annular BOP. The input nodes and transfer CPT of node Annular 1, 2, 9, and 10 are similar to those in Table 2 because these input nodes satisfy the Weibull distribution. Table 3 lists the basic parameters. η and β are two basic parameters of the Weibull distribution. μ is a basic parameter of the exponential distribution when maintenance occurs. Annular 4, 6, and 7 are random events. Parameter “a” in Table 3 represents the probability of random events.

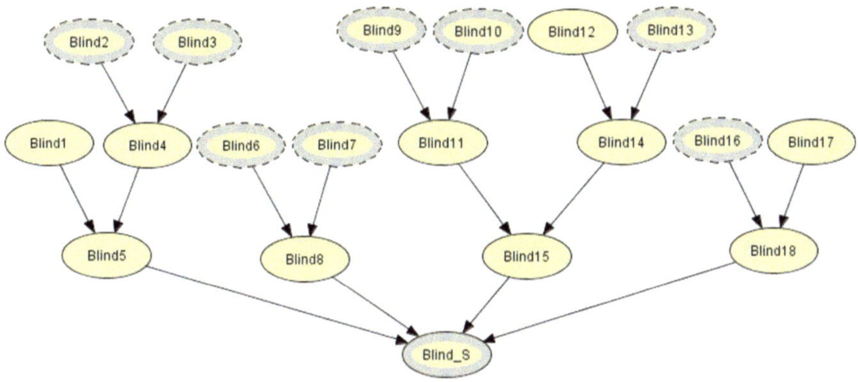

Fig. 12 Static BNs of a shear-blind ram BOP

Table 4 Common failures and several basic parameters of a shear-blind ram BOP

Node	Event	η	β	μ	a
Blind 1	Alignment pin deformation	–	–	–	1.20E−08
Blind 2	Top seal leakage	7000	2.30	1.43E−02	–
Blind 3	Cutter body seal failure	9000	2.30	1.79E−02	–
Blind 4	Seal leakage	–	–	–	–
Blind 5	Untightened sealing of well	–	–	–	–
Blind 6	Middle flange leakage	50,000	2.30	2.00E−03	–
Blind 7	Side door leakage	70,000	2.30	4.55E−03	–
Blind 8	External leakage	–	–	–	–
Blind 9	Hydraulic system leakage	15,000	2.40	3.33E−02	–
Blind 10	Hydraulic block	10,0000	2.50	5.00E−02	–
Blind 11	Hydraulic component failure	–	–	–	–
Blind 12	Wedge deformation	–	–	–	5.60E−09
Blind 13	Locking mechanism hydraulic component failure	30,000	2.30	4.17E−02	–
Blind 14	Locking mechanism failure	–	–	–	–
Blind 15	Switch fault	–	–	–	–
Blind 16	Blade failure	90,000	2.70	5.56E−02	–
Blind 17	Improper operation	–	–	–	5.50E−06
Blind 18	Shear drill pipe failure	–	–	–	–

Figure 12 depicts the static BNs of the shear-blind ram BOP. Node Blind 2, 3, 6, 7, 9, 10, 13, and 16 satisfy the Weibull distribution. These node blinds are considered input nodes because the transfer CPT is similar to that in Table 2. Node Blind 1, 12, and 17 are random events. The basic parameters of the two distributions are presented in Table 4.

Table 5 Common failures and several basic parameters of a pipe ram BOP

Node	Event	η	β	μ	a
Pipe 1	Top seal leakage	7000	2.30	1.43E−02	–
Pipe 2	Front seal leakage	5000	2.30	1.79E−02	–
Pipe 3	Seal leakage	–	–	–	–
Pipe 4	Improper operation	–	–	–	1.00E−05
Pipe 5	Untightened sealing of well	–	–	–	–
Pipe 6	Hydraulic system leakage	15,000	2.40	3.33E−02	–
Pipe 7	Hydraulic block	10,0000	2.50	5.00E−02	–
Pipe 8	Hydraulic component failure	–	–	–	–
Pipe 9	Wedge deformation	–	–	–	5.60E−09
Pipe 10	Locking mechanism hydraulic component failure	30,000	2.30	–	–
Pipe 11	Locking mechanism failure	–	–	–	–
Pipe 12	Switch fault	–	–	–	–
Pipe 13	Middle flange leakage	50,000	2.30	2.00E−03	–
Pipe 14	Side door leakage	70,000	2.30	4.55E−03	–
Pipe 15	External leakage	–	–	–	–

The pipe ram BOP is mainly used for sealing well holes with the tube. The pipe ram BOP can seal higher pressure than the annular BOP. The basic structure of the pipe ram BOP is similar to the shear-blind ram BOP, but the ram structure is different. In Table 5, the common failures of the pipe ram BOP are mainly the top seal leakage, front seal leakage, hydraulic system leakage, and locking mechanism hydraulic component failure.

Figure 13 illustrates the static BNs of a pipe ram BOP. Pipe 1, 2, 6, 7, 10, 13, and 14 satisfy the Weibull distribution. These pipes are considered input nodes because the transfer CPT is similar to that in Table 2. Node Pipe 4 and 9 are random events. The basic parameters of the two distributions are presented in Table 5.

Figure 14a shows the evaluation model of the traditional deepwater BOP system based on the OOBNs. Two annular BOPs in the traditional type of deepwater BOP system are mutually redundant and therefore demonstrate a parallel relationship and are represented by the node Annular. Three pipe ram BOPs are also mutually redundant and hence exhibit a parallel relationship and are represented by the node Pipe. Therefore, the annular, pipe ram, and shear-blind ram BOP groups are in series. The whole system is represented by the node BOP-S. Figure 14b depicts the evaluation model of a modern common deepwater BOP system. The main difference between the two types of BOP is that a modern deepwater BOP has additional sets of shear-blind and test pipe ram BOPs. The shear-blind ram BOP causes redundancy with the existing type of BOP, whereas the test pipe ram BOP is used to detect the entire system. The shear-blind ram BOP can only be used to seal pressure from the upper

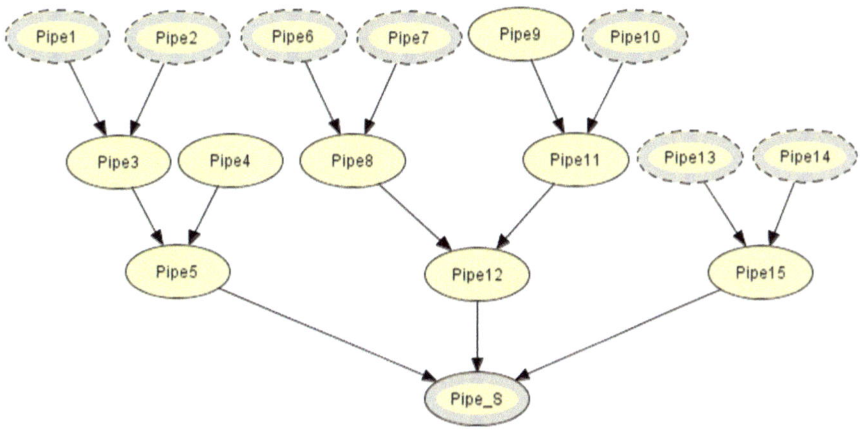

Fig. 13 Static BNs of a pipe ram BOP

part but not the oil pressure from the bottom part. Its main purpose is to increase the efficiency of detection.

Auxiliary nodes, such as Annular 1I1 and 1O1, can be added through the method proposed in Sect. 2 to satisfy the Weibull distribution (Annular 1I1, 1O1, and 1 are a one-to-one correspondence). Based on DOOBNs, Fig. 15 presents a deepwater BOP system model that extended three time slices by adding auxiliary node Annular 1M1 between two time slices. The node can be transferred by changing the relationship between the previous time slice and the intermediate node Annular 1O1. Annular 1M1 and 1I1, in which the nodes of the latter time slice, also have a one-to-one correspondence.

C. Study on the reliability and availability of deepwater BOP

Figure 16 demonstrates the reliability and availability curves of the traditional and modern deepwater BOP systems. The reliability and availability of any type of BOP system decrease gradually with time. On the 36th week, the reliability and availability of the traditional system were reduced to 99.9166191 and 99.9890985%, respectively. The reliability and availability of modern systems decreased to 99.99840 and 99.99997%, correspondingly. In addition, the reliability and availability of the modern BOP system are clearly improved, thereby indicating that the redundant structure of the shear-blind ram BOP significantly increased the reliability and availability of the system. The value of availability is significantly higher than the value of reliability within the same type of BOP, thereby showing that maintenance can significantly improve the performance of the system.

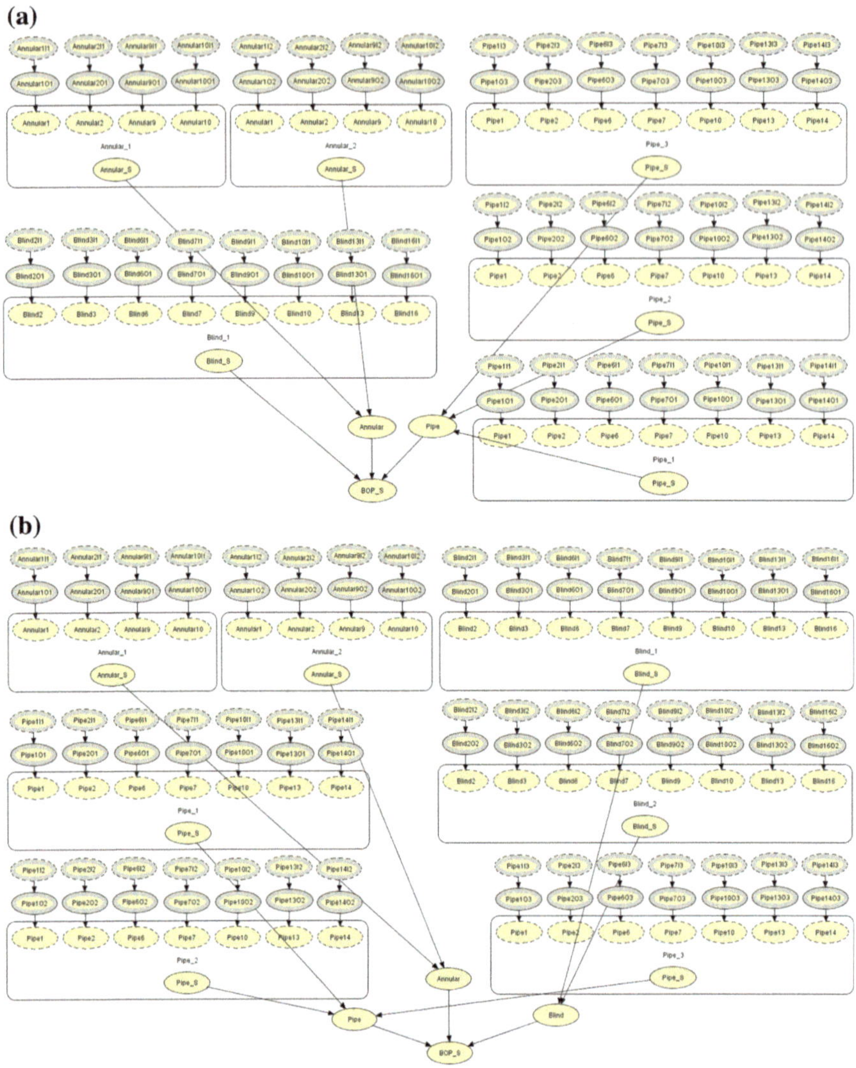

Fig. 14 Model of the deepwater BOP system based on OOBNs, **a** traditional deepwater BOP system and **b** modern deepwater BOP system

5 Conclusion

In this paper, a method for DOOBN-based reliability evaluation of complex dynamic systems is proposed. Six general systems, namely the all-series, all-voting, voting-after-series, series-after-voting, parallel-after-series, and series-after-parallel systems, are used to validate the proposed method. The reliability and availability of

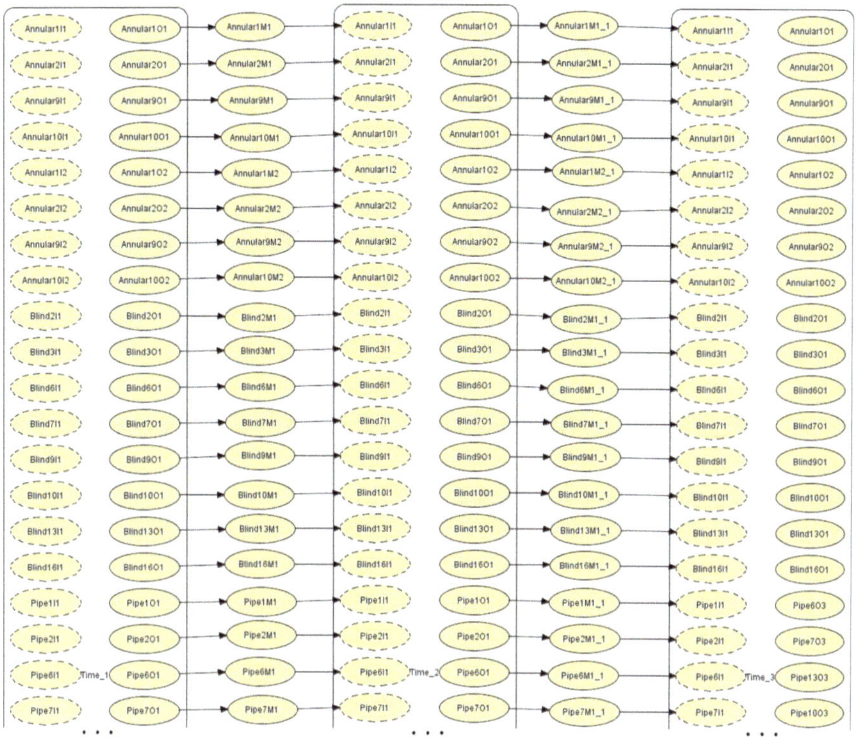

Fig. 15 Model of the deepwater BOP system based on the DOOBNs

a complex mechanical system, such as deepwater BOP system, are also analyzed using the proposed method. Several contributions of this work are summarized.

(1) This reliability modeling methodology consists of two main phases, namely the construction phases of OOBNs and of DOOBNs. Systems with exponential and Weibull distribution can be modeled using the proposed methodology.
(2) The availability curve for the six systems used in exponential distribution decreased rapidly and reached stable levels at different times in various systems, and the reliability curve decreased constantly because no repair actions occur.
(3) For the same systems with Weibull distribution, the availability curves rapidly and more dramatically decreased than the curves of the exponential distribution. The availabilities are not maintained at stable values but continuously decrease. The tendency of the availability curves is slower in the systems with Weibull distribution than in the systems without maintenance.
(4) The shear-blind and testing pipe ram BOPs can significantly improve the reliability and availability by comparing the traditional and modern BOP systems through the DOOBN-based method.

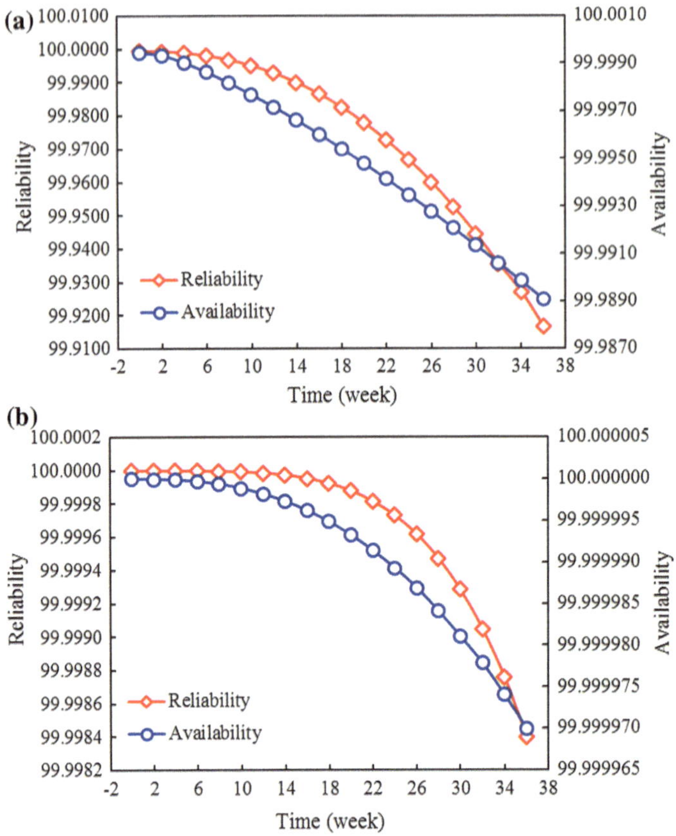

Fig. 16 Model of the deepwater BOP system based on the DOOBNs

References

1. Y. Bai, Q. Bai, Subsea Engineering Handbook, 2012
2. Y. Ma, B. Cai, L. Huang, B. Zhan, X. Yuan, Y. Liu, "A reliability evaluation methodology of complex systems based on dynamic object oriented Bayesian networks," IEEE *Reliability, Maintainability and Safety (ICRMS), 2016 11th International Conference on*, pp. 1–5, Oct. 2016
3. H. Langseth, L. Portinale, Bayesian Networks in reliability. Reliab. Eng. Syst. Safe. **92**(1), 92–108 (2007)
4. A. Darwiche, *Modeling and reasoning with Bayesian Network* (Cambridge University Press, New York, 2009)
5. B. Cai, Y. Liu, Q. Fan, Y. Zhang, Z. Liu, S. Yu, R. Ji, Multi-source information fusion based fault diagnosis of ground-source heat pump using Bayesian Network. Appl. Energy **114**, 1–9 (2014)
6. B. Cai, Y. Zhao, H. Liu, M. Xie, A data-driven fault diagnosis methodology in three-phase inverters for PMSM drive systems. IEEE Trans. Power Electron. **32**(7), 5590–5600 (2017)
7. B. Cai, L. Huang, M. Xie, Bayesian Networks in fault diagnosis. IEEE Trans. Industr. Inf. **13**(5), 2227–2240 (2017)

8. B. Cai, Y. Liu, Z. Liu, X. Tian, X. Dong, S. Yu, "Using Bayesian Networks in reliability evaluation for subsea blowout preventer control system," *Reliab. Eng. Syst. Safe.*, vol. 108, pp. 32–41, 2012. %%25
9. I. Tien, K.A. Der, Algorithms for Bayesian Network modeling and reliability assessment of infrastructure systems. Reliab. Eng. Syst. Safe. **156**, 134–147 (2016)
10. S. Barua, X. Gao, H. Pasman, M.S. Mannan, Bayesian Network based dynamic operational risk assessment. J. Loss Prevent. Proc **41**, 399–410 (2016)
11. T. Daemi, A. Ebrahimi, M. Fotuhi-Firuzabad, Constructing the Bayesian Network for components reliability importance ranking in composite power systems. Int. J. Elec. Power. **43**(1), 474–480 (2012)
12. J. Wu, R. Zhou, S. Xu, Z. Wu, Probabilistic analysis of natural gas pipeline network accident based on Bayesian Network. J. Loss Prevent. Proc. **46**, 126–136 (2017)
13. J. Bhandari, R. Abbassi, V. Garaniya, F. Khan, Risk analysis of deepwater drilling operations using Bayesian Network. J. Loss Prevent. Proc. **38**, 11–23 (2015)
14. B. Cai, Y. Liu, Q. Fan, Y. Zhang, S. Yu, Z. Liu, X. Dong, Performance evaluation of subsea BOP control systems using dynamic Bayesian Networks with imperfect repair and preventive maintenance. Eng. Appl. Artif. Intel. **26**(10), 2661–2672 (2013)
15. B. Cai, Y. Liu, Y. Zhang, Q. Fan, Z. Liu, X. Tian, A Dynamic Bayesian Networks modeling of human factors on offshore blowouts. J. Loss Prevent. Proc. **26**(4), 639–649 (2013)
16. J. Luque, D. Straub, Reliability analysis and updating of deteriorating systems with subset simulation. Struct. Saf. **64**, 20–36 (2016)
17. J. Zhu, M. Collette, A dynamic discretization method for reliability inference in Dynamic Bayesian Networks. Reliab. Eng. Syst. Safe. **138**, 242–252 (2015)
18. B. Cai, Y. Liu, Q. Fan, A multiphase Dynamic Bayesian Networks methodology for the determination of safety integrity levels. Reliab. Eng. Syst. Safe. **150**, 105–115 (2016)
19. N. Khakzad, "Application of Dynamic Bayesian Network to risk analysis of domino effects in chemical infrastructures," *Reliab. Eng. Syst. Safe.*, vol. 138, pp. 263–272
20. J. Foulliaron, L. Bouillaut, A. Barros, P. Aknin, Dynamic Bayesian Networks for reliability analysis: From a Markovian point of view to semi-Markovian approaches. IFAC-PapersOnLine. **28**, 694–700 (2015)
21. D. Koller, A. Pfeffer, "Object-Oriented Bayesian Networks. In Proceedings of the Thirteenth Conference on Uncertainty in Artificial Intelligence," *Morgan Kaufmann Publishers Inc.*, 1997
22. B. Cai, H. Liu, M. Xie, A real-time fault diagnosis methodology of complex systems using Object-Oriented Bayesian Networks. Mech. Syst. Signal Pr. **80**, 31–44 (2016)
23. P. Weber, L. Jouffe, Complex system reliability modelling with Dynamic Object Oriented Bayesian Networks (DOOBN). Reliab. Eng. Syst. Safe. **91**, 149–162 (2006)
24. T.B. Jensen, A.R. Kristensen, N. Toft, N.P. Baadsgaard, S. Østergaard, H. Houe, An Object-Oriented Bayesian Network modeling the causes of leg disorders in finisher herds. Prev. Vet. Med **89**(3–4), 237–248 (2009)
25. G. Weidl, A.L. Madsen, S. Israelson, Applications of Object-Oriented Bayesian Networks for condition monitoring, root cause analysis and decision support on operation of complex continuous processes. Comput. Chem. Eng. **29**(9), 1996–2009 (2005)
26. Q. Liu, F. Pérès, A. Tchangani, Object Oriented Bayesian Network for complex system risk assessment. IFAC-PapersOnLine **49**(28), 31–36 (2016)
27. DNV (DET NORSKE VERITAS), "Offshore reliability data handbook," 5th Edition, 2010. [Online]. Available: http://www.oreda.com/handbook.html
28. B. Cai, Y. Liu, Y. Zhang, Q. Fan, S. Yu, Dynamic Bayesian Networks based performance evaluation of subsea blowout preventers in presence of imperfect repair. Expert Syst. Appl. **40**(18), 7544–7554 (2013)

Operation-Oriented Reliability and Availability Evaluation for Onboard High-Speed Train Control System with Dynamic Bayesian Network

Abstract The reliability and availability of the onboard high-speed train control system are important to guarantee operational efficiency and railway safety. Failures occurring in the onboard system may result in serious accidents. In the analysis of the effects of failure, it is significant to consider the operation of an onboard system. This paper presents a systemic approach to evaluate the reliability and availability for the onboard system based on dynamic Bayesian network, with taking into account dynamic failure behaviors, imperfect coverage factors, and temporal effects in operational phase. The case studies are presented and compared for onboard systems with different redundant strategies, i.e., the triple modular redundancy, hot spare double dual, and cold spare double dual. Dynamic fault trees of the three kinds of onboard system are constructed and mapped into dynamic Bayesian network. The forward and backward inferences are conducted not only to evaluate the reliability and availability, but also recognize the vulnerabilities of the onboard systems. A sensitivity analysis is carried out for evaluating the effects of failure rates subject to uncertainties. To improve the reliability and availability, the recovery mechanism should be paid more attention. Finally, the proposed approach is validated with the field data from one railway bureau in China and some industrial impacts are provided.

Keywords Operation-oriented · Reliability · Availability · Train control system · Dynamic bayesian network

1 Introduction

The main purpose of high-speed railways is to provide safe, efficient, and punctual services to travelers between cities. The train movements are controlled by the train control system, which consists of onboard and trackside subsystems. Recent national and international standards and regulations have introduced new requirements to ensure a higher level of interoperability and flexibility in the exploitation of the railway infrastructure. These requirements have resulted in practical standards for train control system, such as the European Train Control System (ETCS) and Chinese Train Control System (CTCS). The standards allow integration with traditional

B. Cai et al., *Bayesian Networks for Reliability Engineering*,
https://doi.org/10.1007/978-981-13-6516-4_6

systems installed in the railway infrastructure, but also specify a totally new train control philosophy and architecture. For this reason, both CTCS and ETCS are introduced with different levels, depending on the specific implementation selected. The CTCS has five levels, level 0 to level 4, while ETCS has four levels, level 0 to level 3 [1, 2]. Transitions between these levels are performed according to defined functions and procedures. Some of these system levels, such as CTCS level 3 (CTCS-3) and ETCS level 2 (ETCS-2), use wireless communication (GSM-R or LTE-R) to realize continuously bidirectional information transmission between the subsystems.

The CTCS/ETCS systems, both onboard and trackside, are safety-critical. In this paper, the attention is directed mainly to the onboard system, as the higher levels of CTCS and ETCS will have many of the trackside functions transferred to the onboard systems. The onboard system has two main purposes: to avoid too high speed and to prevent violating a red (stop) signal. A failure occurring in the onboard system may have the potential to cause serious accidents.

Achieving a high degree of reliability and availability has obvious advantages in the safety and efficiency for the train control system. For this reason, both active and standby redundancy strategies are introduced for the onboard system. For example, the triple modular redundancy (TMR), hot standby double dual (HDD), and cold standby double dual (CDD) architectures are widely applied. One important feature in a redundancy system is the dependencies among components, which cause the dynamic failure behaviors. It has critical effects on the reliability and availability evaluation and should be modeled when considering the use of spares such as hot spare, warm spare, and cold spare [3]. Another significant feature in a redundant system is the coverage factor, the probability that a single failure entails a complete system failure [4]. The imperfect coverage factor can be used to model the inaccuracy of a recovery mechanism in a redundant system.

Models and approaches for reliability and availability analysis of high-speed train control system have been studied in the past decade. For example, Flammini et al. combined fault tree (FT) and Bayesian network (BN) to evaluate the reliability of ETCS-2 [5]. A multiformalism model is used to perform an availability analysis of an ETCS-2 at early phases of its development cycle [6]. Regarding to the ETCS-2 onboard system, the components are adopted active redundancy strategy and the FT model is utilized to perform a reliability analysis. Di et al. utilized the reliability block diagram (RBD) and Markov model to evaluate the reliability, availability, and maintainability of CTCS-3 onboard system [7]. Su and Che modeled CTCS-3 onboard system by mapping FT to BN and analyzed the multi-state situations [8]. For the CTCS-3 onboard system, the components are adopted standby redundancy strategy. Qiu et al. proposed a simulation approach to model the availability of ETCS-2 as a system of systems [9, 10]. Bernardi et al. presented a model-driven approach for the development of formal maintenance and reliability models for the availability evaluation of a train control system [11]. Some studies formulated non-Markovian models for the moving-block control (CBTC or ETCS level 3), and their works are more detailed in the communication availability [12–14].

For the onboard system, the studies discussed above mainly focus on the reliability and availability analysis in the design phase. Since railways require a long-term and sustainable strategy, performance during operation and maintenance needs to be analyzed. Based on the corrective maintenance, the field data can be used for reliability and availability evaluation. Moreover, the uniqueness of the operation and maintenance of high-speed trains should be taken into account. For example, the onboard system cannot be repaired online in working time considering the safety and efficiency of railway network, and it is only repaired in the train inspection and repair station. The onboard system may degrade on account of the wireless communication timeout. In operational phase, the dynamic failure behaviors and imperfect coverage factors of redundant onboard system also have significant effects on the reliability and availability. Such kind of behaviors should be reflected in an effective reliability and availability analysis model.

BN and DBN have been proved effective in modeling the complexity of industrial system [15–17]. Boudali proposed a discrete-time BN reliability formalism for system reliability prediction and diagnosis [18]. Cai et al. presented a real-time fault diagnosis methodology of complex systems using object-oriented BN [19]. Neil and Marquez utilized a hybrid BN framework to evaluate the effects and corrective maintenance time, logistic delay, and preventive maintenance on the availability of renewable systems [20]. The static BN represents a joint probability distribution at a fixed time slice, while dynamic Bayesian network (DBN) can extend the static BN at different time slices to model dynamics of random variables. For instance, Liang et al. applied a DBN to evaluate the reliability of warship and dealt with the dynamic failure and multistage missions [21]. Cai et al. proposed a DBN model to analyze the reliability and availability of subsea blowout preventer [22]. An availability-based engineering resilience metric was proposed from the perspective of reliability engineering using DBN model [23]. Generally, BN and DBN models can be obtained by mapping the traditional reliability method, such as RBD, FT, and dynamic fault tree (DFT) [24–26].

In this paper, we aim to adopt the DBN approach to evaluate the reliability and availability of onboard high-speed train control systems in the operation phase, with taking into account dynamic failure behaviors, imperfect coverage factors, and multistate. Based on the corrective maintenance, the field data regarding to onboard system are collected to calculate the failure rate of the components. In practice, different redundant strategies of the onboard system (TMR, HDD, and CDD) are adopted. The reliabilities and availabilities of those onboard systems are presented and compared to provide some industrial impacts. The DBN forward, backward, and sensitivity analyses will be conducted to support the maintenance decision making. Finally, the availability obtained is well validated by the field data.

The remainder of the paper is organized as follows: Upcoming section introduces the structure and operation of an onboard system. Then, the corresponding DBN model and model validation are presented. Next, a case study of reliability and availability evaluation of CTCS-3 onboard system will be conducted. Lastly, conclusions occur in the final section.

2 System Description

2.1 Structure of an Onboard System

High-speed train control depends on the bidirectional information transmission between the trackside and onboard subsystem. Through wireless communication system, information about train speed and location can be transmitted to the Radio Block Center (RBC), and meanwhile, the movement authorities, generated by RBC, can be sent to the onboard system. The onboard system consists of the following components: vital computer (VC), radio transmission unit (RTU), balise transmission module (BTM), driver–machine interface (DMI), and so on. Generally, the onboard system can be divided into five functional modules: wireless communication processing module (WP), lineside data processing module (LP), train and driver interface (TD), kernel information processing module (KN), and power and bus (PB), as shown in Fig. 1. The VC is the core component to realize KN function. KN processes the information from WP, LP, and TD, and conduct safe control, protection, and supervision of the train. The other components of CDD, HDD, and TMR onboard system adopted cold standby, hot standby, and parallel structure, respectively.

2.2 Operation of an Onboard System

Any failure in the five function modules can result in the failure of the onboard system. According to the fail-safe concept in railway, the train should be stopped

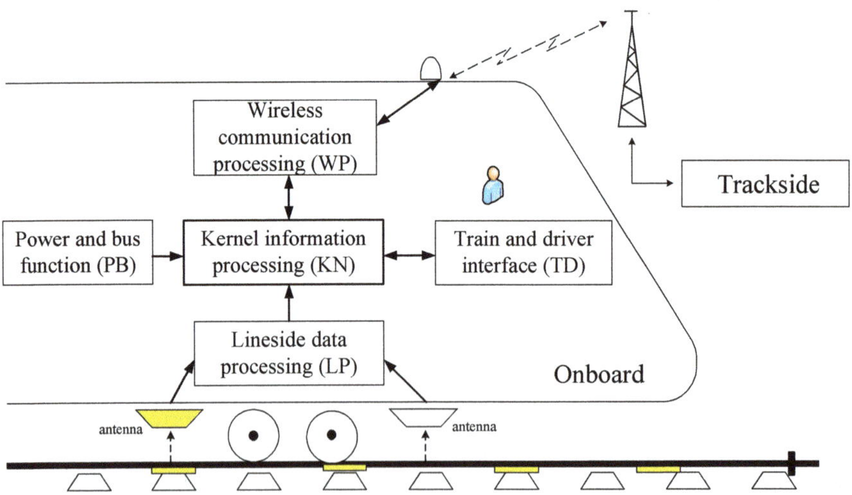

Fig. 1 Function modules of the onboard system

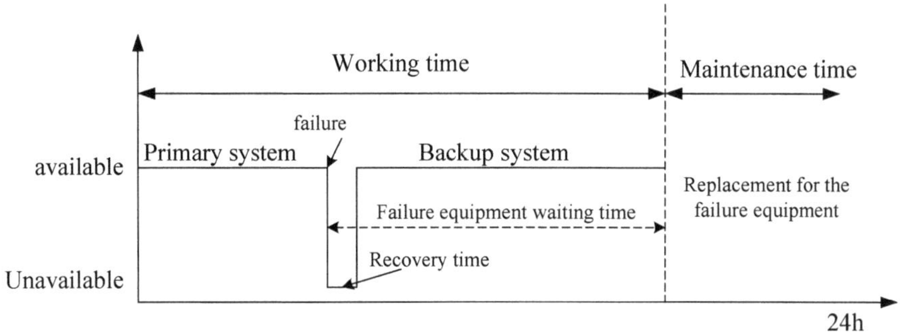

Fig. 2 Working timetable of the onboard system

and then the driver needs to restart the onboard system. The working timetable for onboard system in one day is shown in Fig. 2. Availability means the onboard system is operative in the working hours, and unavailability comes from the recovery time when the primary system suffers a failure. The average working time for a high-speed train is 18 h in one day. For the rest time, the maintenance is required for the train in an inspection and repair station. The onboard system cannot be online repaired, namely it is kept operating during the working hours only by the redundancy strategy. Meanwhile, the onboard system can be fully repaired during the maintenance hours. The recovery time is influenced by some dynamic factors, such as the recovery mechanism and the operational condition.

Field data can be collected, containing failure mode/effect/cause and recovery time. These field data are very useful for the calculation of failure rate and the validation of results.

3 DBN-Based Reliability and Availability Modeling

3.1 Introduction on BN and DBN

A Bayesian network consists of two parts, corresponding a directed acyclic graph (DAG) and a joint probability distribution [27]. Precisely, a Bayesian network is a triplet <(*V*, *E*), *P*>, where (*V*, *E*) are the variables (nodes) and edges (arcs) of a DAG, and *P* is the probability distribution for every *V* [28]. The DAG realizes the qualitative analysis of dependence V. Meanwhile, the Conditional Probabilistic Table (CPT) over *V* accomplishes the quantitative analysis. For discrete random variables $V = \{X_1, X_2 \ldots X_N\}$, the joint probability distribution can be given by:

$$P(V) = P(X_1, X_2, \ldots X_N) = \prod_{X_i \in V} P(X_i|\mathrm{Pa}(X_i)) \quad (1)$$

where N is the number of random variables in the graph, and $\mathrm{Pa}(X_i)$ is the parent of X_i.

DBNs can model the dynamic behavior of random variables by extending the static BN with the temporal dependencies at different time slices. Intra-slice arcs at the same time slice and temporal arcs in different time slices are two types of conditional dependencies between nodes. Generally, two-time slices temporal Bayesian networks (2TBN) are considered in modeling the temporal evolution, assuming that temporal arcs between the consecutive time $t-1$ and t satisfy the first-order Markov process. The 2TBN defines $P(X_t|X_{t-1})$ by means of a DAG as follows: [29]

$$P(X_i|X_{i-1}) = \prod_{i-1}^{N} P(X_t^i|\mathrm{Pa}(X_t^i)) \tag{2}$$

where X_t^i is the ith node in time slice t, and $\mathrm{Pa}(X_t^i)$ indicates the parent of X_t^i which can only stand in slices $t-1$ and t.

Note that the nodes in the first time slice do not associate any parameters, while each node from the second time slice has an associated CPT.

Thereby, the joint distribution can be defined by "unrolling" the 2TBN with T time slices as follows:

$$P(X_{1:T}) = \prod_{t-1}^{T}\prod_{i-1}^{N} P(X_t^i|\mathrm{Pa}(X_t^i)) \tag{3}$$

3.2 DBN Structure Modeling

The structure modeling presents the mapping rules from DFT into DBN. Here, we briefly model the OR gate, AND gate, 2003 voting gate, and spare gate as they will be later used in the case study. For those static logic gates (the OR gate, AND gate, and 2003 voting gate), mapping rules are described in the previous study [24]. As indicated in Fig. 3b, the relationship between C1, C2, and A is linked by intra-slice arcs. Each node involves two states denoted by Working (W) or Failed (F). The significant feature, coverage factor c, in a redundant system is taken into account to model the inaccuracy of a recovery mechanism. The coverage factor c is presented as $c =$ probability {system recovers|fault occurs} [30]. Then, DBN extends the BN by incorporating temporal dependencies at different time slices. For instance, the node C1 (t) is extended to C1 ($t + \Delta t$) with a temporal arc. Similarly, the DBNs of OR gate and 2003 voting gate are shown in Fig. 3a and c, respectively.

Dynamic logic gates are designed to express the time sequence and failure behaviors of the systems. The priority AND (PAND) gate, the functional dependency (FDEP) gate, and the spare gate are commonly used in DFT modeling. The mapping rules of spare gate are described based on the previous study [25, 26]. Generally, a

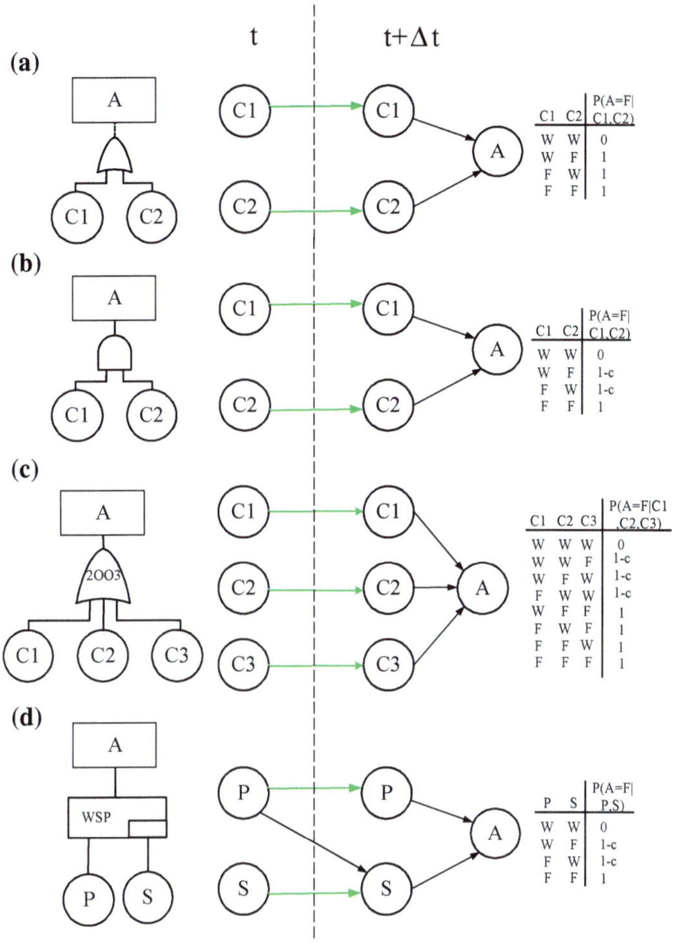

Fig. 3 Mapping the OR gate (**a**), AND gate (**b**), 2003 voting gate (**c**), and spare gate (**d**) to DBNs

spare gate consists of two types of elements: the primary modules and one or multiple redundant modules. For example, in Fig. 3d, the DBN structure is similar to that in Fig. 3a and b, but, the former one demonstrates that component S at $t + \Delta t$ time slice is dependent on both P at t time slice and S at t time slice. Assume that the primary P is active at the t time slice with failure rate λ, and the failure rate of one spare S is λ in active state or $\alpha\lambda$ at inactive state, where α is the dormancy factor. Hot and cold spares can be modeled by setting α equal to 1 and 0, respectively. Whenever the P fails, a replacement is initiated and the S will be powered up to keep the system functional.

3.3 Determination of DBN Parameters

DBN parameters are based on the prior probabilities of root nodes and the CPT of intermediate nodes and leaf nodes. The node C1 in Fig. 3b is demonstrated as an example. Assuming C1 follows the exponential distribution with failure rate λ, it can be obtained:

$$P\{\mathrm{C1}(t + \Delta t) = F|\mathrm{C1}(t) = F\} = 1 - e^{-\lambda t} \tag{4}$$

Considering a repair action, the availability of C1 can also be obtained. If the repair rate of C1 is μ, it can be obtained:

$$P\{\mathrm{C1}(t + \Delta t) = W|\mathrm{C1}(t) = F\} = 1 - e^{-\mu t} \tag{5}$$

The CPT for C1 at $t + \Delta t$ time slice given C1 at t time slice is provided in Tables 1 and 2.

For spare gate shown in Fig. 3d, the CPT of node S without and with repair are given in Tables 3 and 4, where α is the dormancy factor.

Table 1 CPT of C1 at $t + \Delta t$ time slice without repair

C1 at t	C1 at $t + \Delta t$	
	W	F
W	$e^{-\lambda t}$	$1 - e^{-\lambda t}$
F	0	1

Table 2 CPT of C1 at $t + \Delta t$ time slice with repair

C1 at t	C1 at $t + \Delta t$	
	W	F
W	$e^{-\lambda t}$	$1 - e^{-\lambda t}$
F	$1 - e^{-\mu t}$	$e^{-\mu t}$

Table 3 CPT of S at $t + \Delta t$ time slice without repair

P at t	S at t	S at $t + \Delta t$	
		W	F
W	W	$e^{-\alpha\lambda t}$	$1 - e^{-\alpha\lambda t}$
W	F	0	1
F	W	$e^{-\lambda t}$	$1 - e^{-\lambda t}$
F	F	0	1

Table 4 CPT of S at $t + \Delta t$ time slice with repair

P at t	S at t	S at $t + \Delta t$	
		W	F
W	W	$e^{-\alpha\lambda t}$	$1 - e^{-\alpha\lambda t}$
W	F	$1 - e^{-\mu t}$	$e^{-\mu t}$
F	W	$e^{-\lambda t}$	$1 - e^{-\lambda t}$
F	F	$1 - e^{-\mu t}$	$e^{-\mu t}$

3.4 Evaluation and Validation

Through the forward inference, the reliability and availability of different redundancy strategy can be obtained. Meanwhile, the posterior probabilities of each node are generated by the backward inference after an evidence is introduced. A sensitivity analysis is carried out with the assumption that the prior probabilities of five function modules are subject to the uncertainty of 10%. Moreover, the effects of coverage factor on reliability and availability will be calculated.

The validation of the proposed approach is a significant procedure to prove that it is reasonable for the reliability and availability evaluation of the actual system. In this paper, the validations are accomplished in two ways: A partial validation of the model usability should satisfy three axioms proposed by Jones et al. [31]. The availabilities obtained from the proposed approach are validated by analyzing the field data of one railway bureau in China.

4 Case Study

We choose a CTCS-3 onboard system as an example to illustrate the proposed approach. Before the reliability and availability analysis, three assumptions should be made:

1. All components are mutually independent;
2. Failures of all components in the system follow the exponential distributions with a constant failure rate, because they are mainly electronic products;
3. All components are considered "as good as new" after repairs.

4.1 CTCS-3 Onboard System

Figure 4 shows the hierarchical architecture of the CTCS-3 onboard system [32, 33]. According to the system requirements specification, the mean time between failure (MTBF) shall be not less than 10^5h and the availability of CTCS-3 onboard system

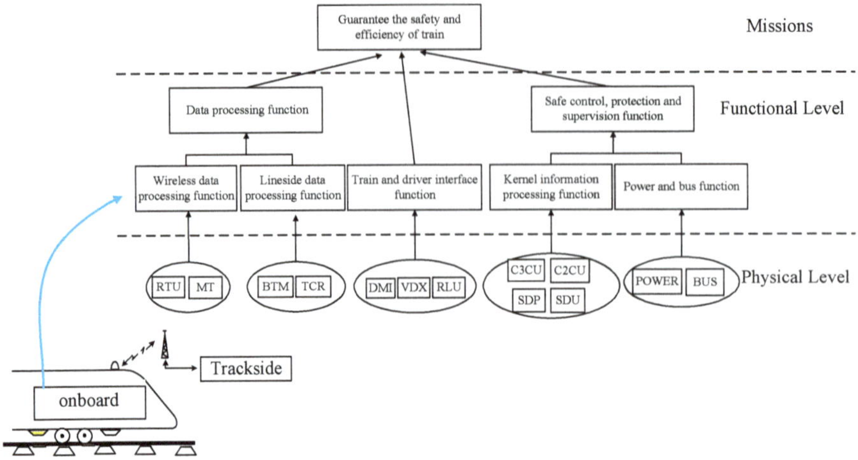

Fig. 4 Architecture of the CTCS-3 onboard system

shall be not less than 0.9999 [32]. Since the transition between CTCS-3 and CTCS-2 is possible, a CTCS-3 onboard system also includes the components of CTCS-2. The five functional modules are realized by the corresponding components shown in the physical layer. It should be noted that:

1. WP function is realized by Mobile Terminal (MT) and RTU. RTU is used for processing the messages, received or transmitted by MT.
2. LP function is conducted by BTM and track circuit reader (TCR). BTM is designed for processing the telegrams, received by BTM antenna. TCR receives the messages from TCR antenna and transmits the messages to VC.
3. TD function is conducted by DMI and train interface unit (TIU). TIU consists of vital digital input/output unit (VDX) and relay unit (RLU).
4. KN function accomplishes safe control, protection, and supervision of the train. CTCS-3 vital computer (C3VC) and CTCS-2 vital computer (C2VC) are the core computing systems for preventing the train from overspeed or overrunning in CTCS-3 and CTCS-2, respectively. Speed and distance processing unit (SDP) measures the speed and distance.
5. PB function is conducted by POWER and BUS.

Field data have been collected from one railway bureau in China. The partial specification of field data is shown in Table 5. It should be mentioned that the field data are related to specific railway lines and operational environment. The failure number of components in a given runtime is obtained from the field data. The failure rate can therefore be expressed as:

$$\lambda = N/N_m/T \tag{6}$$

Table 5 Partial specification of field data

Failure time	Failure mode	Failure effect	Recovery time (minutes)	Failure cause
21/05/2015	Invalid BTM port	Stopped the train	11	BTM1 failure
18/05/2015	Wireless communication timeout	Degraded to CTCS-2	10	MT1 failure
18/05/2015	Error on C3CU	Stopped the train	10	C3CU failure
17/05/2015	Error on train interface	Stopped the train	18	VDX2 failure

where N is the number of component failures, N_m is the total number of the onboard systems, and T is the runtime.

4.2 *The DFT of CTCS-3 Onboard System*

The construction of DFT is constructed based on the system structure and the expert experience. Failure of the CTCS-3 onboard system is considered as the top event. There are five intermediate events, i.e., lineside data processing failure, wireless communication failure, power and bus failure, kernel information processing failure, and the train and driver interface failure. The relationship between events is shown in Fig. 5. The spare gate is either cold or hot spare gate. Specifically, VDX1 and VDX2 are connected by an OR gate because any fault in the VDX1 or VDX2, the train will be stopped unconditionally.

4.3 *Mapping the DFT to DBN*

In this paper, the DBNs for CDD, TMR, and HDD onboard systems are analyzed using GeNIe 2.1 academic software developed at the University of Pittsburgh [34]. To establish the DBNs, the mapping rules from DFTs to DBNs are utilized. The DBN of HDD or CDD onboard system with two-time slices is shown in Fig. 6, whereas, the DBN of TMR onboard system should replace the C3VC and C2VC with 2003 voting architecture. The top event is mapped into the corresponding child node (CTCS-3 onboard), and the basic events are translated into the corresponding parent nodes in DBNs. The failure rates of the components are based on both field data and expert experience, shown in Table 6.

In consideration that the degradation from CTCS-3 to CTCS-2 level is possible, the child node (CTCS-3 onboard) has three states corresponding to working (W), degraded (D) and failed (F), whereas the other nodes only have two states (W, F). It

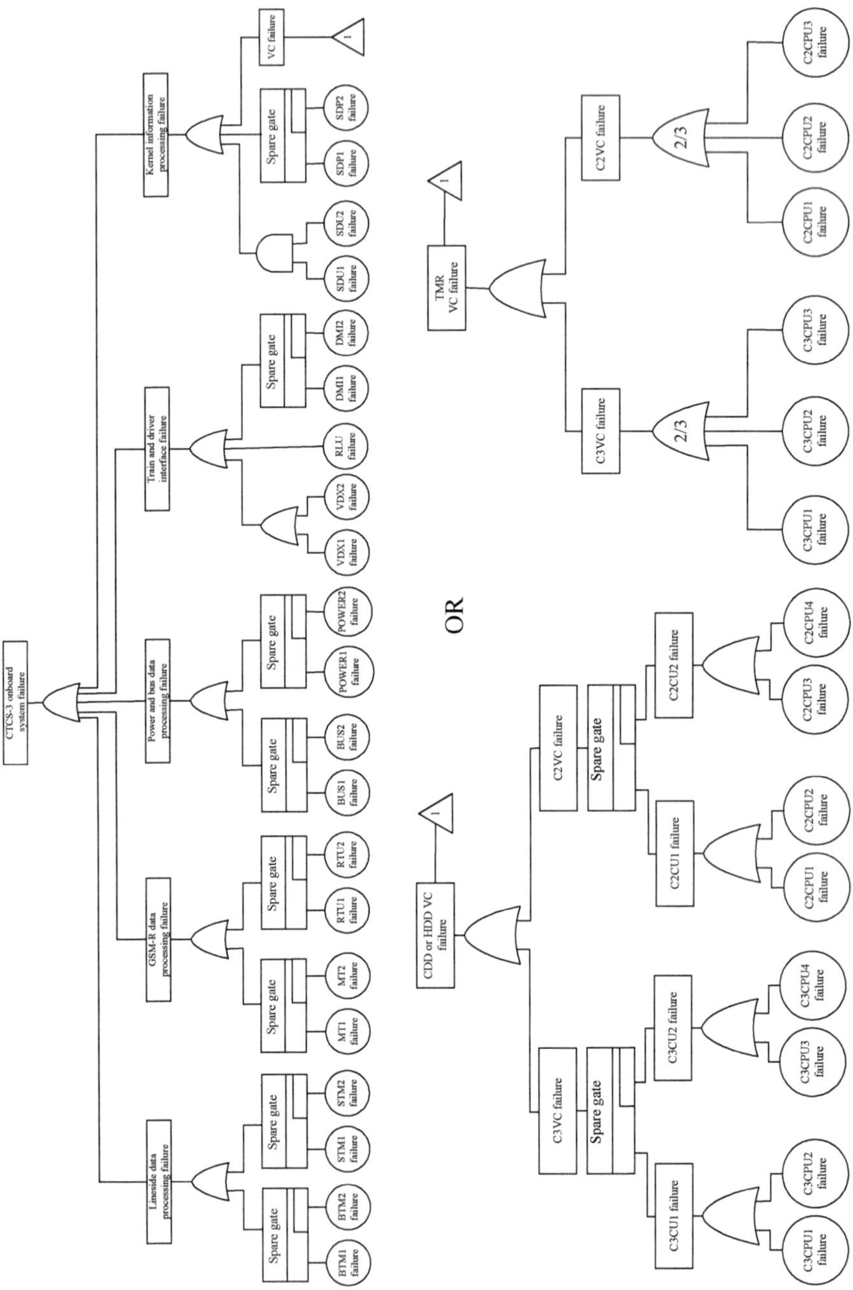

Fig. 5 DFT of CDD, HDD, and TMR onboard systems

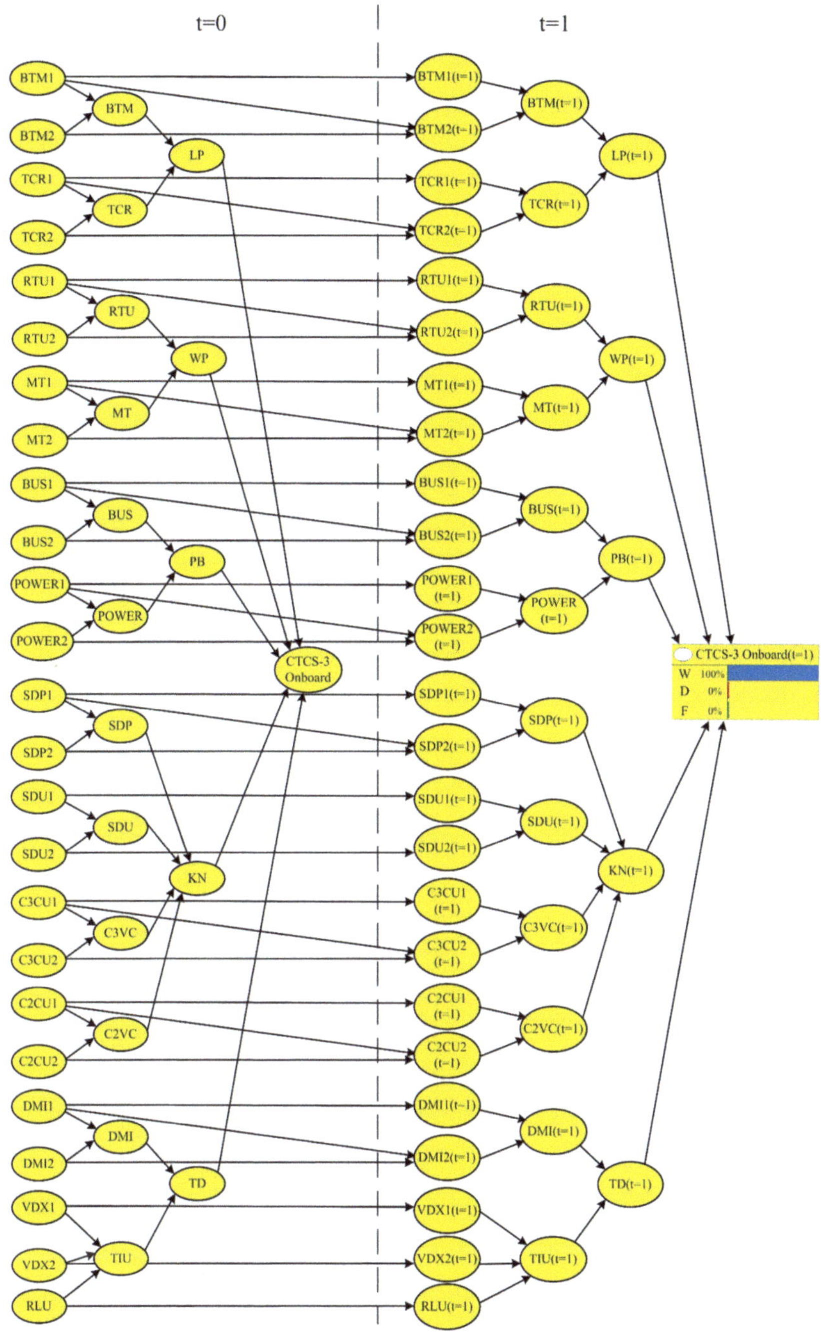

Fig. 6 DBN of HDD or CDD onboard system

Table 6 Failure rates, prior, and posterior probabilities of components

No.	Component	Failure rate λ (/h)	CDD		HDD		TMR	
			Prior probabilities	Posterior probabilities	Prior probabilities	Posterior probabilities	Prior probabilities	Posterior probabilities
1	C3CPU1	7.45E−06	8.56E−02	1.54E−01	8.56E−02	1.61E−01	8.56E−02	1.82E−01
2	C3CPU2		7.71E−03	6.14E−02	8.56E−02	1.61E−01	–	–
3	C2CPU1	6.00E−06	7.02E−02	1.21E−01	7.02E−02	1.25E−01	7.02E−02	1.39E−01
4	C2CPU2		5.12E−03	4.08E−02	7.02E−02	1.25E−01	–	–
5	SDP1	5.00E−07	6.27E−03	8.63E−03	6.27E−03	8.09E−03	6.27E−03	8.30E−03
6	SDP2		2.93E−05	2.33E−04	6.27E−03	8.09E−03	6.27E−03	8.30E−03
7	SDU1	2.50E−07	3.14E−03	4.31E−03	3.14E−03	4.04E−03	3.14E−03	4.15E−03
8	SDU2		1.46E−05	1.17E−04	3.14E−03	4.04E−03	3.14E−03	4.15E−03
9	RTU1	1.80E−06	2.24E−02	2.24E−02	2.24E−02	2.24E−02	2.24E−02	2.24E−02
10	RTU2		2.51E−04	2.51E−04	2.24E−02	2.24E−02	2.51E−04	2.51E−04
11	MT1	6.00E−06	7.28E−02	7.28E−02	7.28E−02	7.28E−02	7.28E−02	7.28E−02
12	MT2		7.28E−02	7.28E−02	7.28E−02	7.28E−02	7.28E−02	7.28E−02
13	BTM1	2.07E−06	2.57E−02	3.66E−02	2.57E−02	3.49E−02	2.57E−02	3.60E−02
14	BTM2		3.31E−04	2.64E−03	2.57E−02	3.49E−02	2.57E−02	3.60E−02
15	TCR1	2.30E−06	2.86E−02	4.09E−02	2.86E−02	3.91E−02	2.86E−02	4.03E−02
16	TCR2		4.08E−04	3.25E−03	2.86E−02	3.91E−02	2.86E−02	4.03E−02
17	POWER1	6.00E−06	7.28E−02	1.28E−01	7.28E−02	1.12E−01	7.28E−02	1.17E−01

(continued)

Table 6 (continued)

No.	Component	Failure rate λ (/h)	CDD		HDD		TMR	
			Prior probabilities	Posterior probabilities	Prior probabilities	Posterior probabilities	Prior probabilities	Posterior probabilities
18	POWER2		7.28E−02	1.28E−01	7.28E−02	1.12E−01	7.28E−02	1.17E−01
19	Bus1	4.00E−06	4.92E−02	7.32E−02	4.92E−02	7.13E−02	4.92E−02	7.39E−02
20	Bus2		1.22E−03	9.68E−03	4.92E−02	7.13E−02	4.92E−02	7.39E−02
21	DMI1	5.00E−06	6.11E−02	9.29E−02	6.11E−02	9.14E−02	6.11E−02	9.49E−02
22	DMI2		1.88E−03	1.50E−02	6.11E−02	9.14E−02	6.11E−02	9.49E−02
23	VDX1	2.00E−06	2.49E−02	1.98E−01	2.49E−02	1.49E−01	2.49E−02	1.64E−01
24	VDX2		2.49E−02	1.98E−01	2.49E−02	1.49E−01	2.49E−02	1.64E−01
25	RLU	1.50E−06	1.87E−02	1.49E−01	1.87E−02	1.12E−01	1.87E−02	1.23E−01

Table 7 CPT of the node (CTCS-3 onboard)

LP		WP		PB		KN		TD		P (system|LP, WP, PB, KN, TD)		
W	F	W	F	W	F	W	F	W	F	W	D	F
1	0	1	0	1	0	1	0	1	0	1	0	0
1	0	0	1	1	0	1	0	1	0	0	1	0
0	1	1	0	1	0	1	0	1	0	0	0	1
0	1	0	1	0	1	0	1	0	1	0	0	1

should be mentioned that the degradation of the node (CTCS-3 onboard) occurs only if the wireless communication failure happens, and the CPT of this node is shown in Table 7.

In the case study, Δt is set to be one week, i.e., 126 h. Since failures follow the exponential distribution, the initial states are in the perfect functioning in 0th week, and the failure probabilities of those nodes are assigned to 0. According to the DBN parameter modeling, the CPT of the nodes in other time slices can be calculated.

Since the component is considered "as good as new" after repair action, the average availability is calculated as MUT/(MUT + MDT), where the MUT is the average working time, and the MDT is the average down time. Considering the operational situation of onboard system shown in Fig. 2, the MDT mainly depends on the mean waiting time of the component in 18 h after it failed. Therefore, the mean waiting time for both primary and spare components in hot spare and 2003 redundancy system is 9 h, whereas the values are 9 and 4.5 h in cold spare redundancy system.

4.4 Results and Discussions

Reliability and availability evaluation The reliabilities and availabilities within 100 weeks are evaluated by the forward inference, shown in Fig. 7. The coverage factors for the redundant system are assigned to 0.95. As indicated in Fig. 7a, the reliabilities decrease with time. The reliability of CDD system is higher than HDD system, whereas the TMR system is between them. Moreover, the reliabilities of CDD, TMR, and HDD systems at 100th week are 0.81, 0.785, and 0.771, respectively. As indicated in Fig. 7b, the occurrence probabilities of degraded state for the CDD, TMR, and HDD systems increase to 0.065, 0.064, and 0.063 at 100th week, respectively.

As shown in Fig. 7c, the availability of CDD, TMR, and HDD systems is 0.999923, 0.999909, and 0.999902, respectively. The availabilities of the three onboard systems approach steady values in 10 weeks. The high availabilities indicate that the onboard systems can recover rapidly when the primary system suffers a failure. Obviously, the availabilities accord with design specification that it should be greater than 0.9999.

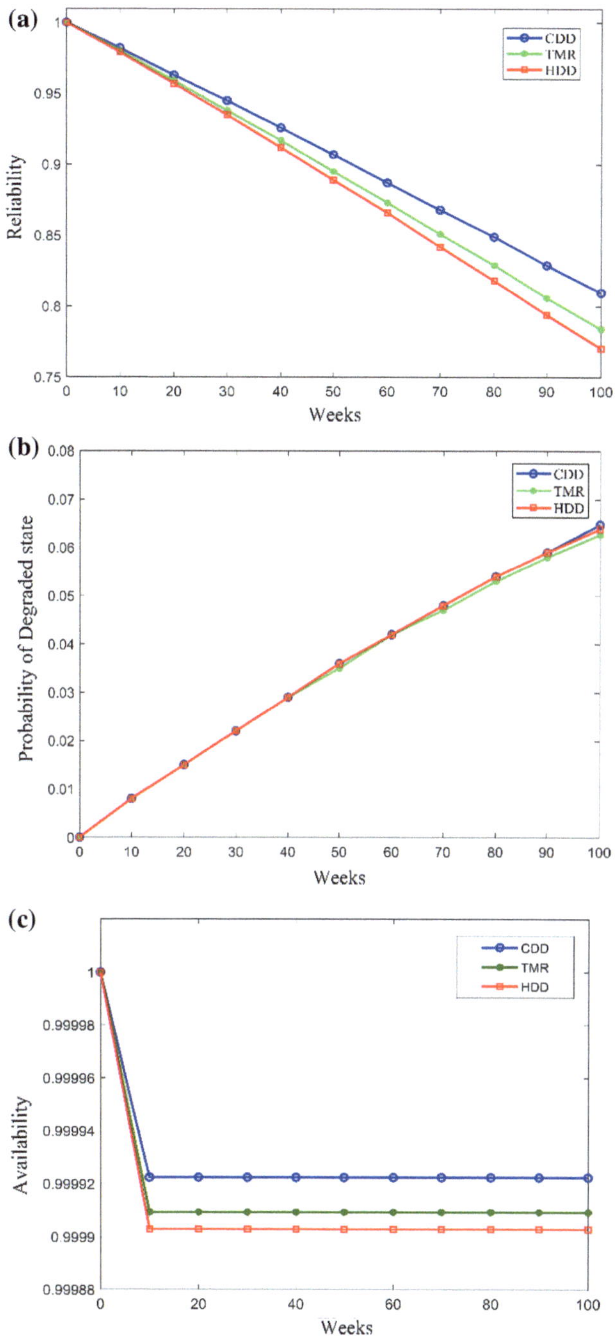

Fig. 7 Reliability (**a**), degraded state (**b**), and availability (**c**) evaluation of three onboard systems

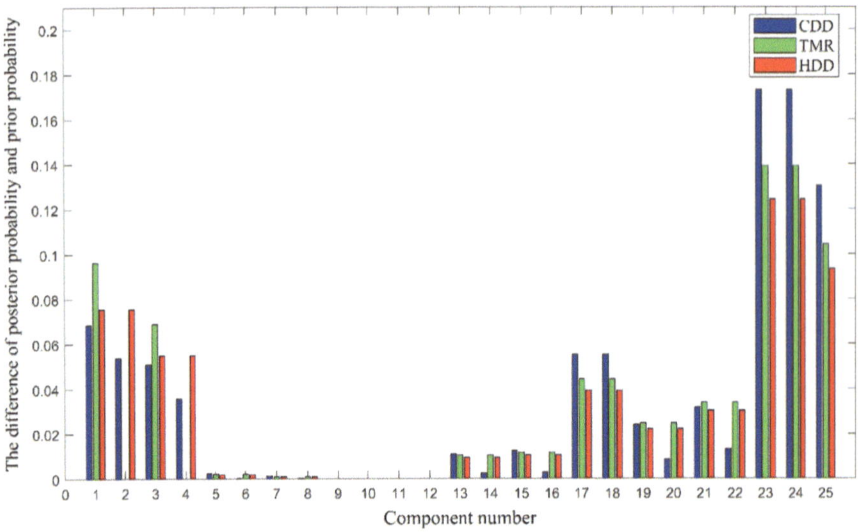

Fig. 8 Difference between posterior probability and prior probability

By setting the failure probability of CTCS-3 onboard node to 1, the posterior probabilities of the component are obtained by backward inference. The prior probability and posterior probability for three onboard systems at 100th week are listed in the 4th–9th columns of Table 6. The difference between posterior probability and prior probability is shown in Fig. 8. It can be seen that the values of VDX, RLU, C3CPU, C2CPU, and POWER are much higher than other components, meaning that the five components should be given more attention to improving the reliability of CTCS-3 onboard system. In addition, the wireless communication failure only leads the system to degraded state so that the values of MT and RTU are 0.

The failure probability for five function modules in different time slices As indicated in Fig. 9, the failure probabilities for five function modules increase with time. WP, KN, and TD have higher failure probabilities than LP and PB. In addition, LP has a negligible effect on system failure. Specifically, WP is mainly responsible for the degraded state of CTCS-3 onboard system, meaning that the higher probability of WP brings the less efficient of the system. It can be concluded that KN and TD are more critical than LP and PB as their failure probabilities are much higher. It should be noted that the failure probability of KN is higher than TD at 100th week for HDD system, whereas the failure probability is lower in CDD and TMR systems.

Sensitivity analysis The variables of failure probabilities for each onboard system at 100th week are calculated with the assumption that the failure rates of each function module change 10%. Effects of changes in each function part are shown in Fig. 10. It can be concluded that the order of effects on the systems failure probabilities are: TD > WP > KN > PB > LP, and KN > WP > TD > PB > LP for CDD and HDD systems, respectively. However, the order is KN > TD > WP > PB > LP for TMR

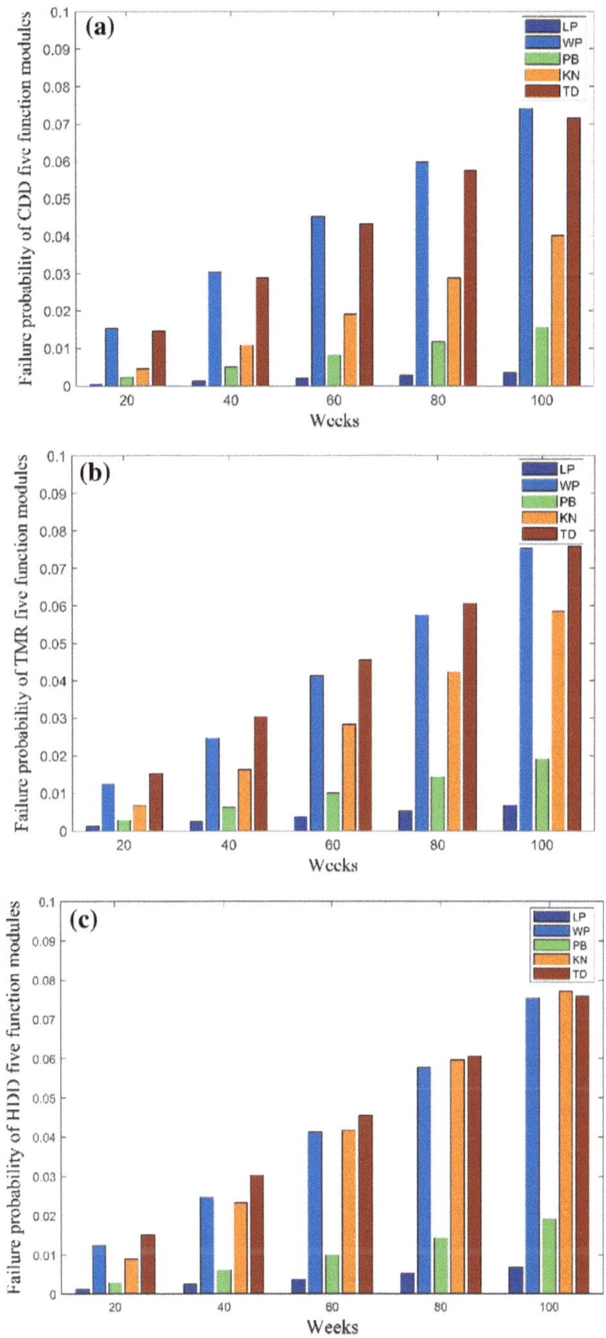

Fig. 9 Failure probability for five function modules in different time slices

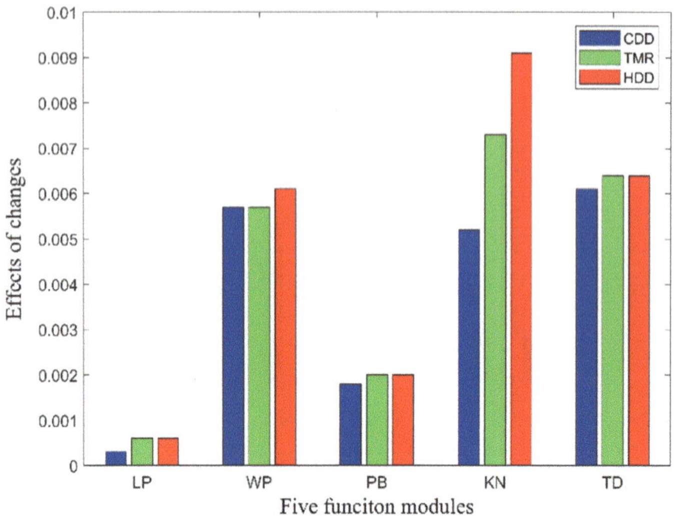

Fig. 10 Effects of changes in each function module

system. Moreover, it shows that the KN has the large fluctuation for the three kinds of CTCS-3 onboard system, whereas others have little fluctuation.

Effects of coverage factor on reliability and availability From the above analysis, the value of coverage factor is 0.95. To analyze the effects of coverage factor, the values are assigned to 0.9, 0.925, 0.95, 0.975, and 1 to calculate the reliability and availability of onboard system at 100th week, as shown in Fig. 11. It can be seen that the reliability and availability increase with the increasing coverage factor, meaning that the recovery mechanism is significant for the three onboard systems. Furthermore, the effects on HDD system are more important than CDD system, whereas the TMR system is between them. The difference of reliability and availability shrinks with the increasing coverage factor. Specifically, the availabilities of the three onboard systems reach the same value, i.e., 0.999949. To achieve high reliability and availability, the recovery mechanism should be paid more attention.

Validation of the model To validate the model usability, the DBNs of HDD system is taken as an example. When the parent node "BTM1" is set to 50 from 0%, the reliability of system decreases to 0.743 from 0.771. When both the parent nodes "BTM1" and "BTM2" are set to 50%, the reliability decreases to 0.55. With the addition of parent node "TCR1" and "TCR2" are set to 50%, the reliability decreases to 0.392. In addition, the sensitivity analysis is also a validation of the model usability. Therefore, the exercise of increasing the influencing node gives a partial validation to the proposed model.

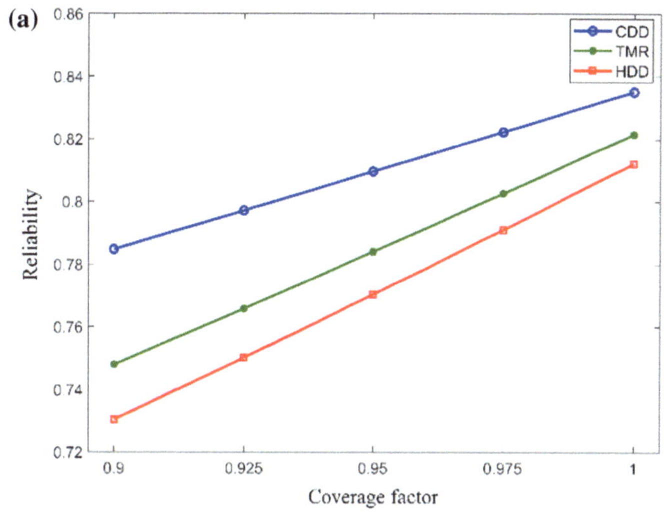

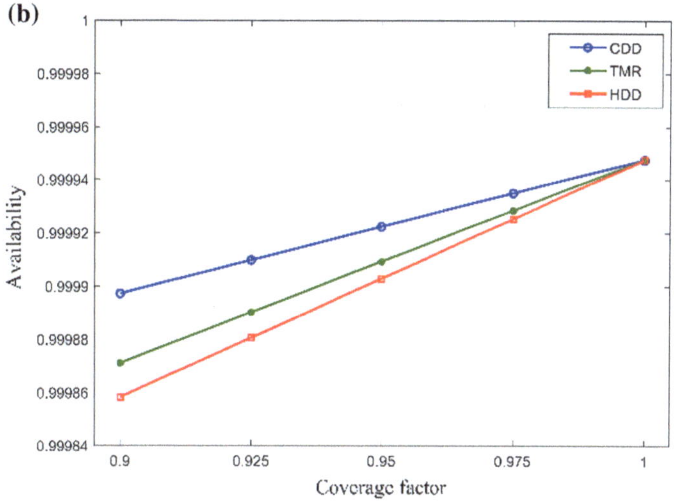

Fig. 11 Effects of coverage factor to reliability and availability

The availabilities obtained can be validated by the field data from a railway bureau in China. Through the description of system, the corrective maintenance data have been analyzed between 2015.5 and 2016.11. The CDD onboard system has been adopted for electrical multiple unit (EMU) and the number of EMU is 63. According to Table 5, the total recovery time is 2683 min and the number of system failure is 155. The starting time for the availability calculation is May 1, 2015. Then, the availabilities can be obtained by A = MUT/(MUT + MDT). For example, the first failure occurs on May 17. The total time is 1088640 min, and the MDT is 18 min. Therefore, the availability is 0.999983. Similarly, all the availabilities are calculated

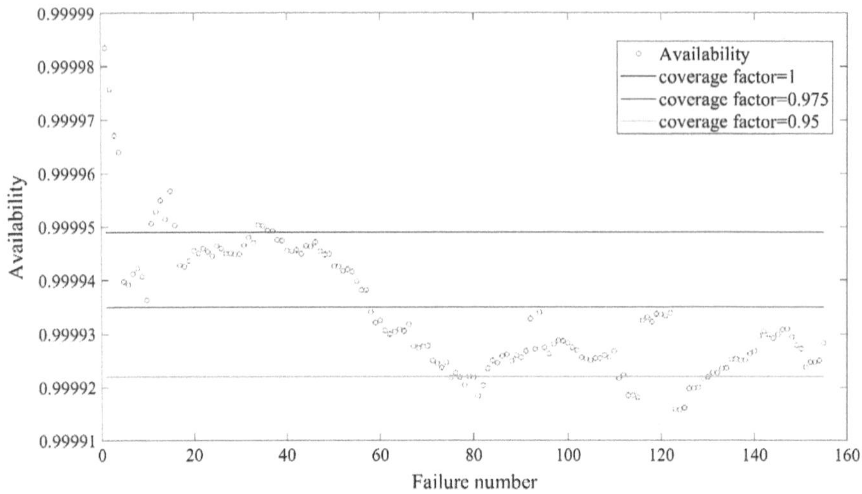

Fig. 12 Availability for the CDD onboard system

and the average availability is 0.999934 during the operational time, as shown in Fig. 12. It can be concluded that most of the availabilities are between 0.999949 (coverage factor $c = 1$) and 0.999923 (coverage factor $c = 0.95$). When the coverage factor c is assigned to 0.975, the availability obtained by the proposed approach is 0.999935, which is approximately equal to the average availability based on the field data. The practical exercise gathering the field data performs a partial validation for the model.

Discussions The CDD, HDD, and TMR onboard high-speed train control systems are adopted in the operational phase, due to the tiny discrepancy of reliability and availability within 100 weeks. Then, the choice of redundancy strategies becomes an additional issue. In the HDD and TMR onboard systems, all the redundant components are affected by operational stresses and the cost of high-reliability components is a real problem. For CDD onboard system, the redundant components are shielded from the operational stresses and they do not fail before operation. Therefore, the cost of CDD is lower than HDD and TMR onboard systems. The choice of CDD onboard system is a reasonable trade-off of reliability and cost, especially when the number of EMU is very large. The CDD onboard system requires a manual switch to activate the redundant component. One failure can cause the train to stop and the driver needs to restart the system, which could affect the efficiency of train operation. However, this feature makes the CDD onboard system satisfy the fail-safe concept.

In this paper, the simplifying assumption of constant failure rates (exponential distribution for components) is made. However, in the operational phase, the failure rate may increase with time, due to the aging and degradation of the components. Future works should try to develop the models in BN with handling of non-constant failure rates.

Another important issue is that this paper mainly focuses on the onboard system, ignoring the relevance of trackside and communication systems. Therefore, future works should focus on the reliability and availability evaluation of the whole high-speed train control system in order to handle the interdependencies and dynamic problems between onboard and trackside system.

5 Conclusions

The reliability and availability of onboard train control system are significant for the performance of high-speed railway network. According to the architecture and operational situation of onboard system, the DFTs and DBNs are constructed. The DBN forward, backward, and sensitivity analyses are conducted to support the maintenance decision making. The DBN-based approach provides a powerful reliability and availability evaluation for onboard system. The main achievements can be summarized as follows:

1. The reliability and availability of CDD, TMR, and HDD onboard systems are evaluated and compared in the operational phase.
2. Based on the system architecture and operation of the onboard system, the problems such as dynamic failure behavior and imperfect coverage factors have been solved.
3. The results of availability are validated by the field data from one railway bureau.
4. To improve the reliability and availability of onboard system, the VDX, RLU, C3CPU, C2CPU, and POWER should be paid more attention. Based on sensitivity analysis, the effects of failure rate have been researched.

References

1. UNISIG. Subset-026 of the ERTMS/ETCS system requirements specification (SRS), 2012
2. CTCS general Rules of technical specification, *Ministry of Railways* (Science and Technology Division, Beijing, China, 2004)
3. J.B. Dugan, S.J. Bavuso, M.A. Boyd, Dynamic fault-tree models for fault-tolerant computer systems. IEEE Trans. Reliab. **41**(3), 363–377 (1992)
4. H. Langseth, L. Portinale, Bayesian networks in reliability. Reliab. Eng. Syst. Saf. **92**, 92–108 (2007)
5. F. Flammini, S. Marrone, N. Mazzocca et al., Modeling system reliability aspects of ERTMS/ETCS by fault trees and Bayesian Network, in *Safety and Reliability for Managing Risk*, ed. by G. Soares, Zio (Taylor & Francis Group, London, 2006), pp. 2675–2683
6. F. Flammini, S. Marrone, M Iacono et al., A multi-formalism modular approach to ERTMS/ETCS failure modeling. Int. J. Reliab. Qual. Safety Eng. **21**(1):1450001-1–1450001-29 (2014)
7. L.Q. Di, X. Yuan, Y.N. Wang, Research on the evaluation method for the RAM goals of CTCS-3. China Railway Sci. **31**(6), 92–97 (2010)

8. H.S. Su, Y.L. Che, Dependability assessment of CTCS-3 on-board subsystem based on Bayesian network. China Railway Sci. **35**(5), 96–104 (2014)
9. S. Qiu, M. Sallak, W. Schön et al., Availability assessment of railway signalling systems with uncertainty analysis using Statecharts. Simul. Modell. Pract. Theory **47**, 1–18 (2014)
10. S. Qiu, M. Sallak, W. Schön et al., Modeling of ERTMS level 2 as an SoS and evaluation of its dependability parameters using state charts. IEEE Syst. J. **8** (4), 1169–1181 (2014)
11. S. Bernardi, F. Flammini, S. Marrone et al. Model-driven availability evaluation of railway control systems, in *International Conference on Computer Safety, Reliability, and Security*, Naples, Italy, 19–22 Sept 2011 (Springer, Berlin), pp. 15–28
12. L. Carnevali, F. Flammini, M. Paolieri et al., Non-markovian performability evaluation of ERTMS/ETCS Level 3, in *European Workshop on Computer Performance Engineering*, Madrid, Spain, 31 Aug–1 Sept 2015 (Springer, Cham), pp. 47–62
13. M. Biagi, L. Carnevali, M. Paolieri et al., Performability evaluation of the ERTMS/ETCS-Level 3. Transp. Res. Part C: Emerg. **82**, 314–336 (2014)
14. G. Neglia, S. Alouf, A. Dandoush et al., Performance evaluation of train moving-block control, in *International Conference on Quantitative Evaluation of Systems*, Quebec City, Canada, 23–25 Aug 2016 (Springer, Cham), pp. 348–363
15. P. Weber, G. Medina-Oliva, C. Simon et al., Overview on Bayesian networks applications for dependability, risk analysis and maintenance areas. Eng. Appl. Artif. Intell. **25**: 671–682 (2012)
16. B.P. Cai, L. Huang, M. Xie, Bayesian networks in fault diagnosis. IEEE Trans. Ind. Inf. **13**(5), 2227–2240 (2017)
17. D. Codetta-Raiteri, L. Portinale, Approaching dynamic reliability with predictive and diagnostic purposes by exploiting dynamic Bayesian networks. Proc. IMechE Part O: J. Risk and Reliab. **228**(5), 488–503 (2014)
18. H. Boudali, J.B. Dugan, A discrete-time Bayesian network reliability modeling and analysis framework. Reliab. Eng. Syst. Saf. **87**, 337–349 (2005)
19. B.P. Cai, H.L. Liu, M. Xie, A real-time fault diagnosis methodology of complex systems using object-oriented Bayesian networks. Mech. Syst. Sig. Process **80**, 31–44 (2016)
20. M. Neil, D. Marquez, Availability modelling of repairable systems using Bayesian networks. Eng. Appl. Artif. Intell. **25**, 698–704 (2012)
21. X.F. Liang, H.D. Wang, H. Yi et al. Warship reliability evaluation based on dynamic bayesian networks and numerical simulation. Ocean Eng. **136**, 129–140 (2017)
22. B.P. Cai, Y.H. Liu, Y.W. Zhang et al. Dynamic Bayesian networks based performance evaluation of subsea blowout preventers in presence of imperfect repair. Expert Syst. Appl. **40**, 7544–7554 (2013)
23. B.P. Cai, M. Xie, Y.H. Liu et al. Availability-based engineering resilience metric and its corresponding evaluation methodology. Reliab. Eng. Syst. Saf. **172**, 216–224 (2018)
24. A. Bobbio, L. Portinale, M. Minichino et al. Improving the analysis of dependable systems by mapping fault trees into Bayesian networks. Reliab. Eng. Syst. Saf. 249–260 (2001)
25. L. Portinale, D. Codetta-Raiteri, S. Montani, Supporting reliability engineers in exploiting the power of dynamic Bayesian networks. Int. J. Approximate Reasoning **51**, 179–195 (2010)
26. S. Montani, L. Portinale, A. Bobbio et al. RADYBAN: a tool for reliability analysis of dynamic fault trees through conversion into dynamic Bayesian networks. Reliab. Eng. Syst. Saf. **93**, 922–932 (2008)
27. J.Y. Zhu, A. Deshmukh, Application of Bayesian decision networks to life cycle engineering in Green design and manufacturing. Eng. Appl. Artif. Intell. **16**, 91–103 (2003)
28. A.G. Wilsona, A.V. Huzurbazar, Bayesian networks for multilevel system reliability. Reliab. Eng. Syst. Saf. **92**(10), 1413–1420 (2007)
29. K.P. Murphy, *Dynamic Bayesian Networks: representation* (Inference and Learning. University of California, Berkeley, 2002)
30. J.B. Dugan, K. Trivedi, Coverage modeling for dependability analysis of fault-tolerant systems. IEEE Trans. Comput. **38**, 775–787 (1989)
31. B. Jones, I. Jenkinson, Z. Yang et al. The use of Bayesian network modelling for maintenance planning in a manufactoring industy. Reliab. Eng. Syst. Saf. 95, 267–277 (2010)

32. CTCS-3 system requirements specification. Ministry of Railways, Science and Technology Division, Beijing, China (2008)
33. CTCS-3 function requirements specification. Ministry of Railways, Science and Technology Division, Beijing, China (2008)
34. University of Pittsburgh, GeNIe and SMILE-Home. https://dslpitt.org/genie/

Failure Probability Analysis for Emergency Disconnect of Deepwater Drilling Riser Using Bayesian Network

Abstract Drilling risers are the crucial connection of subsea wellhead and floating drilling vessel. Emergency disconnect (ED) is the most important protective measure to secure the risers and wellhead under extreme conditions. This paper proposes a methodology for failure probability analysis of ED operations using Bayesian network (BN). The risk factors associated with ED operations and the potential consequences of ED failure were investigated. A systematic ED failure and consequence model was established through fault tree and event sequence diagram (FT-ESD) analyses and, then the FT-ESD model was mapped into BN. Critical root causes of ED failure were inferred by probability updating, and the most probable accident evolution paths as well as the most probable consequence evolution paths of ED failure were figured out. Moreover, the probability adaptation was performed at regular intervals to estimate the probabilities of ED failure, and the occurrence probabilities of consequences caused by ED failure. The practical application of the developed model was demonstrated through a case study. The results showed that the probability variations of ED failure and corresponding consequences depended on the states of critical basic events (BEs). Eventually, some active measures in drilling riser system design, drilling operation, ED test, and operation were proposed for mitigating the probability of ED failure.

Keywords Deepwater drilling riser · Emergency disconnect · FT-ESD model · Bayesian network · Failure probability analysis

1 Introduction

With exploration and development of oil and gas resources moving into deepwater, the demand for drilling vessels capable of drilling in or beyond deepwater is increasing. When drilling operations are conducted from dynamically positioned (DP) drilling rigs, it is necessary to perform ED of the riser system from time to time to avoid serious damage to the drilling riser system and secure the well [1]. The Deepwater Horizon accident, on April 20, 2010, which might be the largest marine catastrophe, was caused partially due to the failure of the blowout preventer (BOP)

B. Cai et al., *Bayesian Networks for Reliability Engineering*,
https://doi.org/10.1007/978-981-13-6516-4_7

and the ED system [2]. ED failure, though rare, is likely to cause blowout which is the most undesired and feared accident that greatly threatens human lives, environment, and assets.

Risk analysis is an effective tool to develop strategies to prevent accident and devise mitigative measures [3]. Quantitative risk and reliability analysis techniques have been widely used to reduce the probability of failure in offshore drilling operations. Some of these techniques include: fault tree (FT), event tree (ET), reliability block diagram, reliability graphs, and the Markov chain. However, nowadays, BN is becoming a popular probabilistic inference technique for reasoning under uncertainty. The main advantage of BN is the ability to perform probability updating and sequential learning, which makes it a superior technique for risk analysis of dynamic systems [3]. Abimbola et al. [4] used BN to conduct safety and risk analysis of managed pressure drilling operation. Khakzad et al. [3] conducted quantitative risk analysis of offshore drilling operations using Bayesian approach. Yang et al. [5] established a systematic corrosion failure model through bow-tie analysis and mapped the bow-tie model into a BN model to conduct failure analysis of subsea pipelines induced by corrosion. Bhandari et al. [6] conducted a dynamic safety analysis of deepwater managed pressure drilling and underbalanced drilling operations using the BN. Cai et al. [2, 7, 8] utilized BN to conduct quantitative risk assessment of subsea BOP operations and reliability evaluation for subsea BOP control system. A BN-based failure evolution model for subsea pipelines was developed by Li et al. [9].

Some hazards related to uncertainty are difficult to model by traditional QRA approaches. Furthermore, historical records of some risk scenarios, particularly extreme hazardous events, are often incomplete and insufficient. Therefore, it may be necessary to carry out a risk assessment based on multiple hazards which are represented in various forms such as probabilistic data, experts' opinions, and linguistic representations. The fuzzy set theory can be used to present subjective, vague, linguistic, and imprecise data and information effectively. In the fuzzy fault tree analysis (FFTA), the probability values of components will be characterized by fuzzy numbers. Using fuzzy set theory, fuzzy number in linguistic term can be transformed into fuzzy failure probability of BEs, and quantitative risk analysis of top events can be conducted by FT method. Lavasani et al. [10–12] applied fuzzy set theory to evaluate the risk of leakage in abandoned oil and natural gas wells and deethanizer failure in petrochemical plant operations. Ren et al. developed an offshore risk analysis method using a fuzzy BN where triangular fuzzy membership functions were used to elicit expert judgments. Ferdous et al. [13] proposed a methodology for computer-aided FFTA. Chen et al. [14] conducted risk assessment of an oxygen-enhanced combustor using a structural model based on the failure mode and effects analysis (FMEA) and FFTA. Shi et al. [15] performed FFTA for fire and explosion accidents for steel oil storage tanks.

In practical operational conditions, various factors, e.g., human, design, operation, time, equipment, and control, are all able to cause the failure of ED which could cause disastrous consequences in deepwater drilling. However, studies for ED operations of deepwater drilling riser from risk perspective can only be found sporadically in the literature. Thus, it is necessary to conduct a comprehensive study to address the failure probability analysis of ED operations for the actual engineering requirements.

The objective of this paper is to propose a failure probability analysis methodology for ED operations of deepwater drilling risers, which could be used to assess the probabilities of ED failure and different failure consequences. In this research, a FT-ESD model was developed to present a systematic accident scenario and accident evolution process caused by ED failure. A BN model was mapped from the developed FT-ESD model to identify the critical events and analyze the most probable paths for ED failure and the most probable paths of the consequences resulting from ED failure by updating the prior probability of BEs. The BN model also aimed to investigate the failure probability of ED by introducing new critical BEs. Finally, some suggestions and measures for ED operations are proposed to reduce the probability of failure accident.

The paper is structured as follows: Sect. 2 introduces the process, analyzes the reasons of ED, and investigates the mechanism of factors influencing the ED operations. In Sect. 3, the failure probability analysis methodology for ED operations of deepwater drilling riser is proposed. Section 4 identifies the hazards and analyzes the accident evolution process of ED failure by FT and ESD. Section 5 is a case study regarding the application of BN in quantitative failure probability analysis of deepwater drilling riser ED operations. Finally, the conclusions are presented in Sect. 6.

2 Background

2.1 Deepwater Drilling Riser System

Deepwater drilling conductor is the first layer of casing installed during the well construction in deepwater drilling, which is generally jetted into the formation without well cementing. After jetting the conductor with low-pressure wellhead (LPW), completing the installation of the casing surface tubular with high pressure wellhead (HPW), and cementing, drilling operation is followed by deployment of riser system and LMRP/BOP by making up the riser joints.

The main components of the riser column include BOP/LMRP stack, lower flex joint (LFJ), slick and buoyancy riser joints, telescopic joint (TJ), and upper flex joint (UFJ). The top end of the riser column is connected to the drilling vessel through the tension system. The TJ consists of inner and outer barrels where the relative motion (stroke) of these barrels can compensate for the length variations of riser column with the motion of the drilling vessel. The LFJ and UFJ can improve the mechanical performance for both ends of the riser column to avoid excessive bending moment and hence damage to the risers [16].

The subsea BOP/LMRP stack includes LMRP and BOP, which is usually equipped with two hydraulic connectors, namely the LMRP connector and wellhead connector.

The LMRP connector is located in the middle of two annular preventers, which is used to connect the LMRP to the BOP, and the wellhead connector is used to connect BOP and HPW [7, 8]. If ED is activated automatically or manually under extreme conditions, the LMRP will disconnect from BOP at the LMRP connector, and the riser column will be lifted up and suspended by the tensioners eventually after the disconnect is completed. If there is drill pipe in the drilling riser, the blind shear rams in BOP will cut through the pipe and seal the well before disconnect.

2.2 *Reasons for Emergency Disconnect*

Generally, there are four reasons for the ED of the drilling riser system that include drift-off, drive-off, storms, and internal solitary waves.

2.2.1 Drift-Off

Drift-off is an event normally caused by loss of power, malfunction in the power system, engine breakdown, or mechanical and human errors. When the DP system can no longer hold the position, the increasing offset of the drilling vessel due to wind, wave, and current will cause large horizontal force and bending moment to the subsea wellhead by drilling riser system, and the ED must be activated to avoid possible accident. If the ED operations cannot be completed successfully in 60 s at most, it may damage the wellhead or break the riser joints. Once the integrity of the well is damaged, the blowout accident will occur inevitably. According to the existing literature, it has been stated that the occurrence probability of drift-off event is 2×10^{-3} per year [17].

Establishing alert offsets for the ED of the vessel-connected riser system through drift-off analysis is used to determine the point of disconnect. Generally, the alert offsets settings are as follows: green region—drilling normally; yellow region—stop drilling and make the preparation for ED while the riser is in the "connected non-drilling mode"; red region—the ED is initiated automatically (it can also be initiated manually in advance) and must be completed before reaching the blue region; blue region—the suspended riser column is in survival mode [18].

2.2.2 Drive-Off

A drive-off is much the same as a drift-off, but it comes from a malfunction in the DP system causing the rig to drive off from its location. This is a very critical event due to the higher velocity of the vessel, and it provides a short available time to activate the ED before the horizontal offset gets too large. The occurrence probability of drive-off event is 1.6×10^{-5} per DP hour [19, 20].

2.2.3 Storm

Generally, the MODU will disconnect from the BOP before a storm is fully developed which is called "planned disconnect." However, if the storm is larger than predicted or if an anticipated rapidly developing seastate happens, an unplanned ED would be needed to secure the drilling risers and wellhead.

2.2.4 Internal Solitary Waves

Internal solitary waves are the nonlinear large amplitude waves existing in the oceanic pycnocline [7, 8]. A large number of measurements and remote-sensing observations have shown that internal solitary waves happen frequently and exist widely in the South China Sea, which have been a fundamental environmental factor that must be taken into account in designing the ocean engineering facilities [21]. On April 6, 2014, when NAIHAI-8 drilling platform was drilling normally in Liuhua oil field in the South China Sea, the internal solitary waves pushed the vessel 137 m away from its original position, and almost reached the red alert offset, causing damage to the ropes of the tensioners.

3 Failure Probability Analysis Technology

3.1 Fuzzy FTA

A fuzzy set takes values from the interval [0,1] and is characterized by a membership function $m(x)$, which represents the relationship among different elements. Fuzzy sets are defined for specific linguistic variables, which can be calculated by triangular fuzzy numbers (TFNs) or trapezoidal fuzzy numbers (ZFNs). The TFNs are denoted by a triplet (a1, a2, and a3), and the ZFNs are denoted by a triplet (a1, a2, and a3) and a quadruple (a1, a2, a3, and a4) and can be defined as follows [14, 22]:

$$\mu(x) = \begin{cases} 0; & x \leq a_1 \\ (x - a_1)/(a_2 - a_1); & a_1 \leq x \leq a_2 \\ (a_3 - x)/(a_3 - a_2); & a_2 \leq x \leq a_3 \\ 0; & x \geq a_3 \end{cases} \tag{1}$$

$$\mu(x) = \begin{cases} 0; & x \leq a_1 \\ (x - a_1)/(a_2 - a_1); & a_1 \leq x \leq a_2 \\ 1; & a_2 \leq x \leq a_3 \\ (a_4 - x)/(a_4 - a_3); & a_3 \leq x \leq a_4 \\ 0; & x \geq a_4 \end{cases} \tag{2}$$

Table 1 Categories and symbols of events [26]

Category	Symbol	Annotation
Initial event	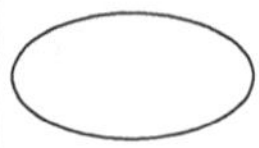	Beginning event of the ESD
Comment event	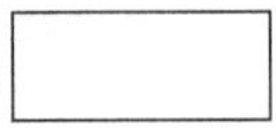	Providing information of the development of event sequence
Termination event	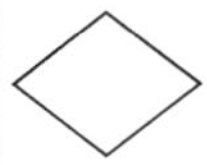	An end state of the ESD

3.2 ESD

ESD is a graphical method for visualizing the sequence of related events. As an effective risk assessment method, ESD has been used in many different fields [23]. The first ESD framework was proposed for risk modeling by NASA in the Cassini space program, and since then it has been employed widely by different researchers [24]. Wu [25] built an ESD model for the driving pump of a spaceship cooling circuit with an initiating event "power failure," and analyzed the related accidents. Zhou et al. [23] applied ESD to evaluate emergency response actions during fire-induced domino effects. To assess the ED failure probability in the present study, ESD was defined based on the work of Swaminathan and Smidts [26].

$$\mathrm{ESD} = (E, C_{\mathrm{d}}, G, \mathrm{Pr}) \tag{3}$$

where E refers to the events which implies any changes from one state to another. Any observable physical phenomenon the analyst chooses to represent in an ESD would be considered as an event. These events could be time-distributed events, demand-based events, non-quantifiable events, or end states. In the present work, events were divided into three categories: (1) "initial event", being the beginning event of an ESD, and starting the potential event sequence; (2) "comment event", describing the development of an event sequence, and (3) "termination event", indicating the termination of the ESD. The symbols used to represent such events and brief definitions are given in Table 1.

C_d indicates conditions which represent the rules controlling the development of an event sequence into different branches. The event sequence will develop in different directions depending on whether the conditions are satisfied or not.

G represents the logic gates, indicating the logical relationships among events. The basic gates are the AND gate and the OR gate, which can be further divided into four types according to event relationships, i.e., output AND gate, input AND gate, output OR gate, and input OR gate. These gates can be used to represent various situations like concurrent processes, synchronization processes, and multiple

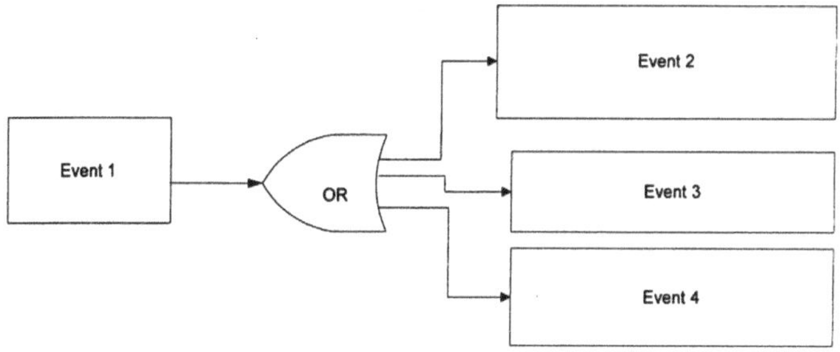

Fig. 1 Output OR gate representing multiple mutually exclusive outcomes

mutually exclusive outcomes. Especially, for the output OR gates, since the outcomes are mutually exclusive, only one of the many possible outcomes will occur. Figure 1 shows an example of an output OR gate. After the occurrence of Event 1, there are three possible scenarios. If *P*2, *P*3, and *P*4 are the probabilities of occurrence of the three events, respectively, then their summation is equal to 1.*Pr* is a set of process parameters, which reflect the states of the system. For example, the above-mentioned occurrence probabilities of the three events are the process parameters, which will influence the evolution of the accident and eventually the probabilities of the termination events (end states).

3.3 Bayesian Networks

BN is a graphical inference probabilistic method widely applied in risk analysis and fault diagnosis [27, 28]. It consists of nodes, arcs, and conditional probability tables which represent a set of random variables and the conditional dependencies among them. Due to its flexible structure and probabilistic reasoning engine, BN is a promising method for risk analysis of large and complex systems.

Considering the conditional dependencies of variables, BN represents the joint probability distribution $P(U)$ of variables $U = \{A1, \ldots, An\}$, expressed as:

$$P(A_i) = \sum_{U \backslash A_i} P(U) \tag{4}$$

where $\mathrm{Pa}(A_i)$ is the parent set of variable A_i. Accordingly, the probability of A_i is calculated by:

$$P(A_i) = \sum_{U \backslash A_i} P(U) \tag{5}$$

where the summation is taken over all the variables except A_i. The main application of BN is in probability updating. BN takes advantage of Bayes' theorem to update the prior probabilities of variables given new observations, called evidence E, rendering the updated or posterior probabilities [29]:

$$P(U|E) = \frac{P(U, E)}{P(E)} = \frac{P(U, E)}{\sum_U P(U, E)} \tag{6}$$

3.4 *Proposed Methodology for Failure Probability Analysis of ED*

In a typical conventional quantitative risk analysis, four steps are involved, i.e., hazard identification, frequency analysis, consequence analysis, and risk quantification [30]. Considering the specialty of ED operations and the simplicity of accident evolution process of ED failure, the proposed methodology framework for failure probability analysis of ED operations involved the following six main steps: (1) defining system and collecting necessary information; (2) reasons analysis and hazard identification; (3) accident scenario modeling, that is, the accident scenarios for ED failure and the accident evolution after ED failure were modeled through FT and ESD analyses, respectively; then, a complete accident model presented after steps 2 and 3 was integrated into FT-ESD and then mapped into BN model; (4) failure probability analysis, that is, the failure probabilities of primary events and end-state consequences were calculated by BN; (5) risk prediction, which means that the failure probability of ED operations was predicted using posterior probability after risk updating and probability adaption [9], and the occurrence probabilities of different consequences caused by ED failure, including blowout, break of the riser, and safe suspension by the vessel, were obtained by BN analysis, respectively. After that, the calculated results and conclusions could provide references for (6) risk decision making of ED operations and proposing preventive measures to reduce the failure probability of ED operations.

4 Methodology Application

4.1 *Disconnect Operation of Deepwater Drilling Riser*

A typical subsea LMRP/BOP system is equipped with two annular preventers and four ram preventers which include a blind shear and three pipe rams. The upper annular preventer is located in LMRP, and the other five preventers are in the BOP [2].

When the vessel is drilling normally with the riser connected, once the vessel moves out of its pre-defined red alert offset due to various reasons, a signal from the DP system will trigger the ED function in the BOP control system. Under such conditions, ED would be initiated automatically to protect the well and riser system (if this action is delayed, the DP operators or the driller can press the ED button manually). Once the ED is initiated, the central control unit (CCU) in the rig will send a signal to the tensioner control system indicating an ED is in progress and will cause the tensioner control system to enter riser anti-recoil mode. On receiving the signal of riser disconnect recoil permission from the tensioner control system, the BOP will complete a series of operations automatically in 30–40 s to hang off the drill string by pipe ram, shear the drill pipe by blind ram, and seal the well.

After the BOP operations are completed, the pre-programmed control and power signals sourced from CCU will be sent to the subsea electronics modules (SEM) by umbilical via umbilical termination. Then, the SEM sends them to motor-operated valves in the actuation modules in LMRP. Meanwhile, power and control signals are also sent to the accumulators, and pressurized fluid in the accumulators is delivered to the motor-operated valves in the actuation modules via hydraulic lines. By electric and hydraulic control, the motor-operated valves are controlled to be open. Under the action of the fluid, the connector between LMRP and BOP can be unlocked, thus the LMRP can be disconnected from the BOP [31].

4.2 *Hazard Identification of ED Failure*

4.2.1 Human Factors

Human factors are seen as a range of issues, including the perceptual, physical, and mental capabilities of people, as well as the interactions of individuals with their jobs and the working environments, the influence of equipment and system design on human performance and, above all, the organizational characteristics that influence safety-related behavior at work [32]. The human factors that affect operation safety may be at organizational and individual levels. At the organizational level, company safety culture, training standards, and system and procedures have the potential to affect the safety of operations. At the individual level, knowledge and skill, distraction, and high work stress affect human performance. In turn, any effect in the performance can influence the operations in the offshore oil and gas industry [2].

4.2.2 Design Factors

As the extreme length of the TJ is also the limit factor for operation alert offset of the deepwater drilling riser system, the system should be configured with telescopic joint in midstroke [33]. Unreasonable configuration of riser column will lead to design defects and will affect the alert offset of the vessel.

As mentioned above, the UFJ and LFJ are configured in the riser column to avoid excessive bending moment, and their extreme rotation angles determine the yellow alert offsets of the vessel which is the position where the ED must be initiated automatically or manually by DP operators or the driller. Since the red alert offset of the vessel at which the LMRP must be disconnected is mainly determined by the bending capacity of LPW and HPW, the bending capacity of LPW and HPW, fabrication defects, yield strength of material, and geometry parameters of conductor will all contribute to the successful completion of ED [18].

4.2.3 Operation Factors

Drilling risers are tensioned structures, and a certain amount of additional overpull (effective tension at the LMRP connector) is needed to keep the riser in tension. A successful ED requires sufficient overpull to lift the LMRP away from the BOP. However, the top tension must be within a particular range to avoid overstress in any components of the riser system, as well as any damage to the equipment caused by recoil due to a sudden upward movement of the riser column [34]. As recoil and recoil control are very complex, they are the subsequent operations of successful ED and were not investigated in the present work. A minimum overpull of 100 kips is required for shallow water, 200 kips for intermediate water depth, and over 300 kips for deepwater and ultra-deepwater [35].

As mentioned earlier, the conductor with LPW is jetted into the formation without well cementing. Since the conductor may not be jetted into the formation in an absolutely vertical direction, and possibly not jetted to the designed depth during jetting operation, the stick-up (height above seabed) of HPW and LPW, the inclination angle of the conductor and LPW both influence the mechanical behavior and the red alert offset of the vessel. Meanwhile, the inclination angle of conductor and LPW makes it difficult to disconnect the LMRP from BOP during ED operations. In addition, since the surface casing should be cemented before the installation of LMRP/BOP, the cementing quality and top level of cement in the annular space will inevitably affect the mechanical performance of wellhead and conductor.

Additionally, during drilling cycle for almost 2–3 months, under low-temperature and high-pressure conditions, the formation of natural gas hydrate around the connectors may freeze the connectors (Fig. 2) and impede the ED operations. Besides, the bonding force between connector and HPW induced by seawater corrosion will also affect the successful completion of ED.

4.2.4 Time Factors

For the DP or moored drilling vessel, internal solitary waves, typhoons, and local rapidly developing storms can lead to large offset in several minutes. Under the action of the increasing vessel offset, the riser's deflected shape changes with time and is significantly affected by the vessel's offset speed. The riser's deflected shape governs

Fig. 2 Hydrate freezing the BOP and HPW

the time at which ED limits are exceeded [36]. Generally, in the event of a drive-off or drift-off, the drilling riser should be quickly disconnected (within 30–45 s) from the BOP [32]. Puccio and Nuttall [37] conducted a riser ED test in which 36 s after initiation of ED, the LMRP connector was released, and the LMRP started to lift up. Since BOP operation is included in ED, the time left for separation of LMRP from BOP will depend on the time required for BOP operation.

Additionally, the successful ED also depends on the point of disconnect during a vessel heave cycle. The ED moment should be analyzed in advance at eight points (from phase angle of 0° to 315° by an increment of 45°) during a heave cycle in order to get the most reasonable disconnect moment [35].

4.2.5 Equipment Factors

Most dynamically positioned vessels are equipped with direct acting tensioners, where the pistons stroke in and out with the heave motion of the vessel, thus the maximal stroke length of the tensioner is also the limit factor of the red alert offset of the vessel [18]. Similarly, the telescopic joint equipped in the rig must have adequate stroke extremity to fully compensate the length change of riser column due to the vessel offset [33]. Additionally, the DP capacity to maintain the position of the vessel plays an important role in ED operations.

4.2.6 Control Factor

As part of the BOP stacks, the LMRP shares the same control system with the BOP. The control system includes electric control system and fluid control system. The fluid control system consists of such components as pumps, valves, accumulators, fluid storage, and mixing equipment and manifold. The electric control system includes CCU, SEM, the connecting umbilical cable, and umbilical termination by which power and control signals are sent, transmitted, and distributed. The CCU is microprocessor-based and typically utilizes triple modular redundancy programmable logic controllers (PLCs) to transmit commands initiated on the surface to the SEM. Two completely independent SEMs, namely subsea blue SEM and subsea yellow SEM, provide fully redundant control of all subsea valve operations and all communications with the CCU. Cai et al. [2, 7, 8] analyzed the control system of LMRP thoroughly and developed a reliability evaluation model for subsea BOP control system. In the present work, the effects of control panel, accumulator, motor actuated valve and leakage in fluid control system, as well as the effects of umbilical and termination, PLC, SEM and control software in electric control system on ED operations are considered.

4.3 FT Model of ED Failure

A FT shown in Fig. 3 was developed to identify the potential hazard factors that may cause ED failure.

There are 40 BEs for ED failure for which the details are shown in Table 2. In order to perform probabilistic analysis, most prior probabilities of all BEs were obtained from the estimated values presented in the literature relating to FFTA [10–14]; Kumar and Yadav [22]. The other source for obtaining probability values was reference reviews.

4.4 ESD of ED Failure

The ESD of ED failure is shown in Fig. 4.

Under extreme conditions, when an ED failure occurs, with the increasing offset of the drilling vessel, either the connected riser system components or the wellhead and conductor will fail inevitably. The failure of wellhead and conductor will lead to the blowout accident directly without any barrier. For the riser column, the failure may occur at the top of the riser system (just below the rotary table) or at the bottom of the riser system (just above the LMRP). When the riser column breaks at the bottom of riser column, being a tensioned structure, the riser will recoil due to the sudden release from the LMRP/BOP. Anti-recoil system will be initiated automatically to reduce the tension and to lift up the riser column in a controlled manner. After the anti-recoil operation is completed, the riser column will be in soft hang-off mode [19, 20]. The riser column will be suspended safely by the vessel or will break during suspension due to the motion of the vessel, and will sink to the seabed. When failure of the riser column occurs just below the rotary table, forced by its wet weight and lateral load caused by waves and currents, the riser column will break either above the LMRP or below the BOP. The former will cause the broken riser to sink to the seabed, while the latter will induce blowout directly. The end states shown in Fig. 4 consist of three conditions: (A) blowout, (B) riser column sinking and laying on the seabed, and (C) safe suspension by the vessel.

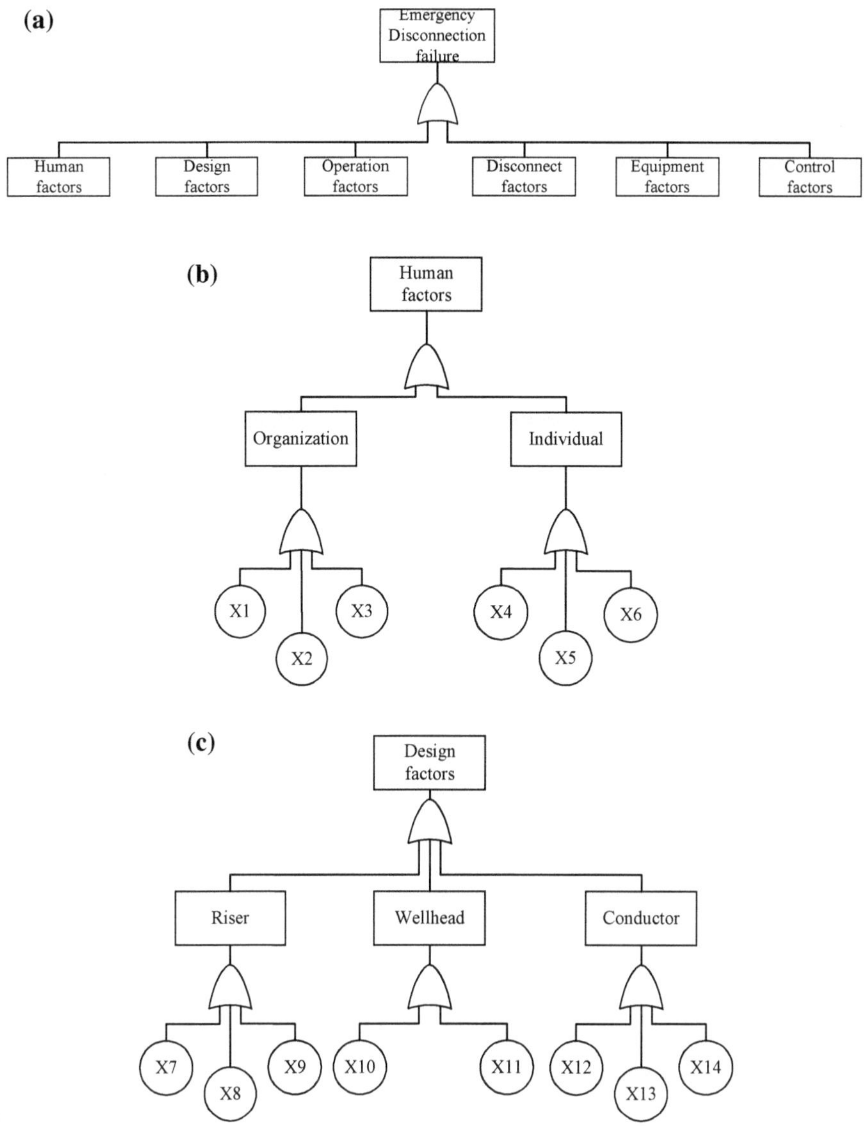

Fig. 3 **a** FT for the ED failure, **b** sub-FT for human factors in ED failure, **c** sub-FT for design factors in ED failure, **d** sub-FT for operation factors in ED failure, **e** sub-FT for disconnect factors in ED failure, **f** sub-FT for equipment factors in ED failure, **g** sub-FT for control factors in ED failure

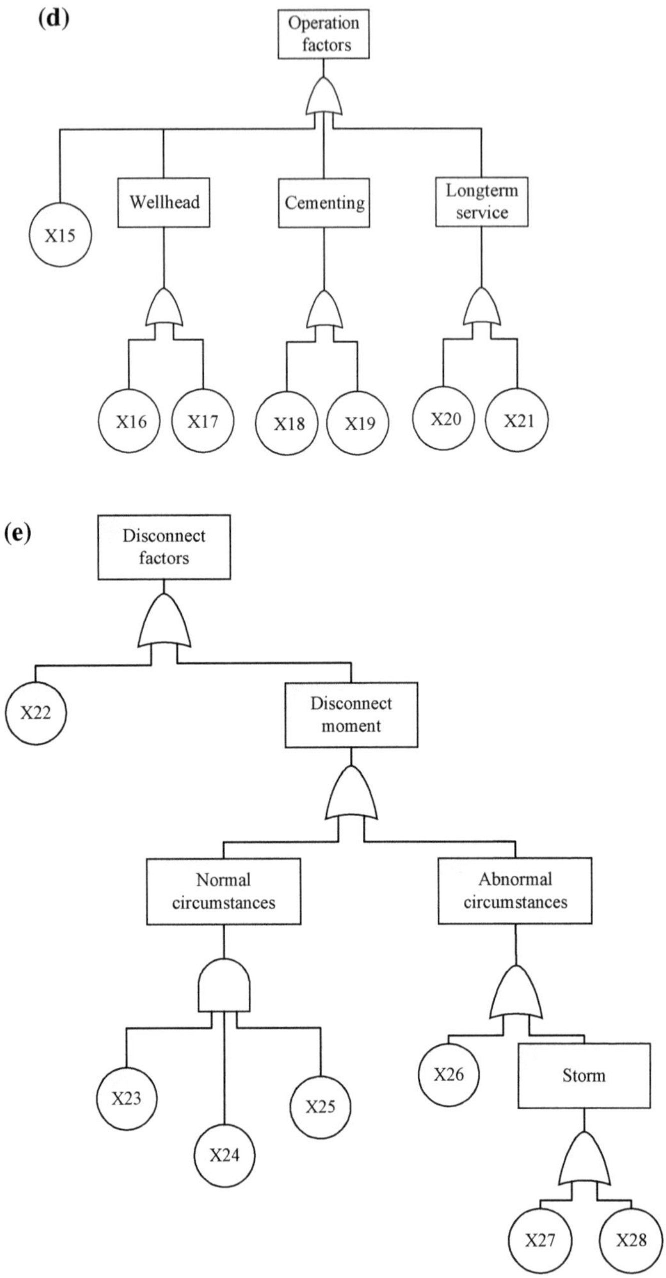

Fig. 3 (continued)

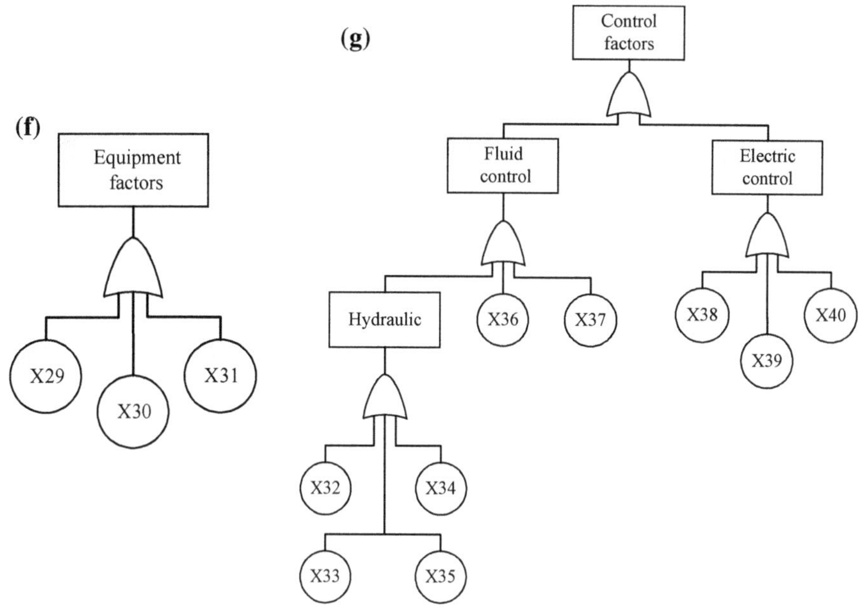

Fig. 3 (continued)

5 Bayesian Network of Emergency Disconnect

5.1 Mapping of FT-ESD Model to BN

To conduct a case-specific failure probability analysis, a BN of ED failure was developed through mapping the FT-ESD model shown in Fig. 5. The left part of the FT-ESD model is FT; mapping from the FT into the BN includes a graphical and numerical translation [38]. In the graphical step, the structure of BN is developed from the FT such that primary events, intermediate events, and the top events of the FT are represented as root nodes, intermediate nodes, and the leaf nodes in the equivalent BN, respectively. The nodes of BN are connected in the same way as the corresponding events in the FT. In the numerical step, occurrence probabilities of the primary events are assigned to the corresponding root nodes as prior probability. For each intermediate node as well as the leaf node, a CPT is assigned. CPTs illustrate how intermediate nodes are related to precedent intermediate or root nodes.

The right part of the FT-ESD model is ESD; mapping from the ESD into BN model is just like that of a FT being mapped into BN model, which also includes a graphical and numerical translation. Similarly, in the graphical step, the structure of BN is developed from the ESD such that initial event, comment events, and the termination events of the ESD are represented as top node, intermediate nodes, and the consequence nodes in the equivalent BN, respectively. In the numerical step,

Table 2 BEs of ED failure and their probabilities

Number	Description of BEs	Prior probability	Posterior probability
X1	Poor company safety culture	7.45E−05	1.12E−04
X2	Poor training standards	5.07E−04	7.66E−04
X3	Unreasonable system and procedures	4.12E−04	6.75E−04
X4	Inadequate knowledge and skill	5.32E−04	8.71E−04
X5	Distraction	4.09E−03	4.08E−03
X6	High work stress	5.07E−04	9.07E−04
X7	Inadequate extremity rotation angles of UFJ	6.32E−05	1.46E−03
X8	Inadequate extremity rotation angles of LFJ	7.43E−05	1.71E−03
X9	Midstroke of TJ is not configured	5.18E−04	1.50E−02
X10	Inadequate bending capacity of HPW	7.28E−05	1.73E−03
X11	Inadequate bending capacity of LPW	6.83E−05	1.62E−03
X12	Fabrication defects	9.40E−05	1.17E−03
X13	Unreasonable geometric parameters	7.88E−05	1.82E−03
X14	Inadequate yield strength of material	7.68E−05	1.77E−03
X15	Inadequate overpull	1.32E−02	2.48E−01
X16	Large stick-up	8.90E−04	2.09E−02
X17	Large inclination angle of wellhead	1.32E−02	1.35E−01
X18	Poor cementing quality	9.54E−05	9.72E−03
X19	Inadequate top level of cement	8.33E−05	8.55E−03
X20	Formation of natural gas hydrate	8.96E−03	8.32E−02
X21	Bonding force induced by corrosion	3.07E−04	3.12E−03
X22	Unreasonable disconnect moment	1.89E−02	2.37E−01
X23	Strong wind	8.10E−03	1.09E−02
X24	Large wave	7.92E−03	1.07E−02
X25	High current	8.35E−03	1.13E−01
X26	Internal solitary waves	8.55E−03	4.82E−02
X27	Typhoon	2.30E−02	7.92E−02
X28	Local rapidly developing storm	8.29E−03	2.93E−02
X29	Inadequate tensioner stroke extremity	5.07E−04	1.24E−02
X30	Inadequate TJ stroke extremity	5.06E−04	1.24E−02
X31	Inadequate DP capacity and accuracy	4.52E−04	8.12E−03
X32	Operation panel failure	1.46E−04	2.23E−03
X33	Leakage of accumulator in BOP	9.88E−04	8.22E−03
X34	Actuator modules failure	9.55E−04	6.13E−03
X35	High unlock pressure in connector induced by leakage of corrosion	4.52E−03	7.62E−02

(continued)

Table 2 (continued)

Number	Description of BEs	Prior probability	Posterior probability
X36	Signal transmission failure	5.07E−04	9.23E−03
X37	Umbilical termination failure	5.33E−04	9.56E−03
X38	Two redundant SEMs failure	7.45E−05	1.82E−03
X39	Triple modular redundant PLCs failure	7.45E−05	1.82E−03
X40	Software failure	5.07E−04	3.24E−03

for each intermediate node as well as the consequence node, a CPT is assigned. CPTs are defined according to the process parameters *Pr* in the ESD model. In the ESD mapping process in this study, the possible events, the conditions controlling the development of an event sequence into different branches, logic relation for evolution of the ED failure accident, and the occurrence probability, which represent the process parameters, were all determined by expert judgments.

Because of the uncertainty and complexity, it is difficult to determine the prior probabilities for BEs and conditional probability tables for different factors. In the present work, the expert judgments were used for this purpose. Five related experts from the oil company as listed in Table 3 were invited to judge the fuzzy number and CPTs based on their experiences.

The expert elicitation is essentially a scientific methodology and is often used in the study of rare events, and various elicitation methods were examined for expert judgments. For the unknown prior probability of BEs of aforementioned human, design, operation, equipment, time, and control factors, the experts were asked to fill out five separate data sheets by linguistic terms, namely occurrence possibility survey tables of BEs. In the next step, using fuzzy set theory, the occurrence possibility of BEs was transformed into fuzzy failure probability (prior probability). The present work used the weighted averages of the five sets of data as the final input data. The weight of each event for the experts was determined based on their qualifications and relevant experiences. The first and the second experts, who were acquainted with human and equipment factors, had a higher weight for human and equipment factors and a lower weight for control factors, whereas the third expert, who was acquainted with time and control factors, had a higher weight for time and control factors and a lower weight for design factors. However, the experts were all acquainted with the ED of the deepwater drilling riser system. With respect to the conditional probability tables of different factors, the five experts discussed and determined the possible data based on their experiences.

The conditional dependencies among elements of BN were assigned in CPT. The logical gate of FT and experience-based judgment were used to determine CPTs in BN model. The logical gate of FT represents deterministic relationship between primary events and intermediate events. For example, if both X10 and X11 succeed, the wellhead would fail inevitably. This relationship is described by a CPT in Table 4. Actually, if both X10 and X11 succeed, the wellhead may not fail. However, although

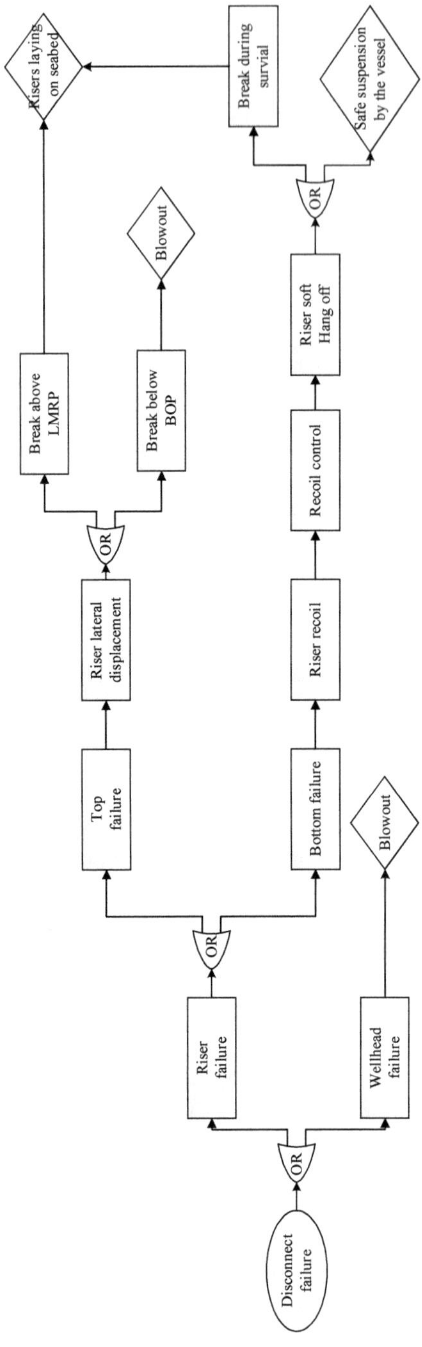

Fig. 4 ESD of ED failure

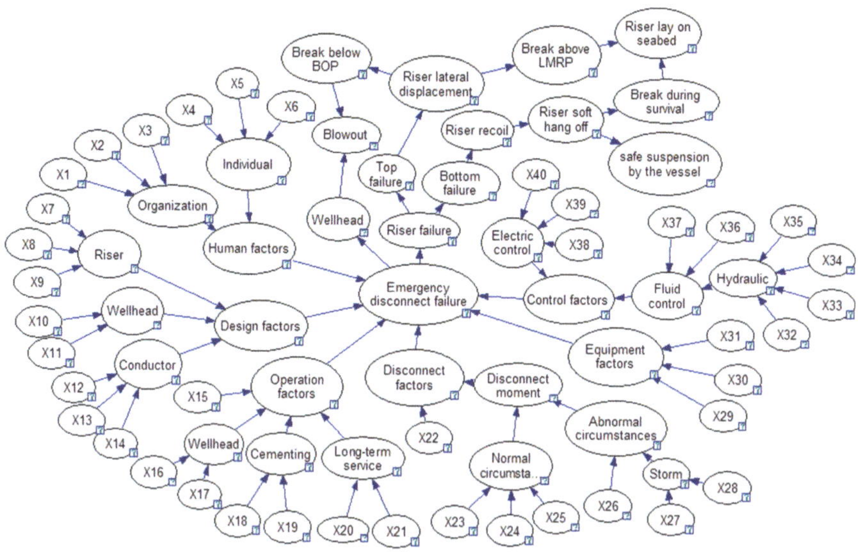

Fig. 5 BN mapped from FT-ESD

Table 3 Experts' information

No.	Professional position	Service time (years)	Educational level
1	Drilling rig manager	16	Bachelor
2	General drilling supervisor	11	Master
3	Deepwater operation manager	12	Bachelor
4	Principal engineer	14	Doctor
5	Senior subsea engineer	10	Master

X10 and X11 do not fail, the failure of wellhead is still possible. Such scenario can be modeled through an amending CPT shown in Table 5. The amending values in CPT were determined by expert judgments. Note that the computed results may be subjected to a margin of error because the input data were obtained from expert judgments and reference reviews, resulting in some possible errors.

The BN of ED failure was developed using graphical network interface (GeNIe) software. Using the BN model of ED failure shown in Fig. 5, with the probabilities listed in Table 2, the probability of ED failure was estimated to be 4.91E−02, and the probabilities of three end states were also calculated, which were 4.17E−3, 1.89E−2,

Table 4 CPT corresponding to OR gate in FT

X10		Success		Failure	
X11		Success	Failure	Success	Failure
Wellhead failure	Success	1	1	1	0
	Failure	0	0	0	1

Table 5 Amending CPT in BN

X10		Success		Failure	
X11		Success	Failure	Success	Failure
Wellhead failure	Success	0.98	0.96	0.96	0
	Failure	0.02	0.04	0.04	1

and 2.37E−2 for blowout, sinking to the seabed, and safe suspension by the vessel, respectively. It is worth noting that the blowout probability is far lower than other consequences, as in no case could the weakest point of the drilling riser system be located below LMRP, and it should be verified by weak point analysis in design stage [39]. The possibility of safe suspension by the vessel and break and sinking to the seabed depends on the seastate and the motion response of the vessel.

According to the analyses results, the order of importance of the influencing factors of ED failure would be as operation factors > time factors > control factors > equipment factors > design factors > human factors. On account of the fact that the corresponding BEs within the operation and time factors may lead to ED failure directly, they play the most significant role in ED operations. Design and equipment factors have certain effects on ED failure because the design level and equipment are usually reliable. Control factors play a minor role in the failure of ED operations because of the use of the redundant control logics, PLC, SEM, and control software. Human factors can hardly contribute to the ED failure because the ED will be initiated automatically in case that the vessel reaches the red alert offset.

5.2 *Risk Updating*

In addition to offering a flexible structure and a robust reasoning engine, the main application of BN is in risk updating. In risk updating, the probability of an accident scenario is updated. This updating is performed in terms of posterior probability of event *xi* given a new evidence. This also helps to identify the critical basic (the most probable) cause of events leading to the evidence [3, 40]. The most common type of evidence used in probability updating is the knowledge about the top event or consequences.

In the present study, the posterior probabilities of BEs given a ED failure, i.e., P (xi|consequence = ED failure), are shown in the last column of Table 2. By com-

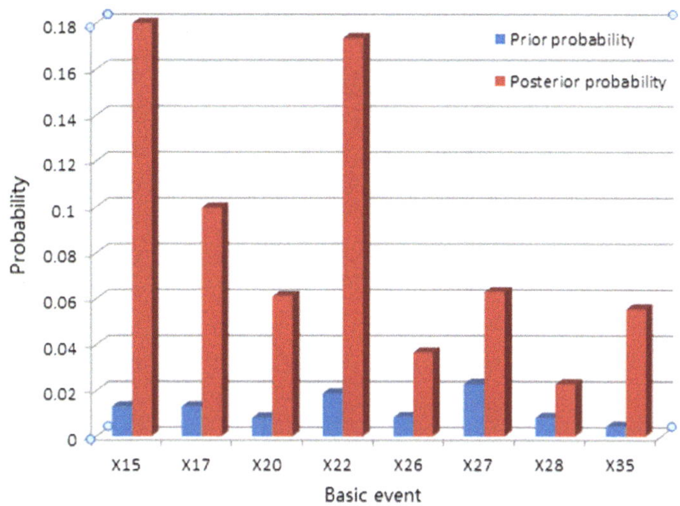

Fig. 6 Comparison between prior and posterior probabilities for critical BEs

paring the posterior probabilities with the prior probabilities of BEs, the critical BEs could be identified. The critical BEs, which are the BEs with a high posterior probability and a high increasing probability, would provide meaningful information for ED operations and preventive actions to avoid ED failure [5]. In Table 2, it is observed that X15, X17, X20, X22, X26, X27, X28, and X35 have the highest increasing probability and significant posterior probability values (Fig. 6). Therefore, the critical events in ED failure are X15 (inadequate overpull), X17 (large inclination angle of wellhead), X20 (formation of natural gas hydrate at LMRP connector), X22 (unreasonable disconnect moment), X26 (internal solitary waves), X27 (typhoon), X28 (local rapidly developing storm), and X35 (high unlock pressure in connector induced by leakage of corrosion). More attention should be paid to all of these critical events during drilling riser design and ED operations.

GeNIe software can also implement strength of influence, through which the probable development paths are found (U.O. Pittsburgh, 2014). The most probable accident evolution paths for ED failure were as follows: (i) X15 (inadequate overpull) and X17 (large inclination angles of conductor), and X20 (formation of natural gas hydrate at the connector) → adverse operation factors → ED failure; (ii) X22 (unreasonable disconnect moment during the heave cycle of the vessel) and X27 (typhoon) → adverse time factors → ED failure; (iii) X35 (high unlock pressure in connector induced by leakage of corrosion) → malfunction of control factors → ED failure. The most probable consequence evolution paths of the ED failure were as follows: (i) ED failure → riser break just above the LMRP → safe suspension by the vessel (after anti-recoil operation); (ii) ED failure → riser break just below the rotary table → riser break above LMRP → sinking to the seabed.

Table 6 Occurrence record of critical BEs in 90 days (Jan–Mar)

BEs	Day					
	0–15	16–30	31–45	46–60	61–75	76–90
X15	0	0	0	1	1	1
X17	1	1	1	1	1	1
X20	0	0	0	0	0	1
X22	0	0	0	0	0	0
X26	0	0	0	0	0	0
X27	0	0	0	0	0	0
X28	0	0	0	0	0	0
X35	0	0	0	0	1	1

5.3 Probability Adaption

Probability adaption, also known as sequence learning, is another important application of BN. It is used for probability updating based on the new information accumulated over time where the new information could be the occurrences of certain BEs or the accident precursors. The critical events for failure of ED were identified by probability updating, and their occurrences were recorded over a period of time in this study. A hypothetical case, in which the occurrences of identified critical events in the South China Sea have been recorded for a time interval of 15 days over one year (Tables 6, 7, 8, and 9), is included as an example to illustrate how to realize probability adaption. Normally, the average drilling cycle for a well is about 3 months. For this reason, the occurrence record of critical events is incorporated with the seasonal weather considerations in Liuhua oil field in the South China Sea. It is worth noting that the occurrence record of critical BEs represents whether the events occur or not, and it is not the actual occurrence record of the events. This means that the critical events occur once the ED is initiated each time. For example, X9 refers to the critical event of midstroke of TJ being not configured. If this event occurs during the installation of drilling riser system, it will exist during the drilling cycle of the well all along. However, for the basic event of X20 (formation of natural gas hydrate around the LMRP connector), as the hydrate can be removed by ROV once it is observed, the occurrence record may change during the drilling cycle.

The prior probabilities can be adapted after occurrences of these critical events for each well, and the revised prior probabilities P can be calculated as follows [5, 41, 42]:

$$P = \frac{a + s}{n + s} \tag{7}$$

where a and n denote the occurrence records of ED failure and total records of ED operations, respectively, for the past wells, and s represents the occurrence record of the critical events for the ongoing well.

Table 7 Occurrence record of critical BEs in 90 days (Apr–Jun)

BEs	Day					
	0–15	16–30	31–45	46–60	61–75	76–90
X15	0	0	0	0	0	0
X17	0	0	0	0	0	0
X20	0	0	0	0	1	0
X22	0	0	0	0	0	0
X26	1	1	1	1	1	1
X27	0	0	0	0	0	0
X28	0	0	0	0	0	0
X35	0	0	0	0	0	1

Table 8 Occurrence record of critical BEs in 90 days (Jul–Sep)

BEs	Day					
	0–15	16–30	31–45	46–60	61–75	76–90
X15	1	1	1	1	1	1
X17	1	1	1	1	1	1
X20	0	0	0	0	1	0
X22	0	0	0	0	1	0
X26	0	0	0	0	0	0
X27	1	1	1	1	1	1
X28	0	0	0	0	1	0
X35	0	0	0	0	0	0

Table 9 Occurrence record of critical BEs in 90 days (Oct–Dec)

BEs	Day					
	0–15	16–30	31–45	46–60	61–75	76–90
X15	1	1	1	0	0	0
X17	0	0	0	0	0	0
X20	0	0	0	0	0	1
X22	0	0	0	0	0	0
X26	0	0	0	0	0	0
X27	1	0	0	0	0	0
X28	1	0	0	0	0	0
X35	0	0	0	0	0	1

Fig. 7 Failure probability of ED for a well (Jan–Mar and Jul–Sep)

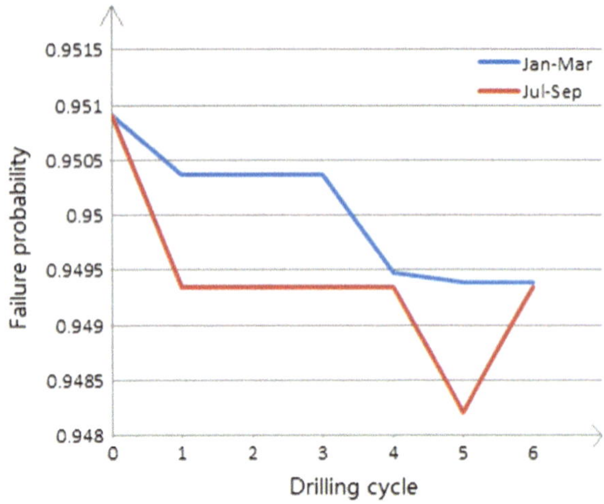

Using the revised probabilities of critical events during a drilling cycle, i.e., January–March and July–September, the probabilities of ED failure were updated (Fig. 7). From January to March, it can be clearly seen from Fig. 7 that the success probability of ED operations decreases slightly from 0.9509 to 0.9504 with the occurrence of critical events X15 (inadequate overpull) and X17 (large inclination angle of wellhead), and keeps unchanged until the occurrence of new events. With the occurrence of more critical events in the subsequent possible ED operations, the success probability of ED operations decreases to 0.9494.

The probability of successful ED operations for the third well is also illustrated in Fig. 7. As the well was drilled from July to September, the frequent typhoons in the South China Sea would affect the safety of ED operations significantly. Similarly, the success probability of ED operations decreases from 0.9509 to 0.9495 because of the simultaneous occurrence of critical events X15 (inadequate overpull), X17 (large inclination angle of wellhead), and X27 (typhoon) for the possible ED operations. With the occurrence of critical event X20 (formation of the natural gas hydrate), the success probability of ED operations continues to decrease to the lowest value. However, the success probability of ED operations would increase to 0.9495 with the removal of the hydrate.

Given that probability of ED failure restore to its initial value after test and maintenance between wells, the probabilities of ED failure for the whole year were calculated. Furthermore, probabilities of successful ED, as well as the probabilities of ED failure consequence (A: blowout) over one year, are illustrated in Figs. 8 and 9, respectively. The calculated results are listed in Tables 10, 11, 12, and 13.

In BN adaptation, by introducing these critical events and analyzing the trends of probability changes shown in Figs. 8 and 9, it can be concluded that the probabilities of ED failure and different consequences caused by it depend on the occurrence of critical events. As can be seen, for the third well, the probability of blowout, which is the most catastrophic consequence of ED failure, increases slightly from

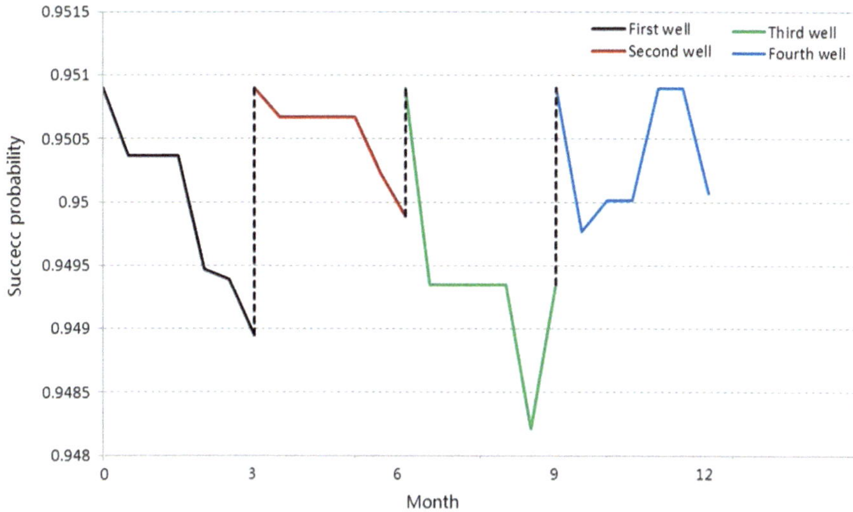

Fig. 8 Probabilities of ED success for the whole year

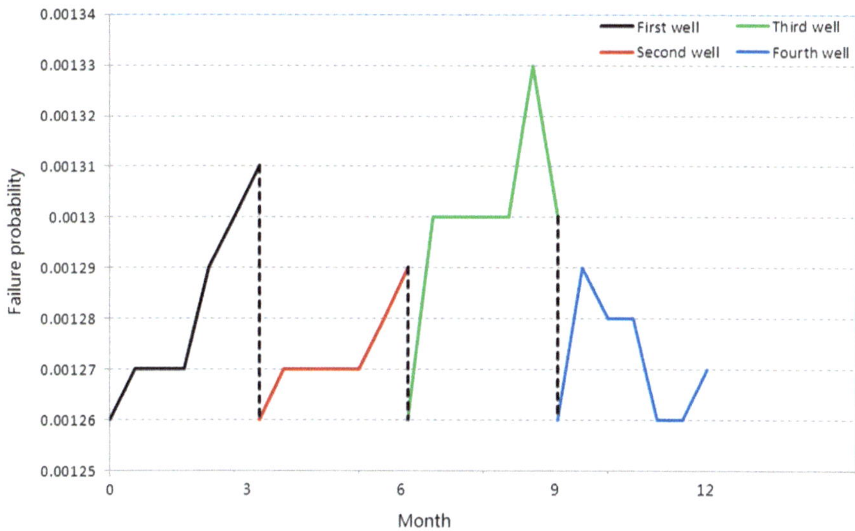

Fig. 9 Probabilities of ED failure consequence (A: blowout) for the whole year

Table 10 Probabilities of ED success for different wells (days)

Well	Initial	0–15	16–30	31–45	46–60	61–75	76–90
1	0.95090	0.95036	0.95036	0.95036	0.94947	0.94939	0.94939
2	0.95090	0.95080	0.95080	0.95080	0.95080	0.95040	0.9507
3	0.95090	0.94935	0.94935	0.94935	0.94935	0.94821	0.94935
4	0.95090	0.94977	0.95002	0.95002	0.95090	0.95090	0.95038

Table 11 Probabilities of ED failure consequence (A) for different wells (days)

Well	Initial	0–15	16–30	31–45	46–60	61–75	76–90
1	0.00126	0.00127	0.00127	0.00128	0.00129	0.00130	0.00131
2	0.00126	0.00127	0.00127	0.00127	0.00127	0.00128	0.00129
3	0.00126	0.00130	0.00130	0.00130	0.00130	0.00133	0.00130
4	0.00126	0.00129	0.00128	0.00128	0.00126	0.00126	0.00127

Table 12 Probabilities of ED failure consequence (B) for different wells (days)

Well	Initial	0–15	16–30	31–45	46–60	61–75	76–90
1	0.02009	0.02031	0.02031	0.02031	0.02067	0.02070	0.02088
2	0.02009	0.02018	0.02018	0.02018	0.02018	0.02036	0.02050
3	0.02009	0.02072	0.02072	0.02072	0.02072	0.02119	0.02072
4	0.02009	0.02055	0.02045	0.02045	0.02009	0.02009	0.02042

Table 13 Probabilities of ED failure consequence (C) for different wells (days)

Well	Initial	0–15	16–30	31–45	46–60	61–75	76–90
1	0.02526	0.02553	0.02553	0.02553	0.02599	0.02603	0.02626
2	0.02526	0.02537	0.02537	0.02537	0.02537	0.02560	0.02578
3	0.02526	0.02605	0.02605	0.02605	0.02605	0.02664	0.02605
4	0.02526	0.02584	0.02571	0.02571	0.02526	0.02526	0.02568

1.26E−3 to 1.33E−3, the sinking probability of the broken drilling risers increases from 2.009E−02 to 2.119E−02, and the safe suspension probability of the drilling risers increases from 2.526E−02 to 2.664E−02. Note that the computed results may be subjected to a margin of error because the conditions in ESD, which represent the rules controlling development of an event sequence into different branches, are obtained from expert judgments, causing some of them to be inaccurate.

As the critical events may induce a major change of probability of ED failure, measures should be taken from all aspects to mitigate the probability of ED failure, including operation, time, control, equipment, and design factors. This means that control of the inclination angles of wellhead during jetting of the conductor, application of adequate overpull over BOP, observing and removing the natural gas hydrate timely, attention and precaution of the abnormal seastate (typhoons, internal solitary waves, and local rapidly developing storms), routine test of the ED, and reasonable design of the drilling riser system will all contribute to a successful ED operation. As an example, for well LW21-1-1, the first ultra-deepwater well in water depth of 2451 m in the South China Sea, drilling operation was postponed after the completion of riserless drilling for the coming typhoon season, and the partial reasons for the delay were the possible ED and the increasing possibility of ED failure.

5.4 *Model Validation*

When a new methodology is developed, it requires a careful validation to ensure its robustness. A sensitivity analysis was carried out in this study to test the proposed model. If the model is robust, the obtained result would be sensitive, but would not show abrupt variations to any minor change of the input parameters [2, 43]. With the assumption of the prior probability of critical BE X15 being increased by 10%, the probability of ED success decreased from 95.09 to 94.97%. When both of the prior probabilities of critical BEs X15 and X17 were increased by 10%, the probability of ED success decreased from 94.97 to 94.91%. When the prior probabilities of critical BEs X15, X17, and X22 were increased by 10% simultaneously, the probability of ED success decreased from 94.91 to 94.81%. As expected, a slight increment of prior probabilities for critical BEs induced the decrement of probability of ED success in a reasonable way, thus giving a validation of the model.

6 Summary and Conclusions

In the present study, four reasons for ED of DP drilling vessel were analyzed which include drift-off, drive-off, storms, and internal solitary waves. Based on the analysis of reasons for ED, the hazards of ED were identified, which include human, design, operation, time, equipment, and control factors, and the influencing mechanism of the hazardous factors on ED was investigated.

Considering the specialty of ED operations and the simplicity of accident evolution process of ED failure, failure probability analysis of drilling riser ED operations was carried out by BN approach. Six categories of influencing factors were modeled and integrated into FT. Accident evolution process and three end states induced by ED failure were modeled by ESD, and then the integrated FT-ESD model was mapped into the BN which could consider polymorphism of BEs as well as conditional dependencies among the primary events of the ED operations. However, BN approach demands more expertise in terms of prior probability, conditional probability, and network construction based on causal relationships between components. A BN model also helps to identify the most probable path of events leading to an ED failure and the most probable paths of consequence caused by ED failure.

The present study indicates that the methodology proposed herein is an alternative approach in failure probability analysis of ED operations for deepwater drilling risers. The study showed that the X15 (inadequate overpull), X17 (large inclination angle of wellhead) and X22 (unreasonable disconnect moment) were the first three critical BEs for the failure of ED operations, and the probabilities of ED failure and potential consequences caused by ED failure varied with the states of critical BEs. Overall, the failure of ED operations is an event with low occurrence probability, and blowout is the consequence with the lowest occurrence probability caused by failure of ED operations. The analysis results obtained from this study could provide reference for risk decision making of ED operations and a better vision of ED safety issues.

References

1. A. Grønevik, *Simulation of Drilling Riser Disconnection-Recoil Analysis* (Norwegian University of Science and Technology, Trondheim, 2013)
2. B. Cai, Y. Liu, Z. Liu, X. Tian, Y. Zhang, R. Ji, Application of Bayesian networks in quantitative risk assessment of subsea blowout preventer operations. Risk Anal. **33**(7), 1293–1311 (2013)
3. N. Khakzad, F. Khan, P. Amyotte, Quantitative risk analysis of offshore drilling operations: a Bayesian approach. Saf. Sci. **57**, 108–117 (2013)
4. M. Abimbola, F. Khan, N. Khakzad, S. Butt, Safety and risk analysis of managed pressure drilling operation using Bayesian network. Saf. Sci. **76**, 133–144 (2015)
5. Y. Yang, F. Khan, P. Thodia, R. Abbass, Corrosion induced failure analysis of subsea pipelines. Reliab. Eng. Syst. Safety **159**, 214–222 (2017)
6. J. Bhandari, R. Abbassi, V. Garaniya, F. Khan, Risk analysis of deepwater drilling operations using Bayesian network. J. Loss Prev. Process Ind. **38**, 11–23 (2015)
7. B. Cai, Y. Liu, Z. Liu, X. Tian, X. Dong, S. Yu, Using Bayesian networks in reliability evaluation for subsea blowout preventer control system. Reliab. Eng. Syst. Safety **108**, 32–41 (2012)
8. S. Cai, J. Xie, J. He, An overview of internal solitary waves in the South China Sea. Surv. Geophys. **33**, 927–943 (2012)
9. X. Li, G. Chen, H. Zhu, Quantitative risk analysis on leakage failure of submarine oil and gas pipelines using Bayesian network. Process Saf. Environ. Prot. **173**, 163–173 (2016)
10. S.M. Lavasani, Z. Yang, J. Finlay, J. Wang, Fuzzy risk assessment of oil and gas offshore wells. Process Saf. Environ. Prot. **89**, 277–294 (2011)
11. S.M. Lavasani, N. Ramzali, F. Sabzalipour, E. Akyuz, Utilisation of fuzzy fault tree analysis (FFTA) for quantified risk analysis of leakage in abandoned oil and natural-gas wells. Ocean Eng. **108**, 729–737 (2015)
12. Lavasani, S.M., Zendegani, A., Celik, M.: An extension to fuzzy fault tree analysis (FFTA) application in petrochemical process industry. Process Saf. Environ. Prot. **93**, 75–88 (2015b)
13. R. Ferdous, F. Khan, B. Veitch, P. Amyotte, Methodology for computer aided fuzzy fault tree analysis. Process Saf. Environ. Prot. **87**, 217–282 (2009)
14. Z. Chen, X. Wu, J. Qin, Risk assessment of an oxygen-enhanced combustor using a structural model based on the FMEA and fuzzy fault tree. J. Loss Prev. Process Ind. **32**, 349–357 (2014)
15. L. Shi, J. Shuai, K. Xu, Fuzzy fault tree assessment based on improved AHP for fire and explosion accidents for steel oil storage tanks. J. Hazard. Mater. **278**, 529–538 (2014)
16. Chang, Y., Design approach and its application for deepwater drilling risers. Dongying: University of Petroleum (East China) (2008)
17. T. Olsen, *Safe Disconnect During Drive-Off/Drift-Off When Drilling on DP* (IADC, Stavanger, Norway, 2001)
18. S. Ju, Y. Chang, G. Chen, X. Liu, L. Xu, R. Wang, Envelopes for connected operation of the deepwater drilling riser. Petrol. Explor. Dev **39**(1), 105–110 (2012)
19. B.D. Ambrose, F. Grealish, K. Whooley, *Soft hangoff method for drilling risers in ultra deepwater* (OTC, Huston, Texas, 2001), p. 13816
20. B.D. Ambrose, M.S. Childs, S.A. Leppard, R.L. Krohn, *Application of a deepwater riser risk analysis to drilling operations and riser design* (OTC, Houston, Texas, 2001), p. 12954
21. W. Huang, *The Investigation on Load and Dynamic Response Characteristics of Deep-Sea Floating Structures in Internal Solitary Waves* (Shanghai Jiao Tong University, Shanghai, 2013)
22. M. Kumar, S.P. Yadav, The weakest t-norm based intuitionistic fuzzy fault-tree analysis to evaluate system reliability. ISA Trans. **51**, 531–538 (2012). https://doi.org/10.1016/j.isatra.2012.01.004
23. J. Zhou, G. Reniers, N. Khakzad, Application of event sequence diagram to evaluate emergency response actions during fire-induced domino effects. Reliab. Eng. Syst. Safety **150**, 202–209 (2016)
24. Y. Liu, Z. Liu, B. Cai, X. Tian, R. Ji, Reliability research on subsea wellhead connector of blowout preventer stack. J. China Univ. Petrol. **37**(5), 140–144 (2013)

25. Q. Wu, *The ESD method and software of astronautics carrier system's risk assessment* (National University of Defense Technology, Changsha, 2005)
26. S. Swaminathan, C. Smidts, The event sequence diagram framework for dynamic probabilistic risk assessment. Reliab. Eng. Syst. Safety **63**, 73–90 (1999)
27. B. Cai, Y. Liu, Q. Fan, Y. Zhang, Z. Liu, S. Yu, R. Ji, Multi-source information fusion based fault diagnosis of ground-source heat pump using Bayesian network. Appl. Energy **114**, 1–9 (2014)
28. B. Cai, H. Liu, M. Xie, A real-time fault diagnosis methodology of complex systems using object-oriented Bayesian networks. Mech. Syst. Signal Process. **80**, 31–44 (2016)
29. X. Li, H. Zhu, G. Chen, R. Zhang, Optimal maintenance strategy for corroded subsea pipelines. J. Loss Prev. Process Ind. **49**, 145–154 (2017)
30. M. Rausand, Risk Assessment: Theory, Methods, and Applications. Wiley (2013)
31. Fenton, S. P., *Riser Emergency Disconnect Control System*. US 0067589 A1 (2012)
32. J.E. Skogdalen, J.E. Vinnem, Quantitative risk analysis offshore-human and organizational factors. Reliab. Eng. Syst. Safety **96**, 468–479 (2011)
33. API RP 16Q, Recommended Practice for Design, Selection, Operation and Maintenance of Marine Drilling Riser Systems, 1st Edition (1993)
34. D.W. Lang, J. Real, M. Lane, *Recent developments in drilling riser disconnect and recoil analysis for deepwater application* (OMAE, Honolulu, Hawaii, 2009), p. 79427
35. P. Ma, J. Pyke, A. Vankadari, A. Whooley, *Ensuring safe riser emergency disconnect in harsh environments: experience and design requirements* (ISOPE, Alaska, USA, 2013)
36. J.N. Brekke, *Key elements in ultra-deep water drilling riser management* (SPE/IADC, Amsterdam, The Netherlands, 2001), p. 67812
37. W.F. Puccio, R.V. Nuttall, *Riser Recoil During Unscheduled Lower Marine Riser Package Disconnects* (SPE, Dallas, Texas, 1998), p. 39296
38. N. Khakzad, F. Khan, P. Amyotte, Dynamic safety analysis of process systems by mapping bow-tie into Bayesian network. Process Saf. Environ. Prot. **91**, 46–53 (2013)
39. K. Kavanagh, M. Dib, E. Balch, P. Stanton, *New Revision of Drilling Riser Recommended Practice (API RP 16Q)* (OTC, Houston, Texas, 2002), p. 14263
40. A. Bobbio, L. Portinale, M. Minichino, E. Ciancamerla, Improving the analysis of dependable systems by mapping fault trees into Bayesian networks. Reliab. Eng. Syst. Safety **71**, 249–260 (2001)
41. A. Meel, W.D. Seider, Plant-specific dynamic failure assessment using Bayesian theory. Chem. Eng. Sci. **61**, 7036–7056 (2006)
42. Q. Tan, G. Chen, L. Zhang, J. Fu, Z. Li, Dynamic accident modeling for high-sulfur natural gas gathering station. Process Saf. Environ. Prot. **92**, 565–576 (2014)
43. Z. Yang, S. Bonsall, A. Wall, J. Wang, M. Usman, A modified CREAM to human reliability quantification in marine engineering. Ocean Eng. **58**, 293–303 (2013)
44. GeNIe. *Decision Systems Laboratory*, 1998–2015. Available at: https://dslpitt.org/genie/
45. P. Luo, Y. Hu, System risk evolution analysis and risk critical event identification based on event sequence diagram. Reliab. Eng. Syst. Safety **114**, 36–44 (2013)

Risk Analysis of Subsea Blowout Preventer by Mapping GO Models into Bayesian Networks

Abstract Bayesian network (BN) is commonly used in probabilistic risk quantification due to its powerful capacity in uncertain knowledge representation and uncertainty reasoning. For the formalization of BN models, this paper presents a novel approach on constructing a BN from GO model. The equivalent BNs of the seventeen basic operators in GO methodology are developed. Therefore, the existing GO model can be mapped into an equivalent BN on basis of these developed BNs of the operators. Subsea blowout preventer (BOP) system plays an important role in providing safety during the subsea drilling activities. A case of closing the subsea BOP in the presence of pump failures is used to illustrate the mapping process. First, its GO model is presented according to the flowchart of the case. Then, BN is obtained based on the presented GO model. The developed BN relaxes the limitations of GO model and is capable of probability updating and probability adapting. Sensitivity analysis is performed to find the key influencing factor. The three-axiom-based analysis method is used to validate the developed BN.

Keywords Risk analysis · Subsea blowout preventer · GO methodology · Bayesian network

1 Introduction

Bayesian networks (BNs) can describe the dependencies between variables both qualitatively and quantitatively [1]. It is a powerful tool in uncertain knowledge representation and uncertainty reasoning. BN is able to predict the probability of unknown variables by forward reasoning or update the probability of known variables given some new information by backward reasoning [2]. Due to this ability, BN is widely used for safety analysis and risk assessment in various fields, such as natural gas pipeline network accident [3], gas explosion accidents [4], offloading process in floating liquefied natural gas platform [5], maritime transportation systems [6], ship recycling sector [7], offshore drilling operations [8, 9], human factor analysis [10–12], managed pressure drilling operation [13], and so on.

B. Cai et al., *Bayesian Networks for Reliability Engineering*,
https://doi.org/10.1007/978-981-13-6516-4_8

A BN is a directed acyclic graph composed of nodes and arcs, which is a graphical and qualitative illustration of relationships among different nodes using directed arcs. Nodes represent random variables, and directed arcs between pairs of nodes denote dependencies between the variables [14, 15]. Conditional probability table (CPT) is specified at each node that has parents, while prior probability is specified at node that has no parents. A BN can be obtained by machine learning using data sets or deducing from expert knowledge [16]. These two methods can be used individually or jointly. A BN has higher uncertain inference capacity in dealing with multisources of information like expert knowledge, empirical data, model output, and so on [17]. In order to make use of the powerful representation in uncertainty, BNs can be developed by converting the other reliability models, such as fault tree [18], dynamic fault tree [19, 20], event tree [21], bond graph model [22], reliability block diagram [23], bow-tie [24], and so on.

GO methodology was originally developed to analyze the safety and reliability of nuclear systems [25]. It is a success-oriented system analysis technique and becomes an effective technique for system reliability analysis [26]. With application of GO model, a new reliability analysis approach for repairable systems with multiple-input and multifunction component is presented [27]. A new method for reliability of vehicle systems by taking into account of typical characteristics based on GO methodology is proposed [28]. A supplemental algorithm for the repair system in GO methodology is developed [29]. The development of GO method is based on decision tree theory, and its basic modeling way is to translate the schematic diagram, flowchart, or engineering chart into GO chart according to some rules. Therefore, GO model is able to reflect the system structure, the relationship, and effects among the components. Unlike fault tree analysis, GO methodology can be used to model the system with multiple states and time-sequential signals. A GO model is created by representing the elements and logical features of a particular system based on the seventeen types of operators. The quantification calculation can be performed one by one along the sequence of the signal flows from the input operator to the final output signal of the system. However, GO method has its limitations. Too many operators make the modeling process not easy. Besides, GO method is unable to construct hierarchical charts. Further, it is hard to describe the effects of uncertainty [30].

In this paper, a novel method on constructing BNs from GO models is presented. The proposed method can relax the mentioned limitations of GO model and enrich the ways of developing BN. Besides, constructing BN from GO model will have some improvements. The new model will be able to update the probability of known variables given some new evidences. Besides, various kinds of dependencies among system components can be accommodated. BN helps to incorporate the modeling aspects of handling multistate variables, dependent failures, functional uncertainty, and expert knowledge, which are common in safety analysis. Given the new evidences, BN can update the probabilities and then it will better reflect system safety characteristics [2]. The corresponding BNs of classical operators in GO methodology are developed firstly. Then a case study of closing subsea blowout preventer (BOP) control system in the presence of pump failures is given to illustrate the proposed method. The reminder of this paper is organized as follows. In Sect. 2, the equivalent

BNs of operators in GO method are developed. Section 3 illustrates the method by a case study. Section 4 discusses the case study. Section 5 summarizes the paper.

2 Proposed Methodology

This section introduces the seventeen basic operators in GO methodology. The equivalent BN of each operator is developed. Netica software is used to build the BNs. Because numbers are not allowed to be state names in Netica, state values of the signals such as 0, 1, 2 in GO methodology correspond to $s0$, $s1$, $s2$ in BNs, respectively.

Type-1 operator is called two-state component [31]. It has one input signal S and one output signal R. It is easily and commonly used, which simulates the element with two states (success and failure). "Success" denotes the signal can get through the operator, while "Failure" means the signal fails to pass. Type-1 operator can simulate switches, pipes, and so on. Define V_S, V_O, and V_R as the state values of input signal, operator and output signal, respectively. For V_O, two values are optional, where 1 denotes success state and 2 denotes failure state. V_S and V_R have N states. The logic rules are shown in Table 1.

For the operator in GO, the output signal R can be regarded as the consequence of input signal S_i and operator O. To map an operator into an equivalent BN, input of the operator S_i will be the parent node and output of the operator R will be the child node. Besides, if the logic rules of the operator are related to the state values of the operator O, node O will also be the father parent node. Once the structure of the BN is established, the parameters need to specify. Prior probability of node S_i is defined according to the state values of the input signal. Similarly, if node O is present, its prior probability is specified based on the state values of operator. Therefore, the equivalent BN of type-1 operator is shown in Fig. 1. Nodes S, O, R represent the input signal, operator, and output signal, respectively. As output signal is related to the input signal and operator, nodes S and O are the parent nodes of node R. Prior probability of nodes S and O needs to be specified. Conditional probability of node R will be established based on the logic rules listed in Table 1.

A case of type-1 operator is given to illustrate the mapping process. Assuming the input signal has two states, namely 1 and 2, so its equivalent BN is shown in Fig. 2. In the BN, $s1$ denotes state 1 and $s2$ means state 2. The probabilities of all the states are shown beside the state names in the BN. For nodes S and O, the probabilities are prior probabilities. According to the logic rules shown in Table 1, CPT of node R is

Table 1 Logic rules of type-1 operator

V_S	V_O	V_R
0, …, N − 1	1	0, …, N − 1
N	1	N
0, …, N	2	N

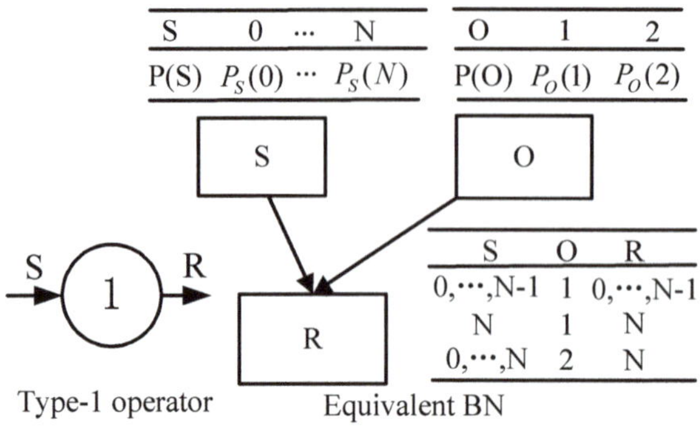

Fig. 1 Type-1 operator and its equivalent BN

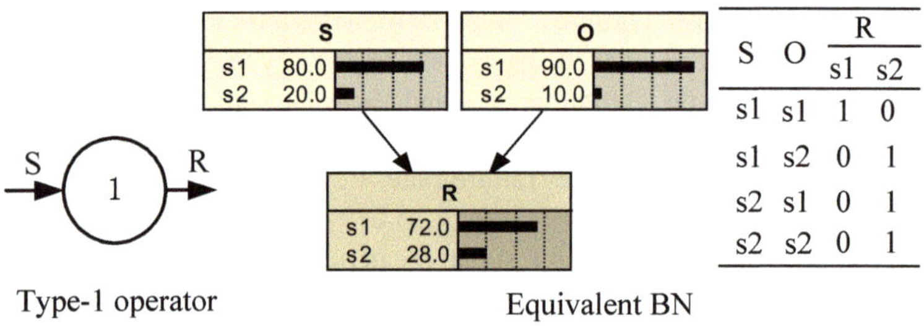

Fig. 2 A case of type-1 operator and its equivalent BN

defined. Figure 2 shows that the probability of the output signal R $P_R(s1) = 0.72$ in the case. The probability of the output signal R can be calculated by

$$P_R(s1) = \sum P(S, O) \cdot P(R = s1|S, O) \tag{1}$$

Based on the Bayesian inference, $P_R(s1) = P_S(s1) \cdot P_O(s1)$, which is the same as value computed by the GO methodology.

Type-2 models an OR gate, which has more M input signal lines. The output signal is only related to the input signals. The output signal is determined by the minimum state value of the M input signals. The equivalent BN of type-2 operator is shown in Fig. 3. The input signal $S_i(i = 0, 1, \ldots, N)$ node is the parent node of the output signal node. Prior probabilities of the input signal nodes need to be defined. The state value of the input signal is denoted by V_{S_i}. Based on the logic rule of the output signal, CPT of R is shown in the figure. Assuming there are two input signals $S1$ and $S2$ with two states, its equivalent BN is shown in Fig. 4. Prior probabilities of the input nodes are given. CPT of R can be defined as shown in the figure. It demonstrates that the probability of the output node R is correct according to the OR logic gate.

Type-3 operator is a triggered generator. It has one input signal and one output signal. It simulates an element with three states (premature, success, and failure).

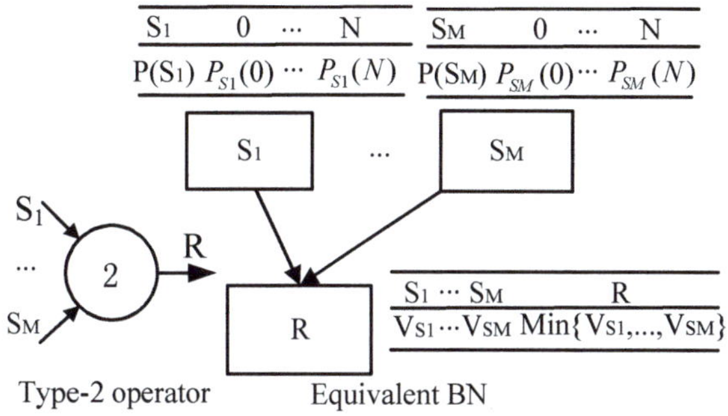

Fig. 3 Type-2 operator and its equivalent BN

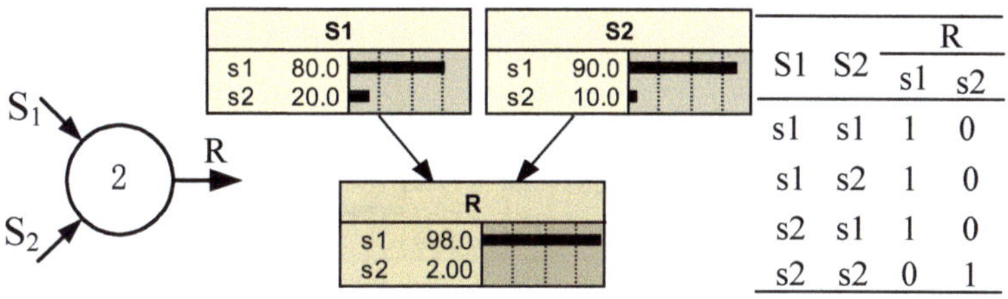

Fig. 4 A case of type-2 operator and its equivalent BN

Table 2 Logic rules of type-3 operator

V_S	V_O	V_R
0, …, N – 1	0	0
0, …, N – 1	1	0, …, N – 1
N	1	N
0, …, N	2	N

In addition, the states "success" and "failure" have the same meanings with type-1 operator. The state "premature" means that the output signal is possible to be present even without an input signal. This state describes the output signal caused by inappropriate actions or unexpected external trigger. Premature, success, and failure of operator are denoted by 0, 1, and 2, respectively. Assuming input signal has N states, the logic rules of type-3 are shown in Table 2. Figure 5 shows the type-3 operator and its equivalent BN. Nodes S, O, and R represent input signal, operator, and output signal, respectively. Node R is the child node of node S and O. CPT of node R can be defined based on the logic rules listed in Table 2.

Figure 6 shows a case of type-3 operator and its equivalent BN. Node S has two states, while nodes O and R have three states. Based on the logic rules of type-3 operator, CPT of node R is listed in Fig. 6. Therefore, all the probabilities of the states of the nodes are shown in the figure. According to Eq. (1), $P_R(s1) =$

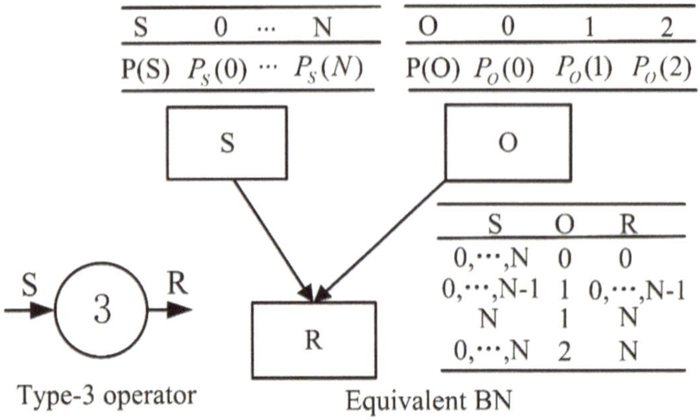

Fig. 5 Type-3 operator and its equivalent BN

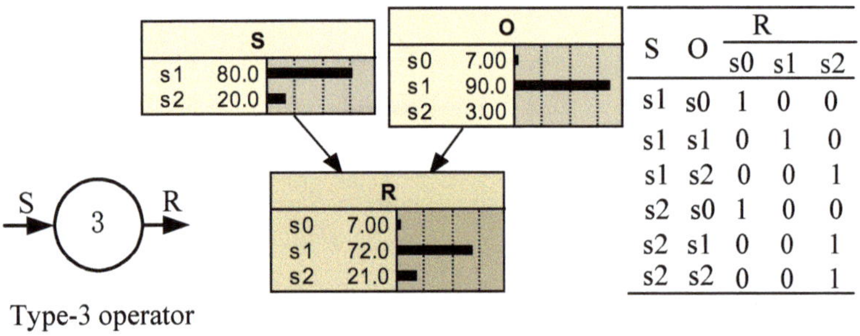

Fig. 6 A case of Type-3 operator and its equivalent BN

$P_S(s1) \cdot P_O(s1) = 0.72$. Similarly, the probability of premature state $P_R(s0)$ can be obtained by

$$P_R(s0) = \sum P(S, O) \cdot P(R = s0, |S, O) \tag{2}$$

Based on the Bayesian inference, $P_R(s0) = P_S(s1) + P_O(s0) + P_S(s2) \cdot P_O(s0) = 0.07$. $P_R(s1)$ and $P_R(s0)$ are identical to the values calculated by GO methodology.

Type-4 is a multiple-signal generator. It has only output signals and no input signals. It generates several dependent and mutually exclusive signals. If the output signals are independent, several type-5 operators can be used to simulate. The state probabilities are denoted by joint probability distribution. If the number of state combination of M input signals is L, the probability of one state combination is $P_i (i = 1, \ldots, L)$, where $\sum_{i=1}^{L} P_i = 1$. M input signals cannot be used alone (Fig. 7). Assuming it generates three mutually exclusive output signals, each output signal has two states. The operator and its equivalent BN are shown in Fig. 8. Nodes $R1$, $R2$, and $R3$ denote the three output signals. Node G has three states, representing the

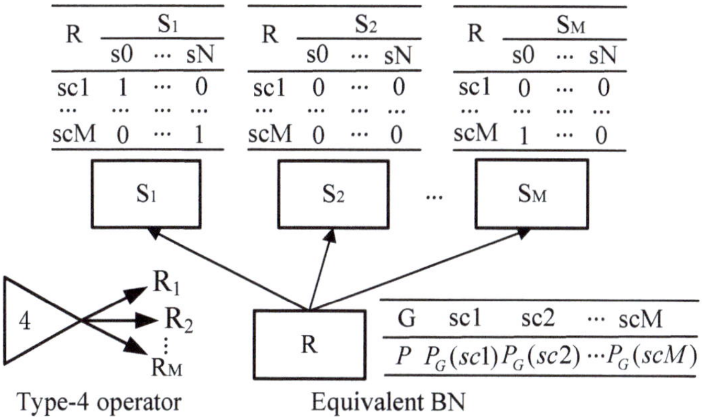

Fig. 7 Type-4 operator and its equivalent BN

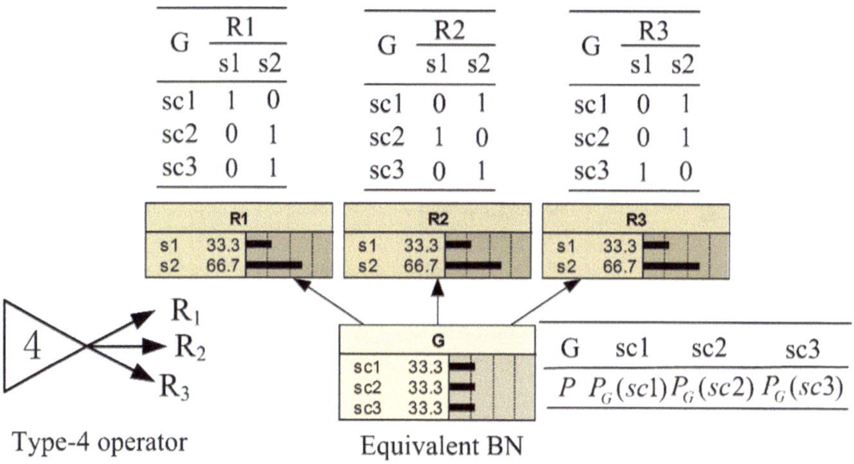

Fig. 8 Type-4 operator and its equivalent BN

three state combinations of output signals. Prior probabilities of node G are the joint probabilities of three output signals. CPT of the child nodes is listed in the figure.

Type-5 is a signal generator. It is a very commonly used input operator. It can be used to simulate generator, power source, water source, etc. The operator and its equivalent BN are shown in Fig. 9. Assuming the output signal R has two states, its CPT is shown in the figure. Prior probabilities of node R denote the probabilities of the states of output signals.

Type-6 operator is a normally open contactor. It simulates the element that produces an output signal in the presence of two input signals. One main input signal $S1$ and one secondary input signal $S2$ used as operation signal are needed. The operator has three states, namely premature, success, and failure. If the input and output signals have two states, its equivalent BN is shown in Fig. 10. Nodes $S1$, $S2$, O, and R

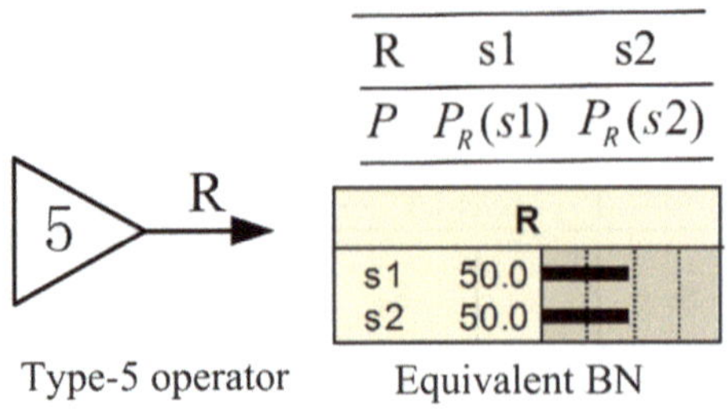

Fig. 9 Type-5 operator and its equivalent BN

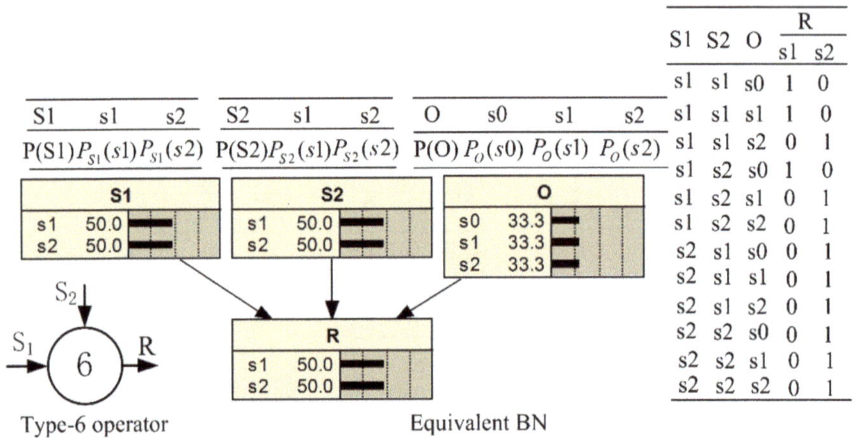

Fig. 10 Type-6 operator and its equivalent BN

denote main input signal, secondary input signal, the operator, and the output signal, respectively. CPT of node R is obtained based on the logic rules. The probability of output signal $P_R(s1)$ is

$$P_R(s1) = \sum P(S1, S2, O) \cdot (R = s1|S1, S2, O) \tag{3}$$

According to Bayesian inference, $P_R(s1) = P_{S1}(s1) \cdot P_O(s0) \cdot P_{S1}(s1) \cdot P_{S2}(s1) \cdot P_O(s1)$, which is identical to the value obtained by GO methodology.

Type-7 operator is a normally closed contactor. It has the opposite function with type-6 operator. Likewise, the operator has three states: premature, success, and failure. When the state of the element is premature, it means that the main input signal $S1$ cannot pass. Besides, the main input signal also fails to pass when there is a secondary signal $S2$ and the state of the element is "success." In other circumstances, the main input signal can pass. Assuming the input and output signals have two states, type-7 operator and its equivalent BN are shown in Fig. 11. CPT of node R is listed in the figure. Based on Eq. (3), $P_R(s1) = P_{S1}(s1) \cdot P_{S2}(s1) \cdot P_O(s2) + P_{S1}(s1) \cdot P_{S2}(s2)$, which is the same as the value calculated by GO methodology.

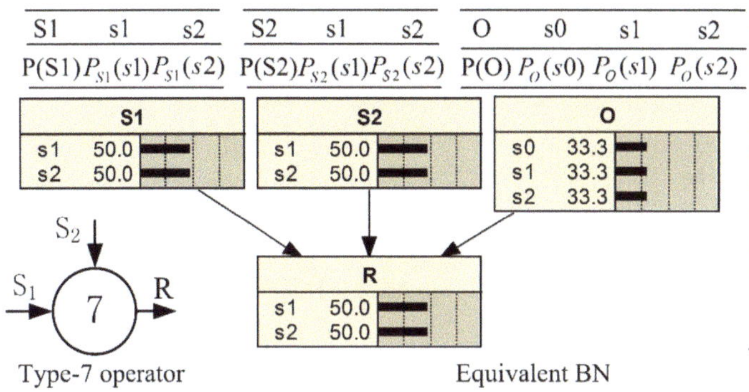

S1	S2	O	R: s1	R: s2
s1	s1	s0	0	1
s1	s1	s1	0	1
s1	s1	s2	1	0
s1	s2	s0	0	1
s1	s2	s1	1	0
s1	s2	s2	1	0
s2	s1	s0	0	1
s2	s1	s1	0	1
s2	s1	s2	0	1
s2	s2	s0	0	1
s2	s2	s1	0	1
s2	s2	s2	0	1

Fig. 11 Type-7 operator and its equivalent BN

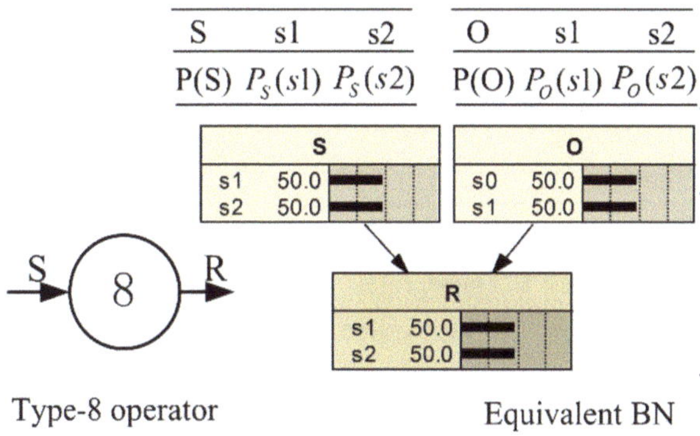

S	O	R: s1	R: s2
s1	s0	1	0
s1	s1	0	1
s2	s0	0	1
s2	s1	0	1

Fig. 12 Type-8 operator and its equivalent BN

Type-8 operator is a delay generator, which simulates an element with delays. It has an input signal S and an output signal R. The state value of output signal is greater than that of input signal. The increment of time points (state values) may denote the corresponding time interval. Assuming the input signal has two states ($s1$, $s2$) and the operator has two delays ($M = 2$), the equivalent BN of type-8 operator is shown in Fig. 12. Node S has two states ($s1$ and $s2$). Node O denotes the operator, and it has two states ($s0$ and $s1$). Based on the logic rules of the operator, the CPT of node R is listed in the figure.

Type-9 operator is a function operator. It has two input signals ($S1$ and $S2$) and one output signal R. R is present when the difference of state values ($V_{S2} - V_{S1}$) is equal to the given value X_j. This operator can be used to simulate a subsystem with cold standby. The state value of output signal is given by

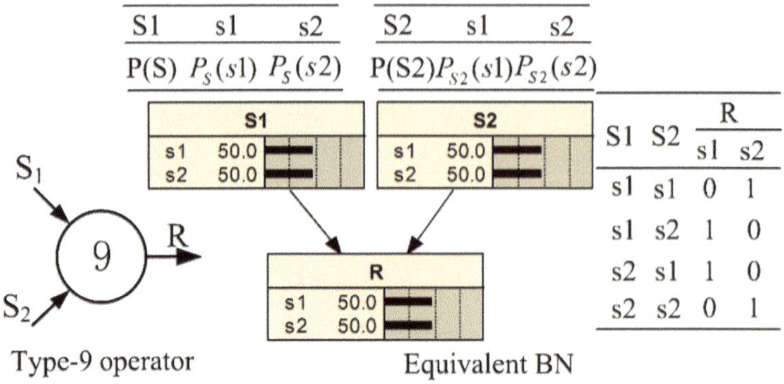

Fig. 13 Type-9 operator and its equivalent BN

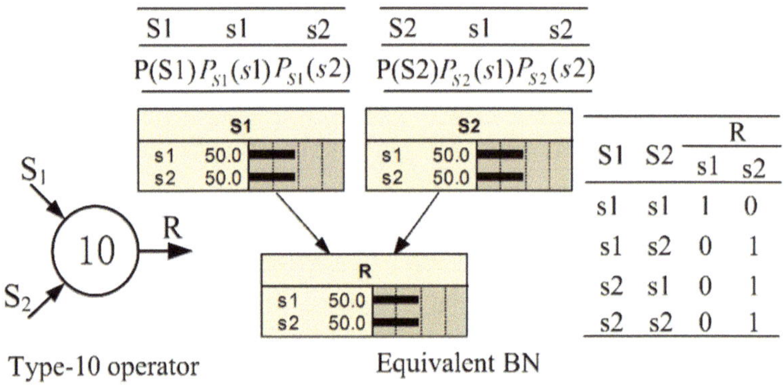

Fig. 14 Type-10 operator and its equivalent BN

$$V_R = \begin{cases} V_{S1} + Y_i, & 0 \le V_{S1} + Y \le N \\ 0, & V_{S1} + Y < 0 \\ N, & V_{S1} + Y > N \end{cases} \tag{4}$$

Type-9 operator is often used as a differential operator. When the difference between the two input signals is great enough, the operator will be triggered then the output signal will be available. Assuming the input signals have two states, 1 denotes success and 2 denotes failure state. The equivalent BN of type-9 operator is shown in Fig. 13. Nodes $S1$, $S2$, and R represent the main input signal, secondary input signal, and the output signal, respectively. CPT of node R can be obtained according to the logic rules.

Type-10 operator is an AND gate. It has several input signals and one output signal. If there are two input signals, the equivalent BN is shown in Fig. 14. Probabilities of the nodes are listed in the figure.

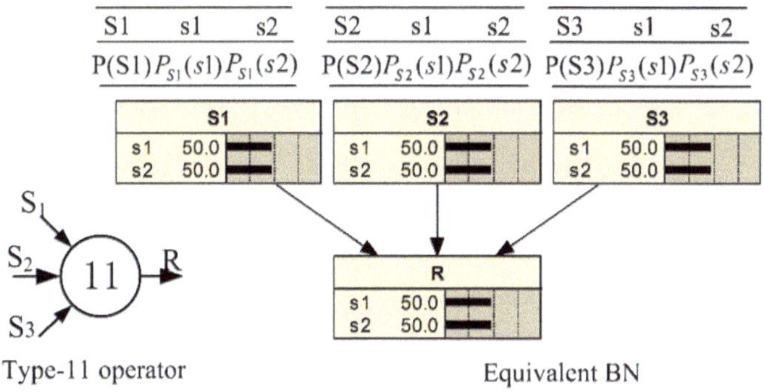

S1	S2	S3	R s1	R s2
s1	s1	s1	1	0
s1	s1	s2	1	0
s1	s2	s1	1	0
s1	s2	s2	0	1
s2	s1	s1	1	0
s2	s1	s2	0	1
s2	s2	s1	0	1
s2	s2	s2	0	1

Fig. 15 Type-11 operator and its equivalent BN

Type-11 operator is a K-out-of-M gate. It has M input signals and one output signal. The state value of output signal is the Kth value in the increasing order of those of input signals. Assuming there is a 2-out-of-3 gate, its equivalent BN is shown in Fig. 15. Based on the logic rules of the operator, CPT of node R is listed in the figure.

Type-12 operator is a path splitter. It has one input signal and M output signals. One path can be selected as the only output while the other paths are closed. All the paths can be closed, which means that it has no outputs. Assuming the operator has three output signals, so its equivalent BN is shown in Fig. 16. Nodes S and O denote the input signal and the operator, respectively. S has two states, and O has four states. Nodes Ri represent the output signal. CPT of node O can be obtained according to the logic rules. Based on Bayesian inference, $P_{R1}(s1) = P_S(s1) \cdot P_O(s1)$, which is identical to the value calculated by GO methodology. So are the probabilities of nodes $R2$ and $R3$.

Type-13 is a multiple-input/output operator. It has K input signals and M output signals. It can simulate the device or system with multiple inputs/outputs and complicated logic relations. Essentially, all the other types of operators can be represented by this operator. However, due to the complexity of the input signals, type-13 operator is seldom used. The equivalent BN is given in Fig. 17. CPT of the output signals is not defined, which can be determined by some certain logic rules.

Type-14 operator is a linear combination generator. It has M input signals and one output signal. The state value of output is decided by the combination of those of input signals. The state value of output signal V_R is obtained by

$$V_R = \begin{cases} N, \; I_1 = I_2 = \cdots = I_M = N \\ N, \; V \geq N \\ V, \; 0 < V < N \\ 0, \; \; V \leq 0 \end{cases} \tag{4}$$

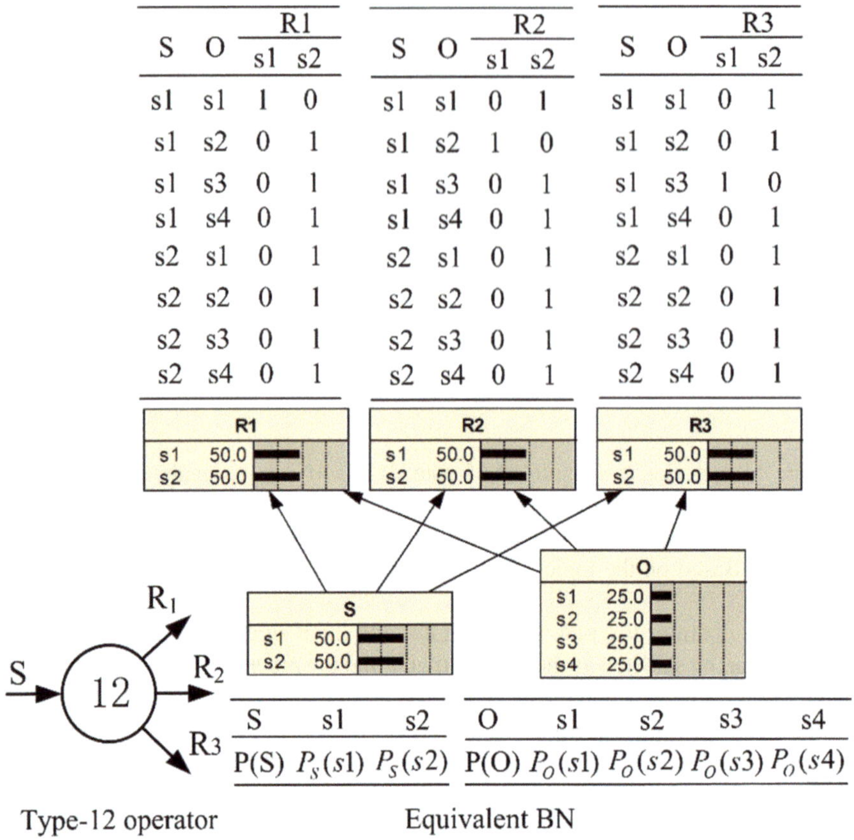

S	O	R1 s1	R1 s2
s1	s1	1	0
s1	s2	0	1
s1	s3	0	1
s1	s4	0	1
s2	s1	0	1
s2	s2	0	1
s2	s3	0	1
s2	s4	0	1

S	O	R2 s1	R2 s2
s1	s1	0	1
s1	s2	1	0
s1	s3	0	1
s1	s4	0	1
s2	s1	0	1
s2	s2	0	1
s2	s3	0	1
s2	s4	0	1

S	O	R3 s1	R3 s2
s1	s1	0	1
s1	s2	0	1
s1	s3	1	0
s1	s4	0	1
s2	s1	0	1
s2	s2	0	1
s2	s3	0	1
s2	s4	0	1

S	s1	s2
P(S)	$P_S(s1)$	$P_S(s2)$

O	s1	s2	s3	s4
P(O)	$P_O(s1)$	$P_O(s2)$	$P_O(s3)$	$P_O(s4)$

Fig. 16 Type-12 operator and its equivalent BN

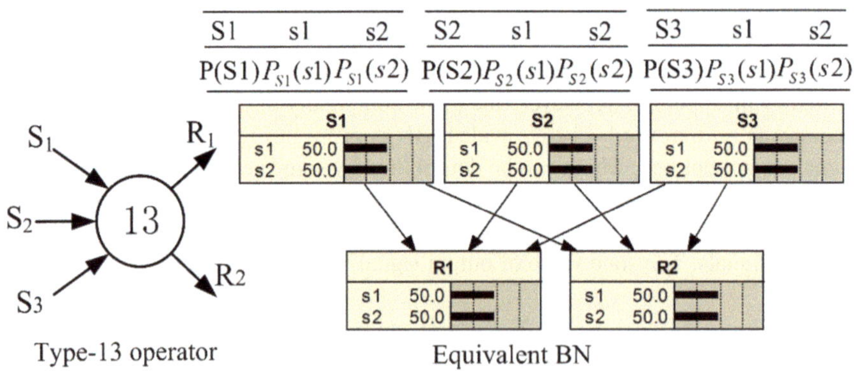

S1	s1	s2
P(S1)	$P_{S1}(s1)$	$P_{S1}(s2)$

S2	s1	s2
P(S2)	$P_{S2}(s1)$	$P_{S2}(s2)$

S3	s1	s2
P(S3)	$P_{S3}(s1)$	$P_{S3}(s2)$

Fig. 17 Type-13 operator and its equivalent BN

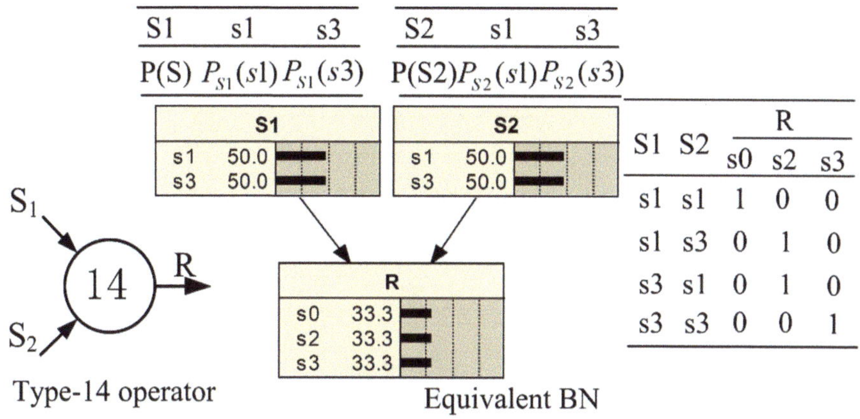

Fig. 18 Type-14 operator and its equivalent BN

where $A = a_0 + a_1 I_1 + a_2 I_2 + \cdots a_M I_M$ and $C = \text{int}(A)$. I_i is the state value of the ith input signal, and a_i is the coefficient of I_i. For example, a logic operation operator has two input signals with two states (1, 3). With $a_0 = -2.0$, $a_1 = 1.0$, $a_2 = 2.0$, $A = -2 + V_{S1} + V_{S2}$. The equivalent BN of type-14 is shown in Fig. 18.

Type-15 operator is a value/probability gate. This operator limits the state value, probability of the input signal, and the thresholds. The state value of the output signal is determined by judging whether the state value and probability of input signal is below the thresholds. V_1 and V_2 are the state values of output signal. Define V_3, V_4 as the thresholds for the state values and P_1, P_2 as the thresholds for the probabilities of the input signal. The state value of the output signal is

$$V_R = \begin{matrix} 0, & I_R \leq 0 \\ I_R, & 0 < I_R < N \\ N, & I_R \geq N \end{matrix} \tag{5}$$

where $I_R = \begin{cases} V_0, V_3 \leq I_s \leq V_4 \text{ and } P_1 \leq P_0 \leq P_2 \\ V_2 \end{cases}$, $V_0 = \begin{cases} V_1, V_1 \neq -1 \\ I_S, V_1 = -1 \end{cases}$.

Assuming $V_1 = 2$, $V_2 = 1$, $V_3 = V_4 = 1$, $P_1 = 0$, $P_2 = 1$, this operator can be used as a NOT gate based on the logic rules. Its equivalent BN is shown in Fig. 19.

Type-16 operator is an actuated normally open contact. It is used to simulate the disconnection of a conductive signal, such as cutting off the electricity or stopping the alarm. The states of input and output signals change from presence to absence. Type-16 operator also has three states: premature, success, and failure. But it has different meanings with type-6 operator. Besides, the output signal has three states. The equivalent BN of type-16 is shown in Fig. 20. Nodes $S1$, $S2$, O, and R denote the main input signal, secondary input signal, operator, and the output signal, respectively. Based on the logic rule, CPT of node R is obtained.

S	s1	s2
P(S)	$P_S(s1)$	$P_S(s2)$

Type-15 operator

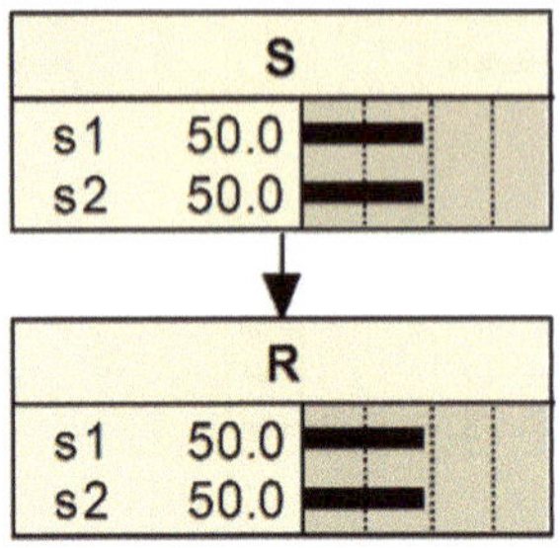

S	R: s1	R: s2
s1	0	1
s2	1	0

Equivalent BN

Fig. 19 Type-15 operator and its equivalent BN

S1	s1	s2
P(S1)	$P_{S1}(s1)$	$P_{S1}(s2)$

S2	s1	s2
P(S2)	$P_{S2}(s1)$	$P_{S2}(s2)$

O	s0	s1	s2
P(O)	$P_O(s0)$	$P_O(s1)$	$P_O(s2)$

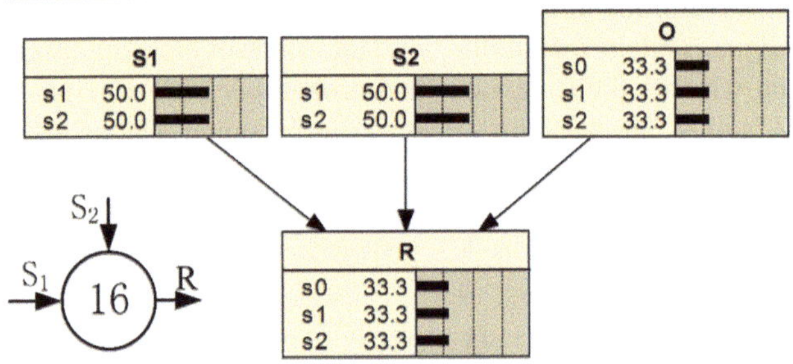

Type-16 operator

Equivalent BN

S1	S2	O	R: s0	R: s1	R: s2
s1	s1	s0	1	0	0
s1	s1	s1	0	1	0
s1	s1	s2	0	1	0
s1	s2	s0	1	0	0
s1	s2	s1	0	1	0
s1	s2	s2	0	1	0
s2	s1	s0	1	0	0
s2	s1	s1	0	1	0
s2	s1	s2	0	0	1
s2	s2	s0	1	0	0
s2	s2	s1	0	0	1
s2	s2	s2	0	0	1

Fig. 20 Type-16 operator and its equivalent BN

Type-17 operator is an actuated normally closed contact. It has the opposite function compared with type-16 operator. It simulates a triggered or closed component, the signal of which is not able to pass at this time. Like type-16 operator, this operator has three states: premature, success, and failure. The equivalent BN of type-17 is shown in Fig. 21. Input signals have two states: $s1$ and $s2$.

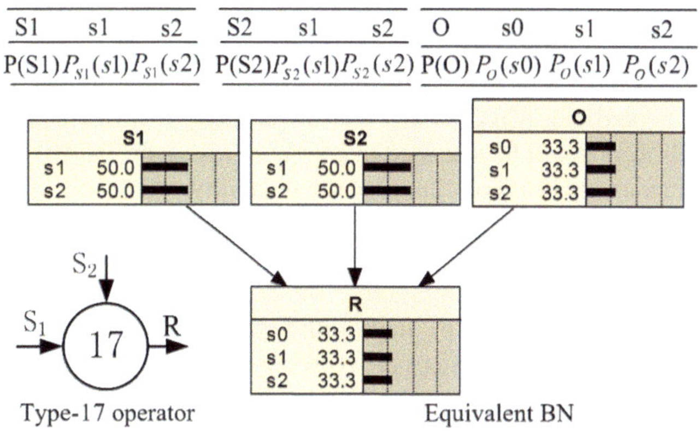

S1	S2	O	R		
			s0	s1	s2
s1	s1	s0	1	0	0
s1	s1	s1	0	0	1
s1	s1	s2	0	0	1
s1	s2	s0	1	0	0
s1	s2	s1	0	0	1
s1	s2	s2	0	0	1
s2	s1	s0	1	0	0
s2	s1	s1	0	1	0
s2	s1	s2	0	0	1
s2	s2	s0	1	0	0
s2	s2	s1	0	0	1
s2	s2	s2	0	0	1

Fig. 21 Type-17 operator and its equivalent BN

3 Cases Study

Offshore drilling for oil and gas is a complex process that depends on careful coordination of hardware and control strategies. A kick is defined as the uncontrolled release of formation fluids into the wellbore [8]. In this situation, a blowout will appear if there are mechanical failures of blowout preventer (BOP) or failures to detect the kick on time [32]. Hence, kick is regarded as an initiating event of a blowout [33]. The first response after detecting a kick should be to isolate the wellbore by activating the BOP and closing the well. Subsea BOP system is developed to cope with the extreme erratic pressures and uncontrolled flow emanating from a well reservoir during drilling processes. Control pod of subsea BOP system serves as the subsea control valve manifold [34]. As the core of the MUX control system, it contains all of the pressure regulators and control valves required to operate the subsea functions. For high reliability, two control pods are used as a redundant system. Each pod can operate all subsea functions, but one pod is active and the other one is hot standby at the same time. The control panels initiated the demand on the subsea control system. The demand signal is multiplexed down the umbilical to the subsea pod, where the signal is decoded and performed. For example, to close a BOP ram, the demand signal will be sent to the subsea control pod and decoded. The decoded signal will open a solenoid electrically and therefore, the proper hydraulic valve will receive a hydraulic pilot signal. This pilot signal will cause the hydraulic valve to shift and send stored and pressurized hydraulic fluid to the BOP ram to be closed.

As mentioned above, blowout could be caused by equipment failure, human operation errors, annular losses, pump failure, swabbing, and so on [8]. This paper takes one managed pressure drilling blowout scenario caused by pump failure as the example to demonstrate the proposed method. A pump failure will be caused by a power or mechanical failure. If the pump failure occurs in managed pressure drilling technology, a kick could be caused by the circulation stop and loss of the annular frictional pressure. At this time, command of closing BOP should be initiated to prevent the kick. Assuming the blue pod is active and yellow pod is standby, the flowchart of closing BOP is shown in Fig. 22. It shows that the command of closing BOP will be initiated once the kick is detected. It is assumed that the function is performed by blue pod firstly. The yellow pod will be activated to continue closing the BOP if the blue pod fails. Two redundant control pods improve the reliability.

Figure 23 shows the GO model of this case. The input of the system is the failures of pump, which are represented by type-5 operators. State "$s1$" means the failure is present, and state "$s2$" denotes there are no failures. The decision nodes are represented by type-12 operators. The processing steps with two inputs are denoted by type-2 operators. All the other processing steps are represented by type-1 operators. In Fig. 23, the first number in the operator symbol denotes the type of operator. The second number denotes the numbering of operator, and the number on the signal line means the numbering of signal.

As type-2 operator is a logical operator, the state probabilities of the other 11 operators are listed in Table 3. Since the case assumes that there are pump failures, probability of both power failure and pump mechanical failure is set 100%. Type-12 operators have three states. For operator E, state values 1, 2, and 3 denote that output signal 6, output signal 9, and none of them are available, respectively. The other operators have two states: Yes and No. The minimal path sets of the developed GO model are $A_1B_1C_1D_1E_1F_1G_1$, $A_1B_1C_1D_1E_2H_1I_1J_1K_1$, $A_1B_1C_1D_1E_1F_1G_1$, $A_1B_2C_1D_1E_1F_1G_1$, $A_1B_2C_1D_1E_2H_1I_1J_1K_1$, $A_1B_2C_1D_1E_1F_2H_1I_1J_1K_1$, $A_2B_1C_1$ $D_1E_1F_1G_1$, $A_2B_1C_1D_1E_2H_1I_1J_1K_1$, $A_2B_1C_1D_1E_1F_2H_1I_1J_1K_1$. Hence, the probability of closing BOP is 86.6%.

Based on the equivalent BNs of the corresponding operators in the above section, BN of the case is developed in Fig. 24. It shows that the successful probability of the system ($P(2-14=s1)$) is 86.6%, which is identical to the value calculated by GO methodology. Therefore, the BN is equivalent to the GO model.

4 Results and Discussion

This case is to investigate the risk of closing the BOP in the presence of pump failures. BN can update the probabilities given the new evidences. In this way, it will better reflect system safety characteristics. With the developed BN model, predictive and diagnostic analysis can be conducted. For example, if the evidence is node *Operator_12 = Output9* as shown in Fig. 25, which means that the valve in blue pod is not activated, predictive analysis will be performed. The probability of closing BOP successfully is 76.6%. As shown in Fig. 26, the evidence is that the BOP is closed successfully, and the diagnostic analysis can be performed. It shows that the close function is most likely to be finished by blue pod (93.1%). Besides, BN can adapt probability if new prior probability or conditional probability distributions are obtained which are collected in the future. Figure 27 shows the new prior probabil-

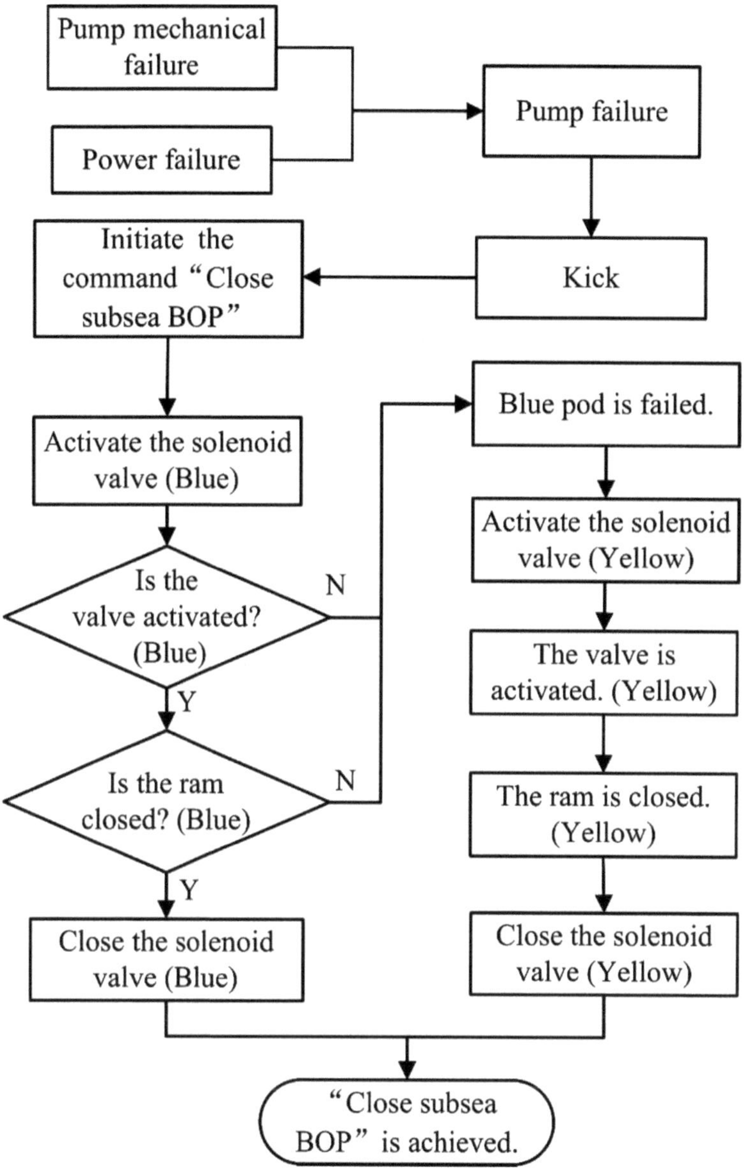

Fig. 22 Flowchart of "close BOP"

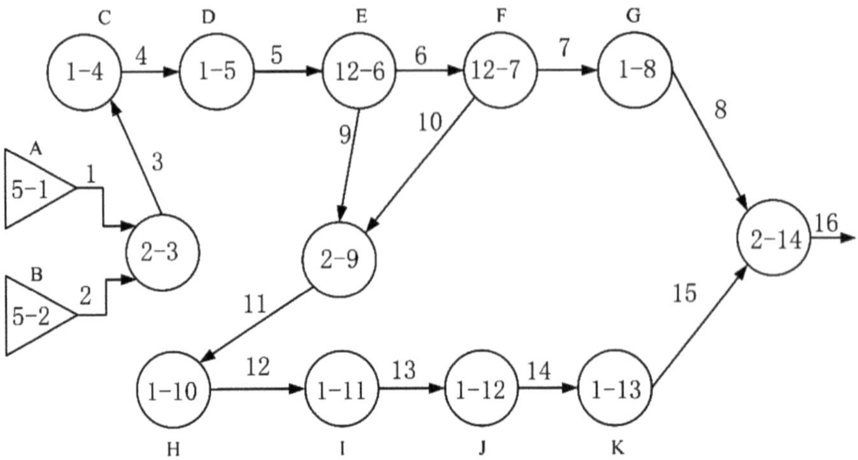

Fig. 23 GO model of the case

Table 3 State probabilities of the operators

No.	Type	Tag	State value	Meanings	Probability
1, 2	5	A, B	1	Yes	1
			2	No	0
4, 5	1	C, D	1	Yes	0.97
			2	No	0.03
6	12	E	1	Output 6 is selected	0.95
			2	Output 9 is selected	0.04
			3	None	0.01
7	12	F	1	Output 7 is selected	0.95
			2	Output 10 is selected	0.04
			3	None	0.01
8, 10, 11, 12, 13	1	G, H, I, J, K	1	Yes	0.95
			2	No	0.05

ity of node Operator_1 which the parent node of node 1–8, and it means that the probability of closing the solenoid valve successfully becomes 96%. Therefore, the probability of closing BOP in the presence of kick improves from 86.8 to 87.5%.

With the developed BN model, sensitivity analysis is performed to find the key factors of influencing closing BOP. Therefore, node 2–14 is selected as the target node, the other nodes are chosen as the findings nodes. The sensitivity analysis can conclude how much entropy reduction of the garget node could be influenced by a single finding at each of the findings nodes in the BN. Assuming the entropy reduction of node 2–14 influenced by itself to be 100%, the entropy reduction of the other important findings nodes are shown in Fig. 28. It shows that node 1–8 has the

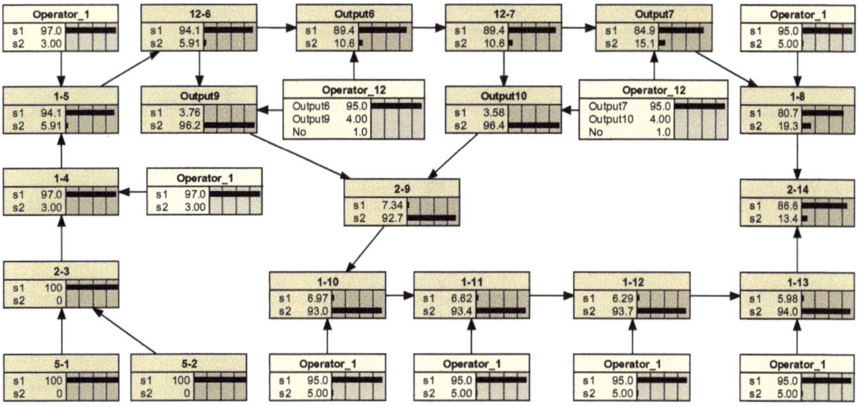

Fig. 24 Equivalent BN of the case

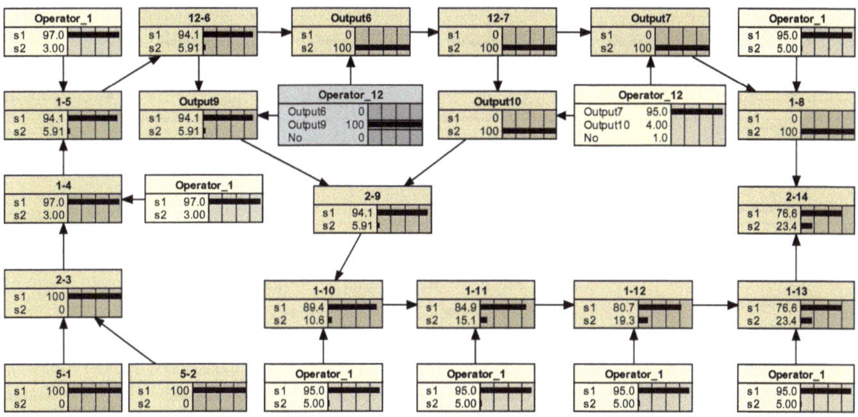

Fig. 25 BN of the case when the valve in blue pod is not activated

greatest influencing degree to the target node 2–14. One of the reasons is that this event is not redundant. If this event fails, the closing BOP function will certainly fails. Therefore, great attention should be paid to the solenoid valve in blue pod in order to improve the probability of closing BOP successfully.

To provide a reasonable amount of confidence to the results of the model, validation is an important part of the proposed BN. In the current work, a three-axiom-based validation method is used for partial validation of the developed BN. The three axioms are as follows [35].

(1) A slight increase/decrease in the prior probabilities of parent nodes must result in the relative increase/decrease of the posterior probabilities of the corresponding child nodes.
(2) The influence degree to the child node caused by the variation of probability distributions of parent nodes should keep consistency.

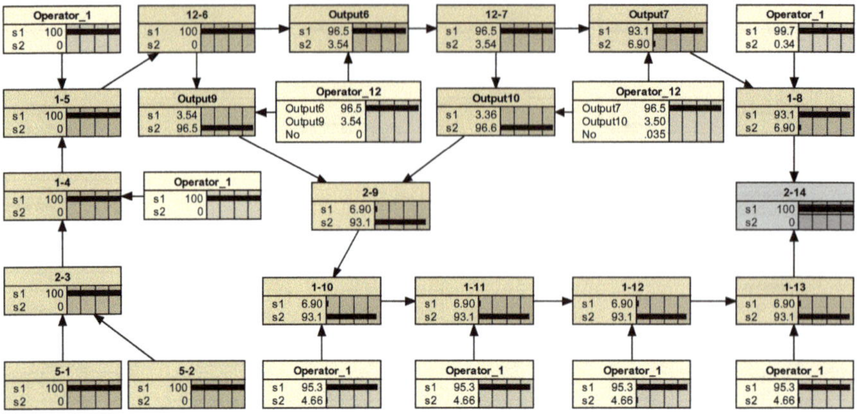

Fig. 26 BN of the case with the evidence that the BOP is successfully closed

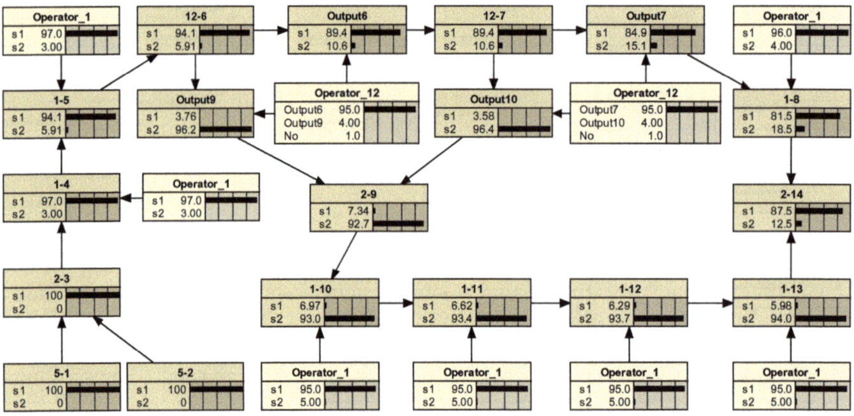

Fig. 27 BN of the case with probability adapting

(3) The total influence magnitudes of the combination of the probability variations from x attributes on the values should always be greater than that from the set of $x - y(y \in x)$.

Take the parent node Operator_1 of node 1–8 as an example, when the prior probability of state "$s1$" is set to be 94%; the probability of closing BOP successfully decreases from 86.8 to 85.8%. Based on this change, when the prior probability of the parent node Operator_1 of node 1–13 is set to be 90%, the probability of closing BOP decreases from 85.8 to 85.5%. The exercise of increasing each influencing node satisfies the axioms, thus giving a partial validation of the model.

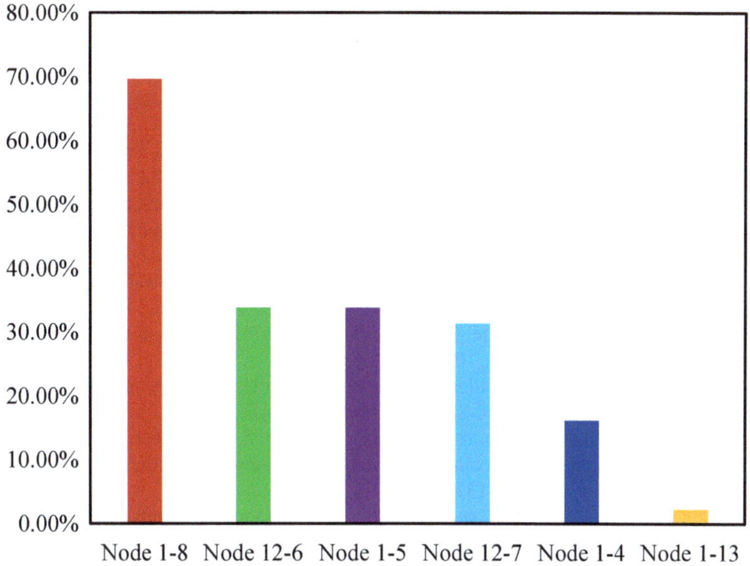

Fig. 28 Entropy reduction of target node influenced by different findings nodes

5 Conclusions

In this paper, a novel method of mapping GO model into equivalent BNs is proposed. Firstly, the equivalent BNs of seventeen operators in the GO methodology are developed. Then a case of closing subsea BOP in the presence of pump failures is used to illustrate the mapping process. Based on the equivalent BNs of basic operators, BN of the case is presented. The proposed method relaxes the existing limitations of GO model and enriches the ways of developing BN. The developed BN is capable of probability adapting and probability updating including predictive analysis and diagnostic analysis. By sensitivity analysis, key influencing factor is determined. It shows that great attention should be paid to the solenoid valve in blue pod in order to improve the probability of closing BOP successfully. Finally, the developed BN is partially validated by the three-axiom-based validation method.

References

1. L. Sanmiquel, J.M. Rossell, C. Vintro, Study of Spanish mining accidents using data mining techniques. Saf. Sci. **75**, 49–55 (2015)
2. N. Khakzad, F. Khan, P. Amyotte, Safety analysis in process facilities: comparison of fault tree and Bayesian network approaches. Reliab. Eng. Syst. Saf. **96**, 925–932 (2011)
3. J.S. Wu, R. Zhou, S.D. Xu, Z.W. Wu, Probabilistic analysis of natural gas pipeline network accident based on Bayesian network. J. Loss Prevent Proc. Ind. **46**, 126–136 (2017)

4. Y.M. Huang, G.W. Ma, J.D. Li, Grid-based risk mapping for gas explosion accidents by using Bayesian network method. J. Loss Prevent. Proc. Ind. **48**, 223–232 (2017)
5. C.T. Yeo, J. Bhandari, R. Abbassi, V. Garaniya, S.H. Chai, B. Shomali, Dynamic risk analysis of offloading process in floating liquefied natural gas (FLNG) platform using Bayesian Network. J. Loss Prevent Proc. Ind. **41**, 259–269 (2016)
6. F. Goerlandt, J. Montewka, A framework for risk analysis of maritime transportation systems: a case for oil spill from tankers in a ship-ship collision. Saf. Sci. **76**, 42–66 (2015)
7. K. Garmer, H. Sjostrom, A.M. Hiremath, A.K. Tilwankar, G. Kinigalakis, S.R. Asolekar, Development and validation of three-step risk assessment method for ship recycling sector. Saf. Sci. **76**, 175–189 (2015)
8. J. Bhandari, R. Abbassi, V. Garaniya, F. Khan, Risk analysis of deepwater drilling operations using Bayesian network. J. Loss Prevent Proc. Ind. **38**, 11–23 (2015)
9. N. Khakzad, F. Khan, P. Amyotte, Quantitative risk analysis of offshore drilling operations: a Bayesian approach. Saf. Sci. **57**, 108–117 (2013)
10. M.J. Akhtar, I.B. Utne, Human fatigue's effect on the risk of maritime groundings—A Bayesian network modeling approach. Saf. Sci. **62**, 427–440 (2014)
11. M. Musharraf, D. Bradbury-Squires, F. Khan, B. Veitch, S. Mackinnon, S. Imtiaz, A virtual experimental technique for data collection for a Bayesian network approach to human reliability analysis. Reliab. Eng. Sys. Saf. **132**, 1–8 (2014)
12. M. Musharraf, J. Hassan, F. Khan, B. Beitch, S. MacKinnon, S. Imtiza, Human reliability assessment during offshore emergency conditions. Saf. Sci. **2013**, 19–27 (2013)
13. M. Abimbola, F. Khan, N. Khakzad, S. Butt, Safety and risk analysis of managed pressure drilling operation using Bayesian network. Saf. Sci. **76**, 133–144 (2015)
14. B. Cai, H. Liu, M. Xie, A real-time fault diagnosis methodology of complex systems using object-oriented Bayesian networks. Mech. Syst. Signal Process. **80**, 31–44 (2016)
15. B. Cai, M. Xie, Y. Liu, Y. Liu, Q. Feng, Availability-based engineering resilience metric and its corresponding evaluation methodology. Reliab. Eng. Syst. Saf. **172**, 216–224 (2018)
16. Y. Zhao, F. Xiao, S. Wang, An intelligent chiller fault detection and diagnosis methodology using Bayesian belief network. Energy Build. **57**, 278–288 (2013)
17. B. Cai, L. Huang, M. Xie, Bayesian networks in fault diagnosis. IEEE Trans. Industr. Inform. **13**, 2227–2240 (2017)
18. A. Bobbio, L. Portinale, M. Minichino, E. Ciancamerla, Improving the analysis of dependable systems by mapping fault trees into Bayesian networks. Reliab. Eng. Sys. Saf. **71**, 249–260 (2001)
19. H. Boudali, J.B. Dugan, A discrete-time Bayesian network reliability modeling and analysis framework. Reliab. Eng. Sys. Saf. **87**, 337–349 (2005)
20. S. Montani, L. Portinale, A. Bobbio, Codetta-Raiteri D. RADYBAN: a tool for reliability analysis of dynamic fault trees through conversion into dynamic Bayesian networks. Reliab. Eng. Sys. Saf. **93**, 922–932 (2008)
21. G. Bearfield, W. Marsh, Generalising event trees using Bayesian networks with a case study of train derailment. Lect. Notes Comput. Sci. **3688**, 52–66 (2005)
22. C.H. Lo, Y.K. Wong, A.B. Rad, Bond graph based Bayesian network for fault diagnosis. Appl. Soft Comput. **11**, 1208–1212 (2011)
23. M.C. Kim, Reliability block diagram with general gates and its application to system reliability analysis. Ann. Nucl. Energy **38**, 2456–2461 (2011)
24. N. Khakzad, F. Khan, P. Amyotte, Dynamic safety analysis of process systems by mapping bow-tie into Bayesian network. Process Saf. Environ. Prot. **91**, 46–53 (2013)
25. R.L. Williams, W.Y. Gateley, Use of GO methodology to directly generate minimal cut set, in *Nuclear system reliability engineering and risk assessment*, ed. by I.B. Fussell (Society for industrial and applied mathematics, Pennsylvania, PA, 1977), pp. 825–849
26. T. Matsuoka, M. Kobayashi, GO-FLOW: a new reliability analysis methodology. Nucl. Sci. Eng. **98**, 64–78 (1988)
27. X.J. Yi, J. Shi, H.N. Mu, H.P. Dong, Z. Zhang, Reliability analysis of repairable system with multiple-input and multi-function component based on goal-oriented methodology. ASME J. Risk Uncertainty Part B **3**, 014501–014501–014501–014506 (2016)

28. X.J. Yi, B.S. Dhillon, J. Shi, H.N. Mu, Z. Zhang, A new reliability analysis method for vehicle systems based on goal-oriented methodology. Proc. Inst. Mech. Eng. D **231**, 1066–1095 (2017)
29. Z.P. Shen, X.J. Dai, X.R. Huang, A supplemental algorithm for the repairable system in the GO methodology. Reliab. Eng. Syst. Saf. **91**, 940–944 (2006)
30. Z. Shen, J. Gao, X. Huang, A new quantification algorithm for the GO methodology. Reliab. Eng. Sys. Saf. **67**, 241–247 (2000)
31. Z.P. Shen, X.R. Huang, *Principle and application of GO methodology: a system reliability analysis methodology* (Tsinghua University Press, Beijing, 2004)
32. M. Abimbola, F. Khan, N. Khakzad, Dynamic safety risk analysis of offshore drilling. J. Loss Prev. Process Ind. **30**, 74–85 (2014)
33. R. Islam, F. Khan, R. Venkatesan, Real time risk analysis of kick detection: testing and validation. Reliab. Eng. Sys. Saf. **161**, 25–37 (2017)
34. P. Fougere, J.R. Roche, Multiplex BOP system utilizes updated technology, decreases riser disconnect time. Oil Gas J. **98**, 86–94 (1997)
35. B. Jones, J. Jenkinson, Z. Yang, J. Wang, The use of Bayesian network modelling for maintenance planning in a manufacturing industry. Reliab. Eng. Syst. Saf. **95**, 267–277 (2010)

Bayesian Network-Based Risk Analysis Methodology: A Case of Atmospheric and Vacuum Distillation Unit

Abstract Chemical and petrochemical accidents, such as fires and explosions, do not happen frequently but have considerable consequences. These accidents compromise not only human safety but also cause significant economic losses and environmental contamination. The increasing complexity of chemical infrastructures increases the requirements of risk prevention. Thus, risk analysis for petrochemical systems is essential in helping analysts find the weakest process in the entire system and be used to strengthen the process and improve safety. Risk analysis has been previously studied; however, traditional methods have limitations. This study proposes a methodology that is based on Bayesian networks by giving a model for system risk analysis. The event is classified into three categories, cause, incident, and accident, according to criticality and thus, the model is analyzed as a three-layered structure. The application of the methodology is demonstrated by analyzing a vacuum distillation and an atmospheric unit. An exact reasoning method is used to infer the causality and probability within the events. After inferring the relationship between causes and accidents, mutual information and variance of beliefs are calculated to find the most sensitive event in an accident. Subsequently, means of strengthening operations to prevent accidents are suggested. This study may help companies decrease the cost of risk reduction.

Keywords Risk analysis · Bayesian networks · Chemical plant · Three-layer hierarchical model

1 Introduction

Petrochemical process industries utilize complex piping and other equipment necessary for processing [1]. An interlock is an essential component in petrochemical processes; it controls various systems and provides better efficiency and reliability. Thousands of interlocks involved in the control system make the process a complex one. In this case, the occurrence of one event triggers another and in turn, leads to a series of events and thus making sure that the entire system is stable with no fault happening is vital. The occurrence of any fault in such a complex control system

B. Cai et al., *Bayesian Networks for Reliability Engineering*,
https://doi.org/10.1007/978-981-13-6516-4_9

may lead to a series of incidents, even accidents. Accidents occurring in chemical plants have low frequency but significant consequences. Historical data show that several accidents with severe consequences have occurred in this industry. Hisken et al. [2] reported an accident that occurred in December 2005 in Buncefield Complex, at the Hertfordshire Oil Storage Terminal, in Hertfordshire, England, where a massive fire and a series of explosions led to 43 injuries and substantial property losses. The primary cause of the fire was leakage of 250,000 L of fuel, and then an explosion in a vapor cloud of evaporated petrol. Li et al. [3] documented one of the most significant accidents reported in history, which happened at the British Petroleum's refinery in Texas City, Texas, the USA on March 23, 2005. A spillage of raffinate led to the evaporation, and a nearby vehicle engine ignited the vapor cloud, resulting in an explosion. The company lost approximately $1.5 billion in monetary terms, and 15 fatalities and 180 injuries were recorded. A fire and explosion caused by the expanded high-pressure methane gas occurred at the Macondo Well, the Gulf of Mexico in the United States on April 20, 2010, which led to 11 workers missing and 17 injuries [4]. This explosion then resulted in a severe oil spillage in the ocean, which lasted for 87 days. This spillage was excessively harmful to the environment and wildlife, which made it the most massive accidental marine oil spill in the history of the petroleum industry.

The severity of these accidents shows that a risk analysis process for petrochemical systems is essential to potentially help identify the weakest link in the system and thus strengthen the process and improve safety. Several methods have been developed for risk analysis. Cai et al. [5] used the Markov method with the multiple-error shock model to evaluate triple and dual modular redundancy systems within subsea blowout preventer control systems. Ramírez-Marengo et al. [6] utilized a stochastic approach that included a Monte Carlo simulation to assess risks of vapor cloud explosions using Analytic Solver Platform. Fu et al. [7] proposed a quantitative risk assessment model for the potential leakage risk of liquefied natural gas (LNG)-fueled vessels; their methods use event tree analysis and computational fluid dynamics simulation. Martins et al. [8] presented a complete quantitative risk analysis for unexpected events that can happen during LNG processing. Fuentes-Bargues et al. [9] used hazard and operability (HAZOP) analysis and risk evaluation to analyze the risk in industrial plants. However, this HAZOP method is not able to provide quantitative analysis like probability or likelihood dimension to give a risk assessment.

Bayesian network (BN) is an important probabilistic graphical model, which can efficiently deal with various uncertainty problems based on probabilistic information representation and inference. In recent years, researchers have been using Bayesian networks (BNs) for quantitative risk analyses to calculate risk models due to the flexibility of the networks. Cai et al. [10] proposed a methodology that uses BNs for the quantitative risk assessment of operations in offshore oil and gas industries by translating a flowchart into five steps. Yeo et al. [11] provided a dynamic safety analysis model for LNG carrier offloading; their model investigates different risk factors during the process using a BN. Bhandari et al. [12] introduced a dynamic safety analysis that uses a BN for deep-water managed pressure and underbalanced drilling operations. Cai et al. [13] presented a quantitative risk assessment model that uses BNs for an offshore blowout; their model focuses on human factors, which were described as human factor barrier failures. This research was extended by Cai et al. [14] by combining root cause analysis and reliability evaluation phases that are based on BNs and dynamic BNs, respectively, to present a real-time reliability evaluation methodology. Several mature, accurate, and approximate inference algorithms have been developed for BNs. Approximate inference algorithms can be used when the number of components is increasing.

This research presents a BN-based risk analysis methodology, which we apply to a critical system in petrochemical processing: an atmospheric and vacuum distillation unit. This component is necessary for petroleum refinery plants and has universal applications. An atmospheric and vacuum distillation unit has an elaborate system comprising many separate structures and control nodes. Equipment and material failures like cracks in the storage vessels, or human errors like operational errors or external perturbation, can have a significant influence on work safety. Therefore, system control and accident reduction methods need to be developed to identify the weakest processes in the system and improve safety. We use BNs for risk analysis in this work.

The remaining parts of this paper are organized as follows; Sect. 2 presents the proposed methodology, based on a BN as a model for system risk analysis. The event is classified into three categories: cause, incident, and accident, by criticality, and the model is analyzed as three-layered. Section 3 discusses the application of the proposed method to the risk analysis of atmospheric and vacuum distillation units. After inferring the relationship between the causes and the accident in the system, the most plausible cause is detected. Then, suggestions for accident prevention are given. Section 4 concludes the work based on the findings. This research can help companies in the petrochemical industry to reduce the cost of risk analysis.

2 Methodology

2.1 BN Overview

A BN is a probabilistic graphical model, which is a statistical model that represents a set of random events denoted by nodes in a graph. The conditional dependencies of these events are represented by edges via a directed acyclic graph. Each node is associated with a probability function as input that selects a particular set of values for the parent variables of the node and gives the probability or probability distribution of the variable represented by the node. Figure 1 shows a typical BN example.

Several basic statistical formulas and theories for quantitative BNs are presented based on conditional independence and the chain rule and by calculating the product of conditional probability tables (CPT), the joint probability of a set of variables $U = \{A_1, A_2, \ldots, A_n\}$ can be given as

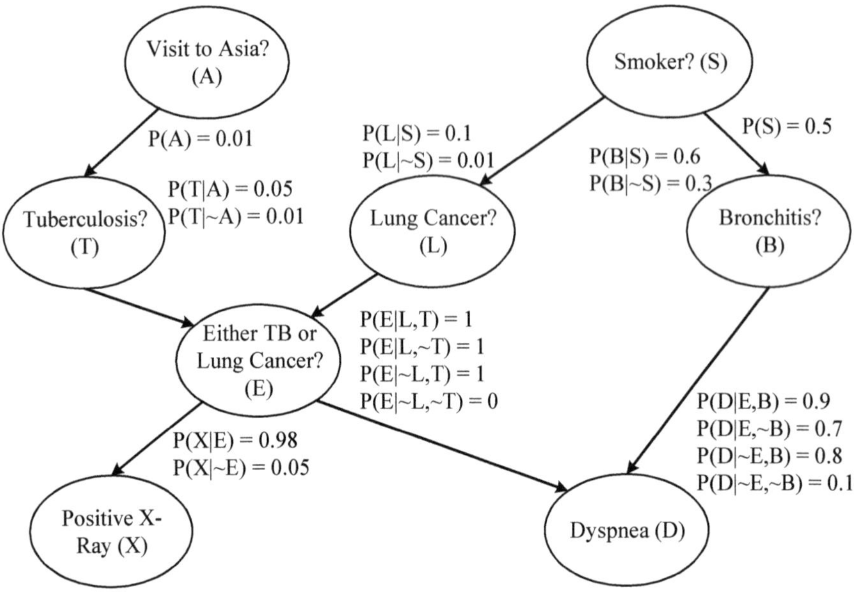

Fig. 1 A typical example of BNs

$$P(U) = \prod_{i=1}^{n} P(A_i|P_a(A_i)), \tag{1}$$

where $Pa(A_i)$ is the parent node of A_i in the BNs and $P(U)$ represents the probabilities [15].

BNs could provide the analyst the capability to perform forward (predictive) and backward (diagnostic) analyses ([16–18]). In diagnostic analysis, a series of evidence E is examined, and the posterior probability distribution can be calculated using various inference algorithms that are based on *Bayes' theorem* as follows [19]:

$$P(U|E) = \frac{P(E|U)P(U)}{P(E)} = \frac{P(E, U)}{\sum_U P(E, U)}. \tag{2}$$

The important degree of the basic event to the system failure can be assessed using Shannon's mutual information (entropy reduction), which is a widely used measurement model for ranking information sources, as shown by Pearl [20]. The uncertainty of a system is assumed to be represented by an entropy function given as

$$H(T) = -\sum_{t} P(t) \log P(t), \tag{3}$$

where $P(t)$ is the probability distribution of the random variable T.

Mutual information is the total uncertainty-reducing potential of X, given the original uncertainty in T before consulting X. The mutual information of T and X is given by

$$I(T,X) = -\sum_{x}\sum_{t} P(t,x)\log\frac{P(t,x)}{P(t)P(x)}, \tag{4}$$

where $P(t,x)$ is the joint probability distribution function of T and X, and $P(t)$ and $P(x)$ represent the marginal probability distribution functions of T and X, respectively.

2.2 BN-Based Risk Analysis Methodology

Many error or non-error causes may happen in chemical plants because they are complex systems with imperfect reliability, and these causes can develop into incidents. An incident is an abnormal event that is not supposed to happen in daily operation. If no response is undertaken when an incident occurs or if self-healing cannot be performed, then an incident may turn into an accident, which is unfavorable.

Cause, incident, and accident illustrate a sequence of increasing level of criticality. The three events are defined as follows. (1) Causes are actions that may cause an abnormal event. (2) Incidents are unusual events that may result in an accident, although they have no severe criticality in and of themselves. (3) Accidents are the events that have severe consequences and threaten to compromise human safety and cause economic loss.

Although several causes or their combinations may cause an accident, humans prefer the non-occurrence of the accident. From previous data, we can give an analysis of the probability of each cause to lead to an accident, if any. Then, from the quantitative analysis, the cause, which has the most significant influence on accidents, can be found. After identifying the most influential causes, we can gain insights into eliminating such accidents.

To find the probability relationship between cause and accident, a hierarchical BN model comprising three layers (cause, incident, and accident) is established in this section. The steps are as follows.

Step 1: Analyzing the system

A system consists of several components and equipment, which have varying reliabilities. Events of different severity levels may happen in the components. The first step is finding the components and possible events in it.

Step 2: Establishing the structural model

This process is qualitative. Figure 2 shows the main idea of this paper: the proposed three-layered hierarchical model that classifies the events into cause, incident, and accident. The different states of an event are variables shown by BN nodes. The arrows show their dependencies and cause–effect relationships.

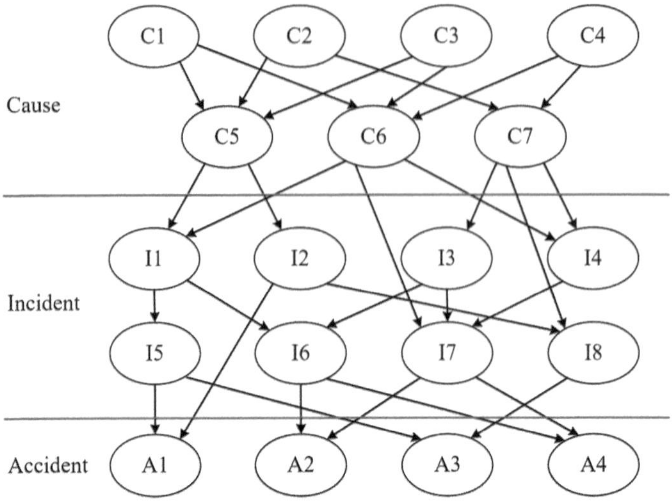

Fig. 2 Three-layered hierarchical model of BNs

Step 3: Establishing the parameter model

This process is quantitative. By analyzing the prior probability of the parent event and establishing CPT (distribution), the quantitative relationships among the related events are represented.

Step 4: Performing inference analysis

By placing the entire established structure and parameter models into Netica, which is a supporting software that is introduced in the succeeding section, the inferential relationship is analyzed, the results are obtained, and the most influential factor is found.

Step 5: Validating the model

Validation can make the model persuasive and is thus a necessary process. To validate the proposed BN model, this study uses a three-axiom-based validation method. The three axioms, which were studied by Jones et al. [21], are as follows.

(1) A slight increase/decrease in the prior subjective probabilities of each parent node should result in a relative increase/decrease of the posterior probabilities of the child node.
(2) Given the variation in the subjective probability distributions of each parent node, the magnitude of influence on the child node values should remain consistent.
(3) The total influence magnitudes of the combination of the probability variations from x attributes on the values should always be higher than those from the set of x-y ($y \in x$) attributes.

3 Risk Analysis of Atmospheric and Vacuum Distillation Unit

3.1 Petrochemical Plant and Equipment

We apply the proposed model in Sect. 2 to an atmospheric and vacuum distillation unit, which is a typical equipment used in the petroleum refining process. Oil refineries use different boiling points of several products to separate them and remove the unwanted product. This process also provides preparations for further processing which utilizes chemical reactions to convert to the desired product. This process yields qualified petroleum products, such as liquefied petroleum gas, jet fuel, fuel oils, and other relevant energy sources. The oil refinery process is divided into primary and secondary processes. During the primary process, distillation separates the crude oil into gasoline, diesel, wax oil, and residual oil, which have different boiling ranges. The secondary process then transfers the heavy products from crude oil distillation to light fractions. Then, refinery light fractions or composite is performed with gas to produce high-octane-rating oil. Refineries depend on distillation units to separate crude oil into fractions. Atmospheric and vacuum distillation units, as necessary units in petroleum refinery plants, have universal applications in the industry. Also, these units are the most common equipment in chemical plants. This study illustrates the proposed method using this unit due to its relative importance in plants.

Atmospheric distillation unit is used to separate petroleum fractions under atmospheric pressure by using temperature. Due to their different boiling point, low boiling fractions vaporize before high boiling fractions. Therefore, the vapor contains more low boiling fractions, and the mixed liquid contains more high boiling fractions. The fractions are separated, and a vacuum distillation unit is used to separate heavy oil fractions come from atmospheric bottoms, into gas oils and asphalt under reduced pressure. Figure 3 shows the external view of the unit. Figure 4 shows the typical flowchart of an atmospheric and vacuum distillation unit. The complex system, which contains many adjunctive systems and control nodes, is shown by the flowchart.

3.2 Application of the BN-Based Methodology

A real petrochemical plant and its atmospheric and vacuum distillation units are analyzed using the proposed methodology. The events and cases are based on previous failure reports and operators' or experts' experience. This application should follow the five steps discussed in the succeeding paragraphs.

Step 1: Analyzing the system

Possible events in the plant and their relationships should be analyzed first. To explain the meaning of "relationships" in the hierarchical BNs model, a common event at the

Fig. 3 External view of atmospheric and vacuum distillation unit

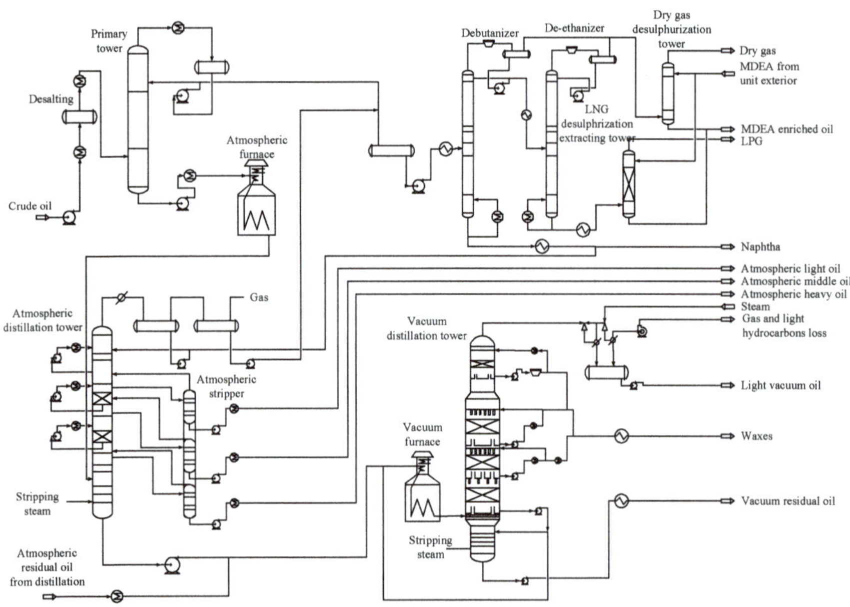

Fig. 4 Flowchart of atmospheric and vacuum distillation unit

plant is used here as an example, excessively low ambient temperature. This event may lead the operator to prefer staying indoors, which may result in the negligence of closing an inlet valve (normally a gate valve) of a tank, then lead to excessive high operation levels, followed by spills, eventually a fire.

Several relationships of possible events are identified and illustrated as follows:

(1) One event happens, leading to another event

For example, sometimes because of fluctuations in production, a busy situation may arise, or there may be an emergency situation, and the system needs to be shut down. In this case, an operator who is doing a shift change may neglect to pass necessary system command to the next crew. The shifted operator does not efficiently execute the command resulting in the command not being completed. Specifically, shift change may cause crew A to forget to pass the command "close the inlet valve of the tank" to crew B, and the inlet valve remaining open will lead to a high level in the tank and eventually resulting in tank overflow.

Another example is of low ambient temperatures that may cause the tank pressure control valve to freeze, leading to a loss of control, creating high pressure in the tank, and finally causing a relief valve to lift.

(2) Several causes concurrently happen, leading to one event

When the ambient temperature is high, the water curtain system is supposed to be triggered to open to cool the equipment. However, if there is a failure of tank water curtain system, it cannot open normally.

The main product of petroleum tank sulfur corrosion is FeS, usually found at the bottom of the tank. At the bottom of the tank containing the petrochemical product, there is a low oxygen content, and some of the FeS partially oxidize leading to the production of free sulfur. This low auto-ignition point sulfur mixed with a loosen structured FeS creates a conducive environment for FeS spontaneous ignition. At the same time, when FeS oxidizes, the process (is exothermic) will release heat resulting in the tank and product to heat up. If the heating is not controlled in time, the rise in temperatures may result in self-ignition. This ignition may burn the rubber seal ring and eventually lead to a fire or explosion.

Another example is of medium volatilization, or heat exchanger, a leakage may cause water to run into the standby pump outlet pipeline. If the operator starts this standby pump without checking it and removing the water, vaporization can occur leading to pump exhaustion due to the difference of water and medium density. The phenomenon of describing when medium cannot be pumped out because system consists of both air and liquid is referred to as "airlock." The outlet pressure and flow may then drop much lower than the normal value. This abnormal value will trigger the interlock to turn off the gas feed causing the furnace to turn off.

(3) Several causes may independently be the reason for one event

For example, an oil leakage, which can lead to a fire, may be caused by any of the following events:

a. High-temperature and high-pressure pumps sealing failure, leading to product leakage.
b. Usually, there is a little water gathering at the bottom of the oil tank. The operator needs to drain and cut off the water then separate it from the oil. When the operator does not identify the water level accuracy, he/she may drain and cut off the water with oil, which may lead to oil leakage.
c. Fracture of outlet pipe fittings such as pipes, elbows, half couplings, nipples, and tees, resulting from corrosion or aging.

Additionally, failure to pump the medium (pump outlet low flow) may be caused by the following events:

a. Insufficient opening level of the inlet valve.
b. Blocked inlet filter.
c. Gas remaining in the pump, because of the low density of the gas, the centrifugal force provided by impeller rotation is not enough to pump out the medium, which may lead to the low outlet flow.

Another example of several causes independently resulting in one event can be seen from the steam turbine; a steam turbine converts thermal energy to mechanical energy. The steam turbine exhaust pressure is one of the principal indicators to measure the work of steam turbine. The lower the exhaust pressure, the better the work. For example, a particular equipment's steam turbine uses 3.5 MPa steam as primary steam supply and 1.0 MPa steam as auxiliary steam supply, to ensure the turbine vacuum. After getting through the turbine, 3.5 MPa steam enters a fixed-tube-sheet heat exchanger which uses seawater as heater tubing medium and cools as condensation water. If there is an abnormal high exhaust pressure of the steam turbine, it may be caused by the following events:

a. Insufficient 1.0 MPa steam pressure.
b. Leakage point in the pipeline.
c. Blocked heat exchanger pipeline, which will cause condensation effectiveness of 3.5 MPa steam get worse.

Table 1 describes other possible events with their name and meaning. Table 2 explains several abbreviations.

Step 2: Establishing the structural model

The identified possible events are then classified into three layers: cause, incident, and accident. On the basis of the identified relationships between the nodes, BNs are established for each layer.

(1) **Cause**: In a petrochemical plant, several common events can cause accidents, three of which are summarized as follows. The first cause is mistakes caused by human error. This cause is the most influential factor because humans refer to subjective operators of the entire unit. Human error can be further divided into organizational, group, and individual errors. Shift change, inadequate knowledge and skills, poor management, high work stress, misunderstandings, and poor attitude may result

Table 1 Events for risk analysis of atmospheric and vacuum distillation unit

	Event	Description	State
1	AbHighP	Abnormal high discharge pressure of compressor	Yes, No
2	AbPiP	Abnormal pipeline pressure	Yes, No
3	Aging	Equipment aging	Yes, No
4	AirLock	Pump airlock	Yes, No
5	AirMonitorFail	Air monitor failure	Yes, No
6	AirT	Air temperature	High, Nor., Low
7	BlockPi	Blocked heat exchanger pipeline	Yes, No
8	ClosingV	Closing lye tank valve	Yes, No
9	Corrosion	Corrosion	Yes, No
10	CVFreezing	Tank pressure control valve freezing	Yes, No
11	Damage	Equipment damage	Yes, No
12	DroponHighT	Liquid drops on high-temperature heat exchanger	Yes, No
13	EquipmentFail	Equipment failure	Yes, No
14	Fatigue	Fatigue operator	Yes, No
15	FeedSpeedFast	Feeding speed too fast	Yes, No
16	FilterB	Inlet filter blocking	Yes, No
17	Fire	Fire	Yes, No
18	GroupErr	Group error	Yes, No
19	HeaterT	Heater temperature	High, Nor., Low
20	HighPiT	High pipeline temperature	Yes, No
21	HighSourWT	High sour wastewater temperature	Yes, No
22	HighTaL	High tank level	Yes, No
23	HighTaP	High tank pressure	Yes, No
24	HumanErr	Human error	Yes, No
25	IndividualErr	Individual error	Yes, No
26	InPuDepletion	Incomplete pump depletion	Yes, No
27	InsExperience	Insufficient experience of operator	Yes, No
28	InsOpening	Insufficient opening	Yes, No
29	InsVOpening	Insufficient inlet valve opening	Yes, No
30	LossC	Loss of control	Yes, No
31	MixOilStain	Cooling water mixed with heavy oil stain	Yes, No
32	NoDecompression	No decompression for pump	Yes, No
33	OilJetting	Oil jetting	Yes, No
34	OilLeakage	Oil leakage	Yes, No
35	OrganizationErr	Organization error	Yes, No

(continued)

Table 1 (continued)

	Event	Description	State
36	PiFracture	Outlet pipeline elbow fracture	Yes, No
37	PiLeakage	Leakage point in pipeline	Yes, No
38	PiW	Water in pump outlet pipeline	Yes, No
39	PMonitorFail	Pressure monitor failure	Yes, No
40	PowerCut	Power cut	Yes, No
41	PuHighTandP	Pump with high-temperature and high-pressure medium	Yes, No
42	PuMediumFail	Pump medium failure	Yes, No
43	PuRemainGas	Remaining gas in pump	Yes, No
44	PuStarting	Pump required to start	Yes, No
45	SafetySFail	Safety and alarm system failure	Yes, No
46	SafetyVFail	Safety valve failure	Yes, No
47	ShiftChange	Shift change	Yes, No
48	Shutdown	Shutdown of unit	Yes, No
49	SpoIgnition	Spontaneous ignition	High, Nor., Low
50	SteamP	Steam pressure	Yes, No
51	TaCrack	Tank crack	High, Nor., Low
52	TaOverflow	Tank overflow	Yes, No
53	TaT	Tank temperature	Yes, No
54	TaWS	Open tank water curtain system	Yes, No
55	ToxicGas	Toxic gas release	Yes, No
56	Vaporization	Vaporization	Yes, No
57	Volatilization	Medium volatilization	Yes, No
58	WOil	Drain and cut off water with oil	Yes, No

in abnormal events. The second cause is equipment failure. All equipment has its reliability that shows that the facility is not entirely reliable. Sometimes, long-time exposure, rain, and time may result in corrosion and aging; device failure or fracture may also happen. When these events happen, the probability of incidents increases. Adequate actions should be implemented immediately to prevent the further development of severe consequences. The third cause is nature. For example, excessive low temperatures may cause medium freezing in the pump. Meanwhile, high ambient temperatures may cause tank temperature rising, and the flammable matter in the tank (e.g., FeS) has the probability to self-ignite. Also, earthquakes, snow, wind, and thunder are also likely to contribute to equipment failure. Figure 5 shows the BNs for the cause layer.

(2) **Incident**: Certain events happening in the plant can lead to more critical results than those causes previously described. Device abnormality can be considered an incident. Four parameters exist as the primary control parameters of the chemical

Table 2 Abbreviations in model

Term	Abbr.
Abnormal	Ab
Blocking	B
Control	C
Error	Err
Failure	Fail
Insufficient	Ins
Level	L
Normal	Nor
Pipeline	Pi
Pressure	P
Pump	Pu
System	S
Tank	Ta
Temperature	T
Valve	V
Water	W

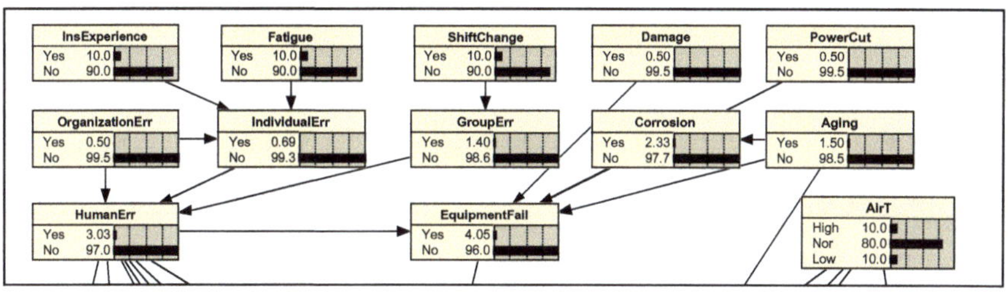

Fig. 5 BNs for cause layer

process, namely level, pressure, temperature, and flow. Individually, their abnormal values are incidents, such as a high level of a tank, which may lead to spill, a high pressure of a pipeline or a vessel, a high temperature of the water tank, or low flow. Figure 6 shows the BNs for the incident layer.

(3) **Accident**: The consequence of an accident is worse than that of others, especially in the chemical plant, although they do not occur frequently. Typical accidents are spills or leaks, fires, explosions, or vessel ruptures. All of them contribute to a substantial economic loss and can sometimes threaten human safety. Figure 7 shows the BNs for the accident layer.

Step 3: Establishing the parameter model

The prior probability of the basic event and child nodes is input by using the supporting software, Netica.

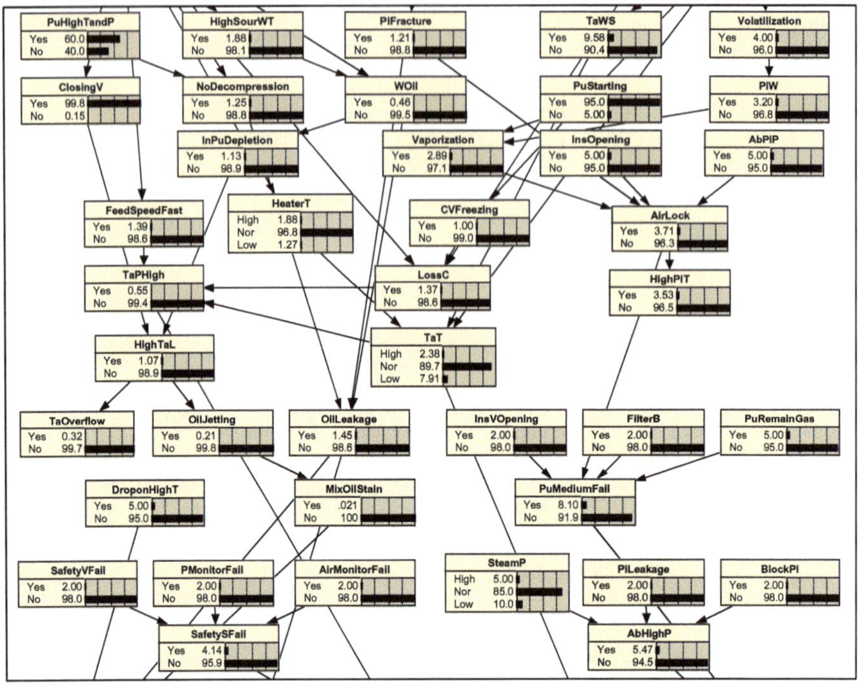

Fig. 6 BNs for incident layer

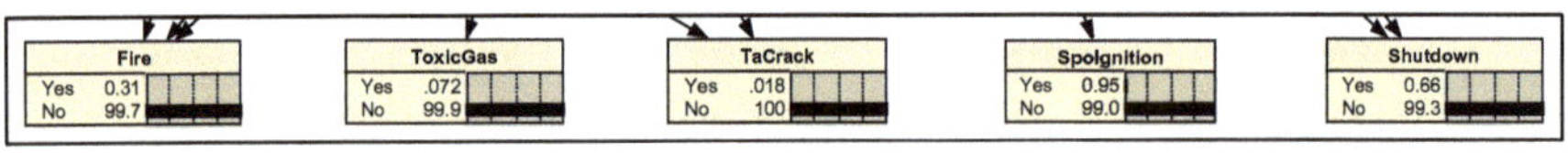

Fig. 7 BNs for accident layer

For a basic event, for example, for the node "Fatigue," the probability of state "Yes" is 10.0%, which means the probability of an operator to become fatigued is 10.0%.

For a child event, n_p (the number of prior probability) of independent entries in the CPT grows exponentially with the number of parents. $n_p = m^n$, where m is the number of its states and n represents the number of parent nodes it has. For example, in Fig. 5, the node "EquipmentFail" has two states ("Yes" and "No"); therefore, $m = 2$. Moreover, this node has four parent nodes; thus, $n = 4$. Thus, for "EquipmentFail," $n_p = m^n = 2^4 = 16$. For the complicated probability, the worst-case scenario can be overcome in the two following ways.

(1) Noisy-OR

Four things can lead to taking the node "PuMediumFail," for example, "InsVOpening," "PuRemainGas," "FilterB," and "AirLock." They are parent nodes of "PuMediumFail". Table 3 shows the CPT of "PuMediumFail". When only one cause hap-

Table 3 CPT (%) of node "PuMediumFail"

InsVOpening	PuRemainGas	FilterB	AirLock	Yes	No
Yes	Yes	Yes	Yes	99.64	0.36
Yes	Yes	Yes	No	96.4	3.6
Yes	Yes	No	Yes	98.2	1.8
Yes	Yes	No	No	82	18
Yes	No	Yes	Yes	99.4	0.6
Yes	No	Yes	No	94	6
Yes	No	No	Yes	97	3
Yes	No	No	No	70	30
No	Yes	Yes	Yes	98.8	1.2
No	Yes	Yes	No	88	12
No	Yes	No	Yes	94	6
No	Yes	No	No	40	60
No	No	Yes	Yes	98	2
No	No	Yes	No	80	20
No	No	No	Yes	90	10
No	No	No	No	0	100

pened, which means only one state of four parent nodes is "Yes" and others are "No," the probability of "PuMediumFail" not happening is an independent failure probability (underlined in Table 3).

The expression formulas of the probabilities are as follows:

$$\mu(\neg \text{PuMediumFail}|\text{VNoEnoL}\&\neg \text{PuRemainGas}\&\neg \text{FilterB}\&\neg \text{AirBinding}) = 0.3$$

$$\mu(\neg \text{PuMediumFail}|\neg \text{VNoEnoL}\&\text{PuRemainGas}\&\neg \text{FilterB}\&\neg \text{AirBinding}) = 0.6$$

$$\mu(\neg \text{PuMediumFail}|\neg \text{VNoEnoL}\&\neg \text{PuRemainGas}\&\text{FilterB}\&\neg \text{AirBinding}) = 0.2$$

$$\mu(\neg \text{PuMediumFail}|\neg \text{VNoEnoL}\&\neg \text{PuRemainGas}\&\neg \text{FilterB}\&\text{AirBinding}) = 0.1$$

Then, other probabilities are calculated by these four independent failure probabilities. For example,

$$\begin{aligned}&\mu(\neg \text{PuMediumFail}|\text{VNoEnoL}\&\text{PuRemainGas}\&\text{FilterB}\&\neg \text{AirBinding})\\&\quad = 0.3 \times 0.6 \times 0.2 \times 0.1 = 0.0036.\end{aligned}$$

Other probabilities follow the same rules.

(2) Experience

The relationship between parents and children is restricted in that conditional independencies exist between the nodes. For instance, if each node has no more than three parents, then a total number of possibility $N < 8n$.

For example, for the node "LossC," three events lead to it: "CVFreezing," "HumanErr," and "Aging." Table 4 shows the CPT of "LossC." According to experience, when "CVFreezing" happens, "LossC" must happen. Therefore, when the state of "CVFreezing" is "Yes," the probability of state "Yes" of "LossC" is 100%. If "CVFreezing" does not happen but "HumanErr" and "Aging" occur, the probability of "LossC" is 15%. If one of "HumanErr" or "Aging" occurs, the probabilities of "LossC" happening are 10 and 5%, respectively. If nothing happens within "CVFreezing," "HumanErr," and "Aging," then the probability of "LossC" is 0.

Step 4: Performing inference analysis

After giving the CPT of each node, Netica is used to calculate the probability of each node by integrating all the probabilities. Figure 8 shows the entire structure. Tables 5, 6, and 7 present the calculated results of each layer.

Step 5: Validating the model

Validation is used to illustrate that the model is a reasonable representation of an actual system. The model is supposed to satisfy the three axioms described in Sect. 2.2. For example, by increasing the probability of state "Yes" of node "HumanErr" to 100%, the probability of state "Yes" of node "Fire" increases from 0.31 to 1.46%. Additionally, the probability of state "Yes" of node "TaCrack" increases from 0.018 to 0.31%. Therefore, if an error occurs because of human operation, then these accidents may happen. Increasing each influencing node satisfies the axioms, thus partially validating the model.

Table 4 CPT (%) of node "LossC"

CVFreezing	HumanErr	Aging	Yes	No
Yes	Yes	Yes	100	0
Yes	Yes	No	100	0
Yes	No	Yes	100	0
Yes	No	No	100	0
No	Yes	Yes	15	85
No	Yes	No	10	90
No	No	Yes	5	95
No	No	No	0	100

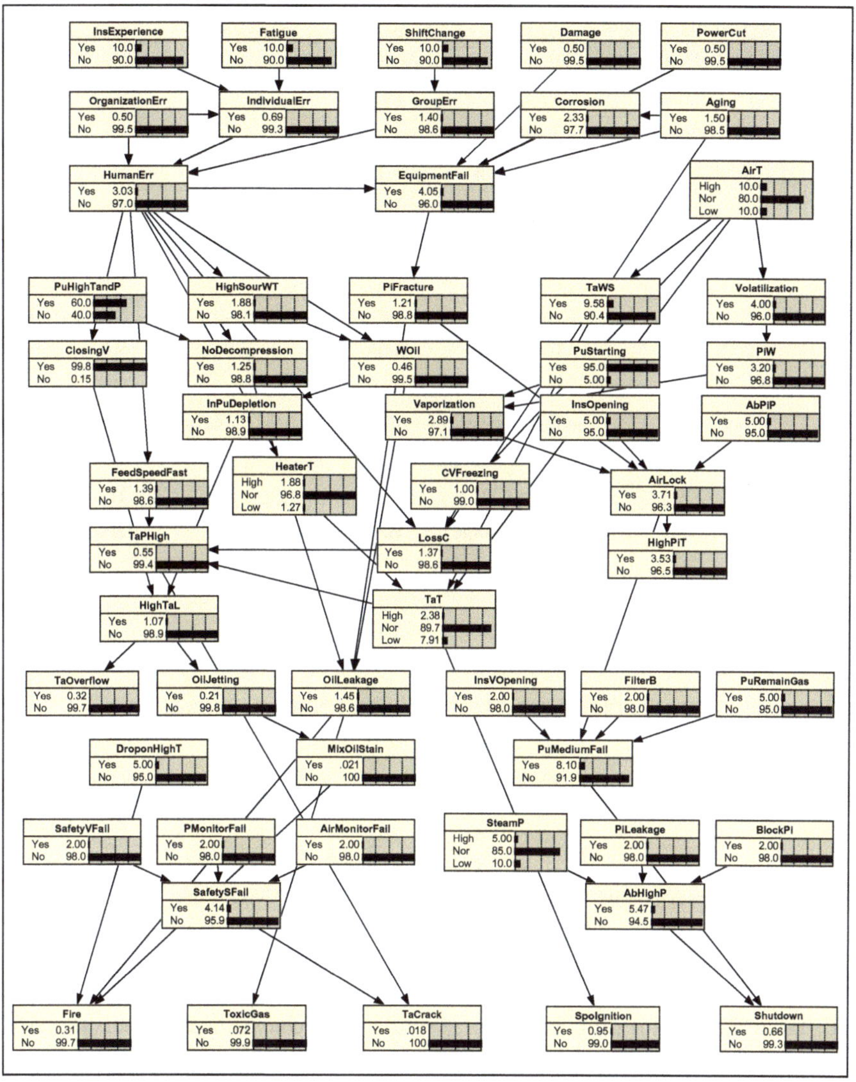

Fig. 8 BNs of the entire unit

Table 5 Probability of cause nodes at state "Yes"

Node	Probability of state "Yes" (%)	Basic node
Aging	10	Yes
AirT	–	Yes
Corrosion	2.33	No
Damage	0.5	Yes
EquipmentFail	4.05	No
Fatigue	10	Yes
GroupErr	1.4	No
HumanErr	3.03	No
IndividualErr	0.69	No
InsExperience	10	Yes
OrganizationErr	0.5	Yes
PowerCut	0.5	Yes
ShiftChange	10	Yes

3.3 *Results and Discussions*

Other results from the model are presented in the following discussion.

3.3.1 Risk Analysis When Finding Evidence

When an accident occurs, its most probable cause must be determined. In the model, the probabilities of state "Yes" of the nodes "Fire," "ToxicGas," and "TaCrack" are changed to 100%, which means they simulate the accidents happening. The probabilities of other nodes change accordingly. Figure 9 shows the probabilities during a fire. Nodes with multiple changes in probabilities changed are circled in red. Table 8 shows the probabilities of several nodes. Underlined numbers represent prior probabilities.

Figure 10 shows the probabilities changing of several basic nodes when finding different evidence. For example, when "Fire" happens, "InsExperience"'s probability increases the most. Thus, it is the most likely cause of the fire. When "ToxicGas" occurs, "InsExperience"'s probability significantly increases. "ShiftChange" also increases considerably and hence can be a possible and primary reason of "ToxicGas." Similarly, when "TaCrack" happens, "InsExperience," "ShiftChange," and "Aging" can be considered first as the possible reasons.

Figure 11 shows the probabilities changing of several child nodes when finding different evidence. When "Fire," "ToxicGas," or "TaCrack" happens, "HumanErr" and "EquipmentFail" are the most responsible. When "ToxicGas" happens, the probability of "EquipmentFail" increases the most. Group error also has a relatively high probability responsible for the happening of "ToxicGas" and "TaCrack."

Table 6 Probability of incident nodes at state "Yes"

Node	Probability of state "Yes" (%)	Basic node
AbHighP	5.47	No
AbPiP	5	Yes
AirLock	3.71	No
AirMonitorFail	2	Yes
BlockPi	2	Yes
ClosingV	99.8	No
CVFreezing	1	No
DroponHighT	5	Yes
FeedSpeedFast	1.39	No
FilterB	2	Yes
Heater	–	No
HighPiT	3.53	No
HighSourWT	1.88	No
HighTaL	1.07	No
HighTaP	0.55	No
LossC	1.37	No
MixOilStain	0.021	No
NoDecompression	1.25	No
InPuDepletion	1.13	No
InsOpening	5	Yes
InsVOpening	2	Yes
OilJetting	0.21	No
OilLeakage	1.45	No
PiFracture	1.21	No
PiLeakage	2	Yes
PiW	3.2	No
PMonitorFail	2	Yes
PuHighTandP	60	Yes
PuMediumFail	8.1	No
PuRemainGas	5	Yes
PuStarting	95	Yes
SafetySFail	4.14	No
SafetyVFail	2	Yes
SteamP	–	Yes
TaOverflow	0.32	No

(continued)

Table 6 (continued)

Node	Probability of state "Yes" (%)	Basic node
TaT	–	No
TaWS	9.58	No
Vaporization	2.89	No
Volatilization	4	No
WOil	0.46	No

Table 7 Probability of accident nodes at state "Yes"

Node	Probability of state "Yes" (%)	Basic node
Fire	0.31	No
Shutdown	0.072	No
SpoIgnition	0.018	No
TaCrack	0.95	No
ToxicGas	0.66	No

Table 8 Probabilities when finding an evidence

Evidence		Probability (%)			
		Origin	Fire	ToxicGas	TaCrack
Parent node	InsExperience	10	11.7	18.5	17.3
	Fatigue	10	10.3	11.7	11.5
	ShiftChange	10	11.3	16.8	15.9
	Damage	0.5	1.72	6.75	0.5
	PowerCut	0.5	1.31	4.66	0.5
	Aging	1.5	3.17	10.1	10.2
Child node	HumanErr	3.03	14.2	59.7	51.9
	IndividualErr	0.69	3.23	13.6	11.8
	GroupErr	1.4	6.54	27.6	24
	OrganizationErr	0.5	2.24	9.38	8.15
	EquipmentFail	4.05	18.5	77.6	46
	Corrosion	2.33	4.71	14.5	10.1
Accident	Fire	0.31	100	4.47	0.92
	ToxicGas	0.072	1.04	100	0.79
	TaCrack	0.018	0.054	0.2	100
	SpoIgnition	0.95	2.09	6.77	7.25
	Shutdown	0.66	0.66	0.66	0.62

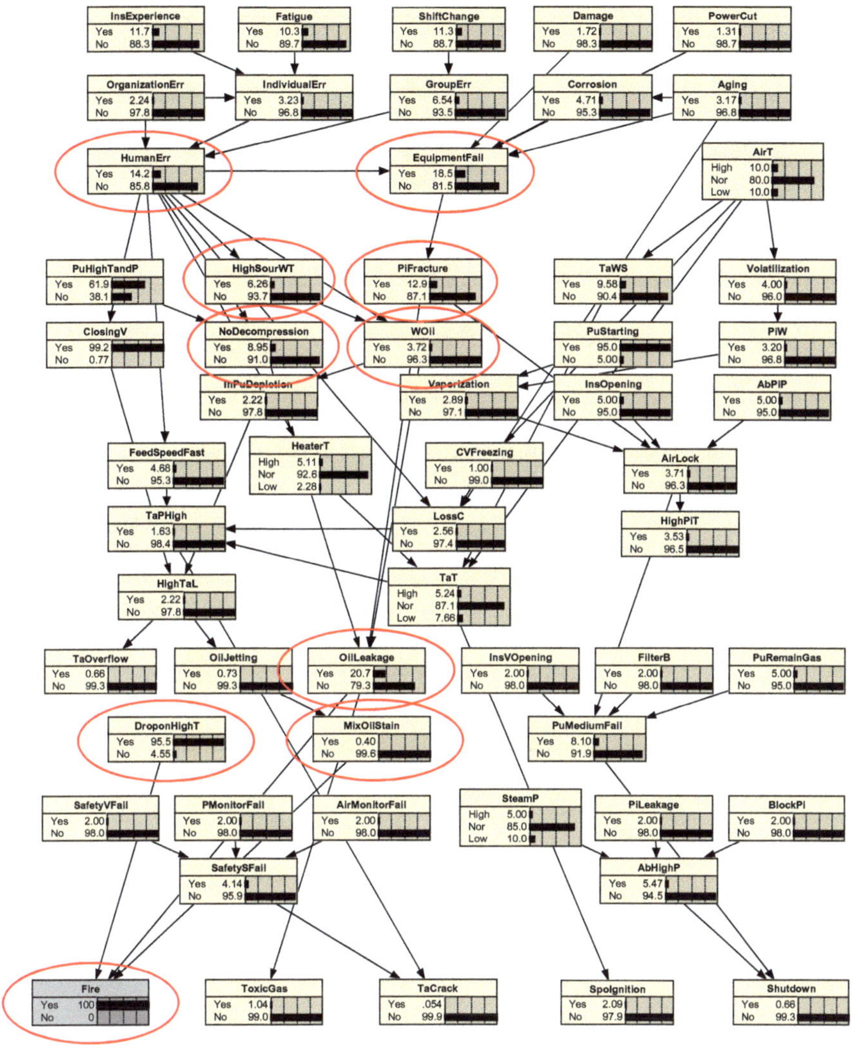

Fig. 9 Probabilities of events when finding "Fire"

3.3.2 Mutual Information and Variance of Beliefs (VB)

Mutual information and VB show the relevance between one node and other nodes. By calculating the mutual information and VB, related events can be found. The calculation can also help find the most probable cause for one accident.

We use Netica to calculate mutual information and VB of "SpoIgnition" with some basic nodes. Table 9 presents the result. Figure 12 shows compared results for mutual information of the basic nodes. Figure 13 shows VB of each basic node. These

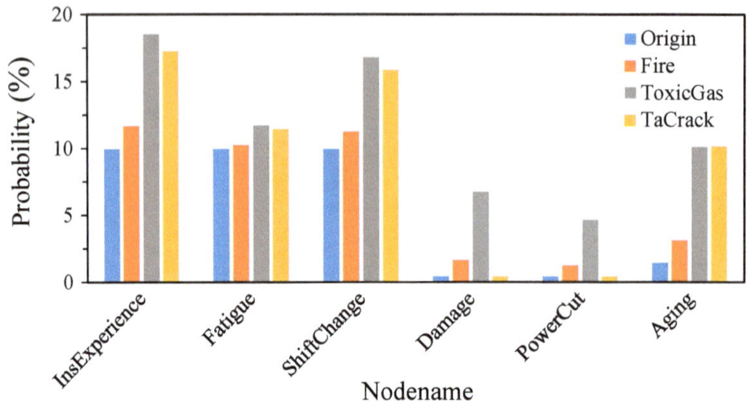

Fig. 10 Compared probabilities of basic nodes when an evidence is found

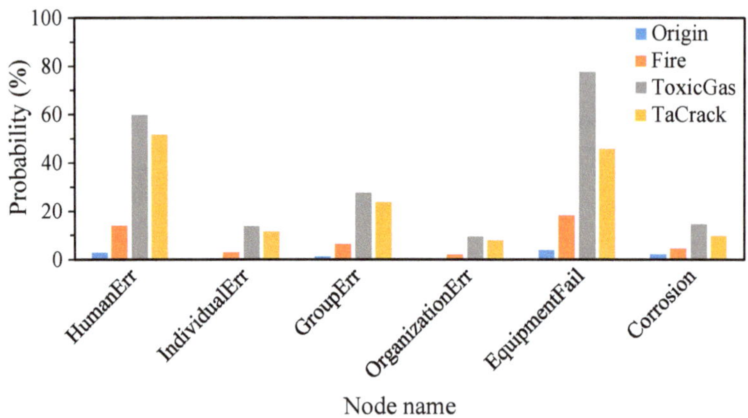

Fig. 11 Compared probabilities of several child nodes when an evidence is found

Table 9 Sensitivity of "SpoIgnition" at other nodes

Basic node	Mutual info	Percent	VB
AirT	0.00182	2.35	0.0000335
InsExperience	0.00015	0.199	0.0000023
ShiftChange	0.0001	0.131	0.0000015
Fatigue	0.00001	0.00896	0.0000001

two figures show that air temperature is the most related event, which means that high ambient temperature is the most probable cause of spontaneous ignition. Therefore, during high-temperature days, operators should remain alert of the weather. In addition, the equipment should be cooled using a drencher system promptly to prevent spontaneous ignition.

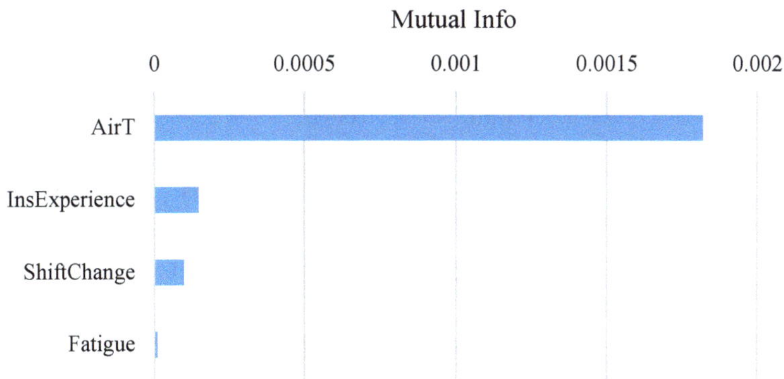

Fig. 12 Mutual information of "SpoIgnition" at other nodes

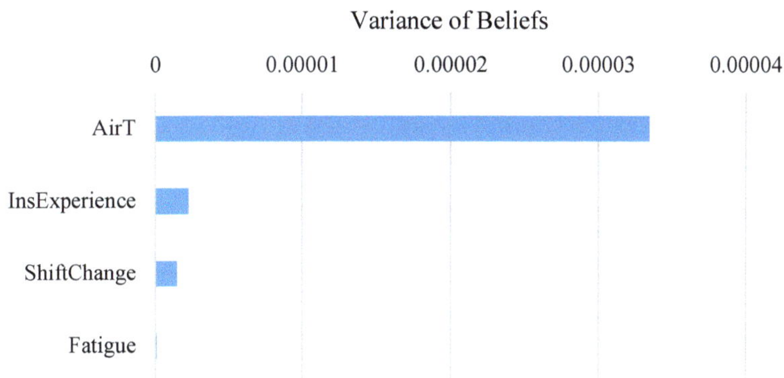

Fig. 13 VB of "SpoIgnition" at other nodes

We use Netica to calculate mutual information and VB of "Shutdown" with some basic nodes. Table 10 presents the result. Figure 14 shows compared results for mutual information of the basic nodes. Figure 15 shows VB of each basic node. The results show that steam pressure is the most related event, which means that high steam pressure easily leads to an abnormally high discharge pressure of the compressor. This event is thus the most probable cause of "Shutdown." Therefore, ensuring steam pressure stability is a relatively efficient way to prevent "Shutdown."

Besides steam pressure, other events influence the occurrence of "Shutdown." Figure 14 shows that by mutual info, "PuRemainGas" (mutual info = 0.00103) has more influence on "Shutdown" than "InsVOpening" (mutual info = 0.00101). On the contrary, Fig. 15 shows that by VB, the value of "InsVOpening" (VB = 0.0000186) is larger than that of "PuRemainGas" (VB = 0.0000149). The results of the two parameters have only a small difference. Therefore, most of them are comparable, and the two parameters should be considered comprehensively.

Table 10 Sensitivity of "Shutdown" at other nodes

Basic node	Mutual info	Percentage	VBs
SteamP	0.00215	3.77	0.0000296
FilterB	0.00126	2.2	0.0000244
PuRemainGas	0.00103	1.8	0.0000149
InsVOpening	0.00101	1.78	0.0000186
AirT	0.00089	1.56	0.0000109
PiLeakage	0.00038	0.661	0.0000056
BlockPi	0.00038	0.661	0.0000056
PuStarting	0.00001	0.013	0.0000001

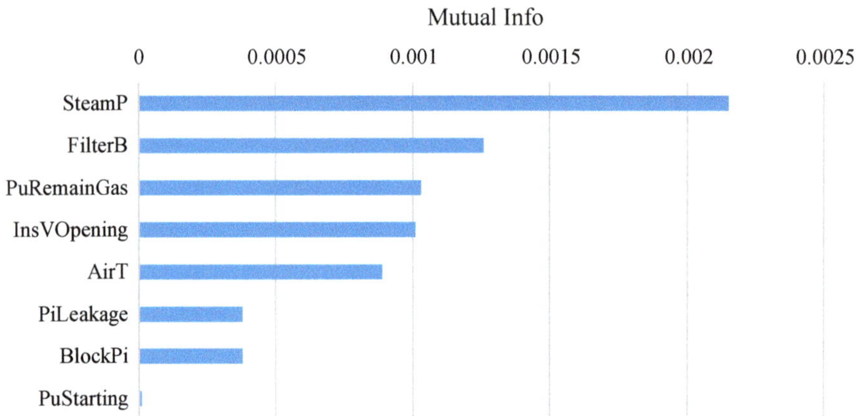

Fig. 14 Mutual information of "Shutdown" at other nodes

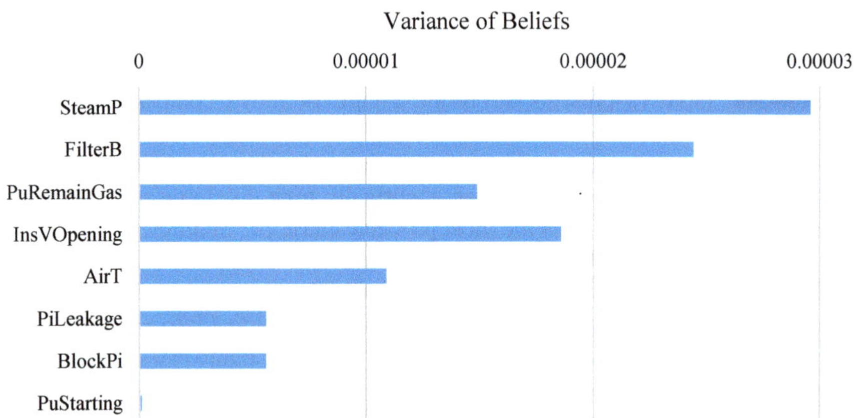

Fig. 15 VB of "Shutdown" at other nodes

Table 11 Probability when finding an error

	Nodes	Probability (%)			
		Nature	HumanErr	OrganizationErr	EquipmentFail
Parent node	InsExperience	10	24.5	10	18.6
	Fatigue	10	12.9	10	11.7
	ShiftChange	10	21.7	10	16.9
	Damage	0.5	0.5	0.5	11.2
	PowerCut	0.5	0.5	0.5	7.6
	Aging	1.5	1.5	1.5	16.1
Child node	HumanErr	3.03	100	95.1	60.2
	IndividualErr	0.69	22.8	0.69	13.7
	GroupErr	1.4	46.2	1.4	27.8
	OrganizationErr	0.5	15.7	100	9.45
	EquipmentFail	4.05	80.3	76.5	100
	Corrosion	2.33	2.33	2.33	23.1
Accident	Fire	0.31	1.46	1.4	1.42
	ToxicGas	0.072	1.42	1.35	1.39
	TaCrack	0.018	0.31	0.3	0.21
	SpoIgnition	0.95	10.9	10.4	6.82
	Shutdown	0.66	0.66	0.66	0.66

3.3.3 Possible Accident When an Error Occurs

Risk is best avoided by increasing reliability. However, mistakes are sometimes inevitable, and when an error occurs, people suffer adverse consequences. Simulation can show the situation in which people know that an error has happened and which accident is most likely to happen. In this case, operators can implement remedial measures to reduce risk. Table 11 gives the probability of several nodes when an error occurs. Underlined numbers represent the prior probabilities of identified causes.

Figures 16 and 17 compare and present the probabilities of these nodes to change with the causes. The different column colors in the bar chart represent various identified causes ("HumanErr," "OrganizationErr," and "EquipmentFail"). The node names are on the horizontal coordinate, whereas the probabilities are on the vertical coordinate. In Fig. 16, the sixth column shows the occurrence of a human error (The probability of state "Yes" of "HumanErr" is 100%.). The probability of operators to become fatigued (state "Yes" of "Fatigue") is 12.9%.

Figure 16 shows the probability changing of certain basic nodes when an error happens. When "HumanErr" happens, "InsExperience" and "ShiftChange" increase the most; this means that they are the most likely reasons to result in human error. When "EquipmentFail" happens, "Damage" and "Aging" increase the most; "PowerCut" also increases. These observations indicate these three events can be possible reasons for equipment failure.

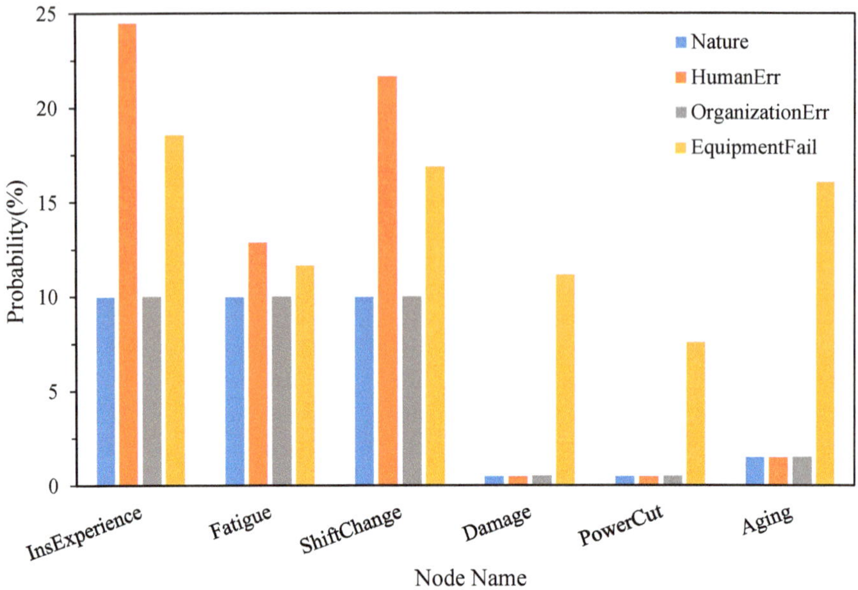

Fig. 16 Compared probabilities of basic nodes when an error is found

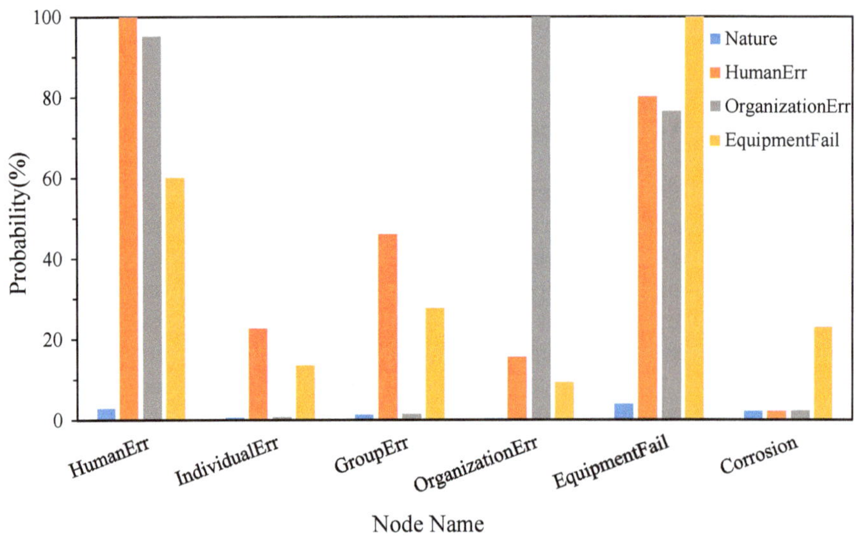

Fig. 17 Compared probabilities of certain child nodes when an error is found

Figure 17 shows the probabilities changing of certain child nodes when an error transpires. When "HumanErr" happens, "EquipmentFail" increases correspondingly. Human error may lead to equipment failure, according to the relationship shown in Fig. 8.

Table 11 shows that when "HumanErr" occurs, the probability of "Fire" increases from 0.31 to 1.46%. "SpoIgnition," which increases from 0.95 to 10.9%, shows the most noticeable change. Therefore, when a human error happens, the temperature should be given special attention to prevent fire and spontaneous ignition.

4 Conclusions

When a petrochemical accident materializes, it is more likely to have devastating consequences than other industries' accidents. A risk analysis model for petrochemical systems should, therefore, be developed to help strengthen the relevant processes and improve safety. This work proposes a BN-based risk analysis methodology, whose core features a three-layered hierarchical model. Then, the methodology is applied to a case study for validation. The conducted analyses lead to the following conclusions.

(1) The case study application indicates the feasibility of the proposed methodology.
(2) When evidence is found, the probability of other events changes accordingly. Insufficient experience is the most responsible for fires and toxic gas releases. For non-basic nodes, human error and equipment failure influence accidents the most.
(3) Mutual information and VB analysis show that air temperature is the most related event within our consideration to spontaneous ignition and steam pressure is the most related event within our consideration to shut down. These parameters sometimes show slightly different results but have similar trends. This research suggests that the two parameters should be considered comprehensively.
(4) When an error occurs, implementing timely and appropriate remedial measures can reduce risk to a certain extent. For example, when a human error happens, the probabilities of fire and spontaneous ignition increase substantially. Therefore, the temperature should be specially monitored to avoid accidents.

Based on the conducted risk analysis, the following recommendations are given;

(1) The reliability of workers should be increased, and the probability of human error should be reduced to prevent accidents because human error is the most influential event in the unit.
(2) On the basis of the five steps in the proposed methodology, plant managers can conduct risk analysis combined with a practical evaluation of the on-site situation, establish structural and parameter models, divide the model into three layers according to criticality, provide a probability for each event, and then calculate. The results can determine the weakest point and the most influential event, thus reducing risk efficiently.

References

1. J. Zhang, B. Cai, Y. Liu, M. Xie, Risk analysis of atmospheric and vacuum distillation unit using Bayesian networks, in *Conference: 2016 11th International Conference on Reliability, Maintainability and Safety (ICRMS)* (2017)
2. H. Hisken, G.A. Enstad, V.D. Narasimhamurthy, Suppression of vortex shedding and its mitigation effect in gas explosions: an experimental study. J. Loss Prevent. Proc. **43**, 242–254 (2016)
3. W. Li, L. Zhang, W. Liang, Job hazard dynamic assessment for non-routine tasks in gas transmission station. J. Loss Prevent. Proc. **44**, 459–464 (2016)
4. J. Feng, J. Fu, P. Chen, J. Luo, Z. Liu, H. Wei, An advanced Driller's Method simulator for deepwater well control. J. Loss Prevent. Proc. **39**, 131–140 (2016)
5. B. Cai, Y. Liu, Z. Liu, X. Tian, H. Li, C. Ren, Reliability analysis of subsea blowout preventer control systems subjected to multiple error shocks. J. Loss Prevent. Proc. **25**, 1044–1054 (2012)
6. C. Ramírez-Marengo, C. Diaz-Ovalle, R. Vázquez-Román, M.S. Mannan, A stochastic approach for risk analysis in vapor cloud explosion. J. Loss Prevent. Proc. **35**, 249–256 (2015)
7. S. Fu, X. Yan, D. Zhang, C. Li, E. Zio, Framework for the quantitative assessment of the risk of leakage from LNG-fueled vessels by an event tree-CFD. J. Loss Prevent. Proc. **43**, 42–52 (2016)
8. M.R. Martins, M.A. Pestana, G.F.M. Souza, A.M. Schleder, Quantitative risk analysis of loading and offloading liquefied natural gas (LNG) on a floating storage and regasification unit (FSRU). J. Loss Prevent. Proc. **43**, 629–653 (2016)
9. J.L. Fuentes-Bargues, C. González-Gaya, M.C. González-Cruz, V. Cabrelles-Ramírez, Risk assessment of a compound feed process based on HAZOP analysis and linguistic terms. J. Loss Prevent. Proc. **44**, 44–52 (2016)
10. B. Cai, Y. Liu, Z. Liu, X. Tian, Y. Zhang, R. Ji, Application of Bayesian networks in quantitative risk assessment of subsea blowout preventer operations. Risk Anal. **33**(7), 1293–1311 (2013)
11. C. Yeo, J. Bhandari, R. Abbassi, V. Garaniya, S. Chai, B. Shomali, Dynamic risk analysis of offloading process in floating liquefied natural gas (FLNG) platform using Bayesian Network. J. Loss Prevent. Proc. **41**, 259–269 (2016)
12. J. Bhandari, R. Abbassi, V. Garaniya, F. Khan, Risk analysis of deepwater drilling operations using Bayesian network. J. Loss Prevent. Proc. **38**, 11–23 (2015)
13. B. Cai, Y. Liu, Y. Zhang, Q. Fan, Z. Liu, X. Tian, A dynamic Bayesian networks modeling of human factors on offshore blowouts. J. Loss Prevent. Proc. **26**, 639–649 (2013)
14. B. Cai, Y. Liu, Y. Ma, Z. Liu, Y. Zhou, J. Sun, Real-time reliability evaluation methodology based on dynamic Bayesian networks: a case study of a subsea pipe ram BOP system. ISA T. **58**, 595–604 (2015)
15. F.V. Jensen, T.D. Nielsen, *Bayesian Networks and Decision Graphs*, Second ed. Information Science and Statistics. Springer (2007)
16. B. Cai, L. Huang, M. Xie, Bayesian networks in fault diagnosis. IEEE Trans. Industr. Inf. **13**(5), 2227–2240 (2017)
17. B. Cai, H. Liu, M. Xie, A real-time fault diagnosis methodology of complex systems using object-oriented Bayesian networks. Mech. Syst. Signal Process. **80**, 31–44 (2016)
18. B. Cai, M. Xie, Y. Liu, Y. Liu, Q. Feng, Availability-based engineering resilience metric and its corresponding evaluation methodology. Reliab. Eng Syst Safety **172**, 216–224 (2017)
19. A. Darwiche, *Modeling and Reasoning with Bayesian networks* (Cambridge University Press, New York, 2009)
20. J. Pearl, *Probabilistic Reasoning in Intelligent Systems: Networks of Plausible Inference* (Morgan Kaufmann Publishers, San Francisco, 1988)
21. B. Jones, I. Jenkinson, Z. Yang, J. Wang, The use of Bayesian network modelling for maintenance planning in a manufacturing industry. Reliab. Eng. Syst. Safe. **95**, 267–277 (2010)

A Multiphase Dynamic Bayesian Network Methodology for the Determination of Safety Integrity Levels

Abstract A novel safety integrity level (SIL) determination methodology based on multiphase dynamic Bayesian networks (MDBNs) for safety instrumented systems is proposed. Proof test interval phase and proof test phase are modeled separately using dynamic Bayesian networks and integrated together to form the MDBNs. The unified structure models of MDBNs for k-out-of-n architectures are constructed, and the procedures of automatic creation of conditional probability tables are developed. The target failure measures, that is, probability of failure on demand, average probability of failure on demand, probability of failing safely, average probability of failing safely, and SIL of safety instrumented systems operating in a low-demand mode, are evaluated using the proposed MDBNs. The effects of time interval of MDBNs, common cause weight, imperfect proof test, and repair on model precision are researched. User-friendly SIL determination software is developed by using MATLAB GUI to assist engineers in determining the SIL value.

Keywords Multiphase dynamic Bayesian networks · Safety integrity level · Safety instrumented system · KooM architecture · KooMD architecture

1 Introduction

Safety has always been an important property in the context of modern industry, which requires designers to address these technical specification challenges to improve reliability, maintainability, and availability constraints. For example, a triple redundant control system is required for subsea blowout preventers to ensure the safety of offshore oil and gas industry [1]. A reliable reconfigurable real-time operating system is used for fault-tolerant traction systems in railway applications [2]. Moreover, redundant controllers, sensors, and actuators are needed for flight-critical distributed systems in avionic applications [3]. Functional safety is a new aspect of safety and is becoming increasingly important, particularly with the introduction of IEC 61508 [4], which is a standard for the functional safety of electrical/electronic/programmable electronic safety-related systems. The standard is regulated in different fields, and various standards such as IEC 61511 in process industry

B. Cai et al., *Bayesian Networks for Reliability Engineering*,
https://doi.org/10.1007/978-981-13-6516-4_10

[5], IEC 61513 in nuclear power plants [6], IEC 62061 in machinery [7], EN 50129 in railway applications [8], and ISO 26262 in road vehicles [9] are issued. The safety integrity level (SIL) of a safety instrumented system (SIS) is determined by the risk reduction factor provided by the SIS to the equipment under control. With the assumption that the SIS prevents all risks, it is a measure of the likelihood of a failure of the SIS. For example, in IEC 61508, SIL is determined by average probability of failure on demand (PFD_{avg}) for low-demand mode and probability of failure per hour (PFH) for high-demand mode; in ISO 26262, SIL is determined by random hardware failure target values; and in EN 50129, SIL is determined by tolerable hazard rate (THR). IEC 61508 is the general and fundamental standard, and it proposes several semi-quantitative and quantitative methods for SIL determination, such as reliability block diagram, fault tree, Markov chain, Petri net, risk graph, layer of protection analysis (LOPA), and hazardous event severity matrix. However, several of these methods are complex and difficult to apply [10], whereas others have important limitations for complex SISs, including binary variable problems [11] and state space explosion problems [12].

Research on SIL determination has attracted considerable attention, and significant results can be found in the literature. Kim et al. [13] proposed an evaluation method for hardware SIL determination by using hazard analysis and risk assessment and failure modes, effects, and diagnostic analysis. Dutuit et al. [14] proposed an evaluation method of PFD in relationship with SILs of SISs by introducing distributions for periodically tested components into fault tree models. Chang et al. [15] proposed a SIL determination procedure for risk graph method, minimum SIL table from OLF 070 and LOPA accounting for uncertainties. Khalil et al. [16] proposed a cascaded fuzzy-LOPA model for SIL determination for certain hazardous scenarios in natural gas industry. Mechri et al. [17] proposed a holistic method for modeling the unavailability of SISs by using switching Markov chain and researching on the influences of several parameters on the performance of SISs, such as common cause failure and imperfect proof testing. Ding et al. [18] proposed an approach for SIL determination based on system degradation by using reliability block diagram. Nait-Said et al. [19, 20] proposed several modified risk graph methods to improve flexibility and reduce the subjective uncertainty. Shu et al. [21] proposed a simplified method for SIL determination for complex SISs by conducting Markov analysis on each channel and combining the results. Sallak et al. [22] proposed a fuzzy probabilistic method for SIL determination of SISs considering the uncertainty of failure rates of SIS components. Jahanian [23] conducted a detailed analysis and derived a generalized form of PFD formula for K-out-of-M (KooM) systems using the same makeup of PFD elements utilized by IEC 61508 standards. Ouache et al. [24] proposed a three-step mathematical model to compute the PFD of SISs and used Bowtie method to conduct the safety analysis of several scenarios by determining the PFD of safeguards. Innal et al. [25] established generic analytical formulations for the assessment of SIS performance in terms of safety integrity and operational integrity.

On another active research frontier, Bayesian networks (BNs) and dynamic Bayesian networks (DBNs) have attracted considerable attention in the field of system reliability, safety, and risk evaluation. In recent years, they have been applied to study the reliability of various systems. Tsilipanos et al. [26] proposed a system of system framework for the reliability evaluation of telecommunication networks based on a combination of hazard analysis techniques along with the BN model and sensitivity analysis. Doguc et al. [27] proposed an automated approach for the reliability assessment of grid systems by using BNs, which require no prior information of the grid system structure. Jiang et al. [28] proposed a BN-based probabilistic model, named hybrid relation model, for the reliability evaluation of programmable logic controller systems. Zhang et al. [29] proposed a fuzzy-BN-based systemic decision support method for safety risk analysis under uncertainty in tunnel construction. Daemi et al. [20, 30] proposed a BN-based reliability evaluation method for composite power systems with emphasis on the importance of degree sequence of components in consideration of load variation and weather conditions. Baraldi et al. [31] applied BNs to handle the uncertainty problems of human reliability analysis and compare it with fuzzy expert system. O'Connor et al. [32] proposed a general dependency model (GDM) that used BNs to model the probabilistic dependencies between components for analysis of common causes and dependent failures in system risk and reliability assessments. Cai et al. [33–35] proposed BN-based reliability evaluation methods in consideration of common cause failure, imperfect coverage, and intermittent faults, and DBN-based real-time reliability evaluation methodology for industrial systems. Ramírez et al. [36] proposed DBN-based evaluation method of life extension for aging repairable systems. Flammini et al. [37] presented both a failure model for KooM systems based on BNs and a maintenance model based on continuous-time Markov chains, which were combined according to a compositional multiformalism modeling approach to analyze the effect of imperfect maintenance on system safety. Moreover, the researchers proposed a BN-based method to evaluate the trustworthiness of 2oo3 decision fusion mechanisms in multisensor applications [38]. Weber et al. [39] applied BNs on circular and linear typical consecutive KooM: F system to estimate its reliability and provide the parameters to distribute the control efforts among the redundant set of actuators.

In SISs, there are proof tests and repairs, which are dynamic processes. DBNs are a long-established extension to ordinary BNs and allow the explicit modeling of changes over time. The technique has an advantage in representing uncertain knowledge in dynamic systems. DBNs can be used to model the dynamic process of SISs. In addition, IEC 61508 standards introduce a multiphase Markovian approach to model the dynamic process of SISs. The main problem with Markov graphs is that the number of states increases exponentially when the number of components of the system under study increases [4]. DBNs can be used to resolve the problem, because the relationship between the size of the models and the number of components of the system is linear but not exponential like Markov processes. Therefore, the work proposes a novel methodology called multiphase dynamic Bayesian networks (MDBNs) to determine the SILs of SISs.

This study is organized as follows: Section 2 introduces the proposed MDBNs for SIL determination. Section 3 investigates the effects of parameters, including time interval, common cause weight, and imperfect proof test on the performance of SISs. Section 4 develops a MDBN-based software package for SIL determination by using MATLAB. Section 5 summarizes the study.

2 Proposed MDBNs for SIL Determination

2.1 *Structure of the Proposed MDBNs*

In SIS, a proof test is required to verify that the specified SIL is met and kept during the life cycle. The test can detect hidden and potentially dangerous failures that cannot be detected through traditional monitoring methods. Thus, a repair can restore the system to an "as new" condition or as close as practical to this condition. Proof test interval is the period of time between adjacent tests. Furthermore, proof test interval varies for each SIS and is dependent on the technology, system architecture, and target SIL. Therefore, the proposed MDBN model for SIL determination is composed of two phases, that is, proof test interval phase and proof test phase. Self-diagnosis, failure, and repair in proof test interval phase can be modeled using DBNs because the proof test interval is generally a long period of time. Similarly, the operation in proof test phase can also be modeled using DBNs according to the actual conditions. The DBNs for proof test interval phase and DBNs for proof test phase are integrated to form the MDBNs, as shown in Fig. 1.

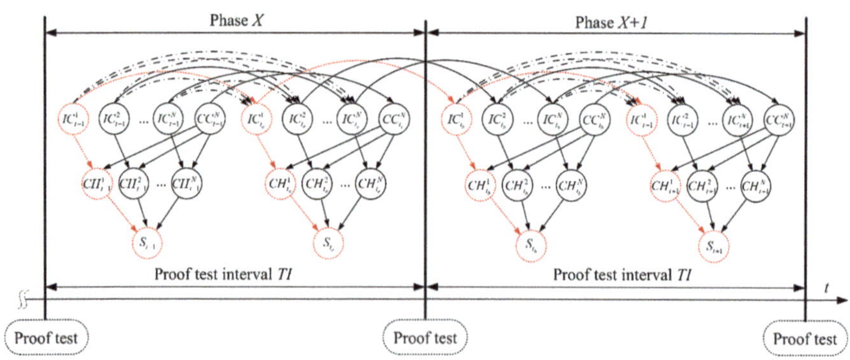

Fig. 1 Schematic diagram of MDBNs for SIL determination

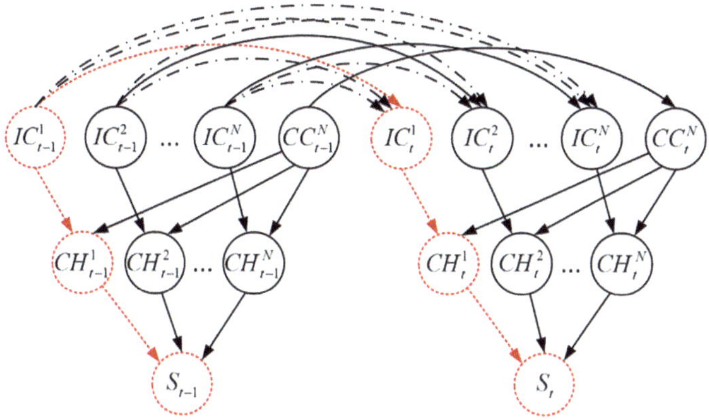

Fig. 2 DBNs for proof test interval phase

2.1.1 DBNs for Proof Test Interval Phase

The proof test interval is a long period of time; the DBNs include many time slices, denoted by different lower indexes, such as t–1 and t. A unified DBN structure model for proof test interval phase is constructed, as shown in Fig. 2. The DBNs contains three layers from the top down, that is, failure cause, single channel state, and system state layers. The node IC in failure cause layer denotes the independent cause failure for a single channel, and the node CC denotes the common cause failure for multiple channels. Common cause failure can be referred to as common mode failure and dependent failure. Moreover, common cause failure is the result of an event. These events with dependencies cause a coincidence of failure states of components in two or more separate channels of a redundancy system, leading to the failure of defined systems to perform its intended function. The node CH in single channel state layer denotes the state of a single channel simultaneously affected by independent cause failure and common cause failure. Therefore, the IC nodes are connected to corresponding CH nodes and the CC node is connected to all of the CH nodes via arcs. The node S in system state layer denotes the state of an entire system consisting of all channels. The nodes IC, CC, and CH have five states, that is, normal state (NS), safe detected failure (SD), safe undetected failure (SU), dangerous detected failure (DD), and dangerous undetected failure (DU). For node S, the states SD and SU are combined to form a safety state (SS). Thus, node S has four states, that is, NS, SS, DD, and DU. In adjacent time slices, the causal relationships of IC nodes are illustrated by connecting the corresponding nodes via inter-slice arcs (black solid line or red dash line) and other nodes via auxiliary inter-slice arcs (black dot dash line).

The DBNs with different architectures of KooM and KooMD systems are determined by the number of channels M. An M equal to 1 represents the architectures of 1oo1 and 1oo1D. The entire DBNs are reduced to the models in red dashed line in

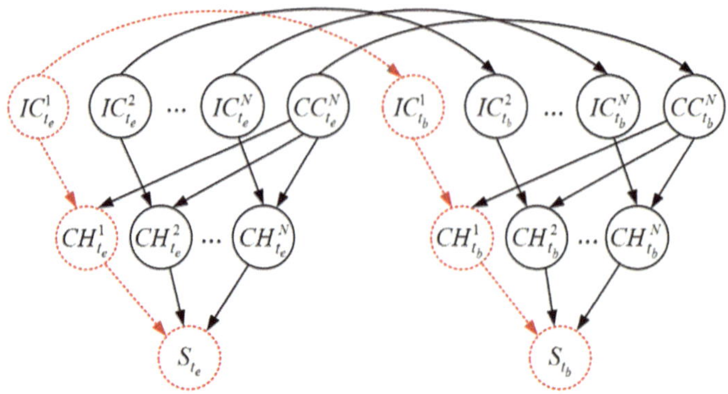

Fig. 3 DBNs for proof test phase

Fig. 2 because the common cause failure for the two architectures is meaningless. An M equal to 2 can represent the architectures of 1oo2, 2oo2, 1oo2D, and 2oo2D. An M equal to 3 can represent the architectures of 1oo3, 2oo3, and others. The DBN structures for systems with same channels are identical; however, the parameter model must be different, which means that DBNs for different architectures with same number of channels are determined by conditional probability tables (CPTs) of nodes.

2.1.2 DBNs for Proof Test Phase

Compared with the proof test interval, the proof test is a short period of time; therefore, the DBNs for proof test phase include only two time slices, as shown in Fig. 3. Actually, the two time slices are the last and the first time slices of DBNs of two adjacent proof test intervals, denoted by lower indexes t_{e} (e for ending) and t_{b} (b for beginning). The causal relationships of each channel at adjacent time are illustrated by connecting the corresponding nodes using inter-slice arcs.

2.2 *Parameter of the Proposed MDBN*

DBN parameter models are the CPTs of nodes in proof test interval phase and proof test phase. For different kinds of nodes, the CPTs are totally different. Therefore, the CPTs of five typical nodes are introduced, including CPTs of IC node, CC node, CH node, and S node in proof test interval phase, and CPTs of IC/CC node in proof test phase.

2.2.1 CPTs of IC Node in Proof Test Interval Phase

The CPTs of IC node of KooM and KooMD systems are illustrated using a flowchart, indicating the time-dependent rules of failure and repair, as provided in Fig. 4. The rules are defined according to practical condition of engineering. When the current state of IC node is NS, the IC node degrades to SD, SU, DD, and DU states exponentially. When the current state of IC node is SD or DD, and if it causes the safe failures of the system, the IC node translates into NS state exponentially with the repair rate μ_{SR}; otherwise, the IC node translates into NS state exponentially with the repair rate μ_{TR}. When the current state of IC node is SU, and if the parent nodes of IC cause the safe failure of the system, IC node translates into NS state exponentially with the repair rate μ_{SR}; if the number of detected failure of parent nodes of IC is equal or larger than 1, the IC node translates into NS state exponentially with the repair rate μ_{TR}; otherwise, it maintains SU state. When the current state of IC node is DU, and if the parent nodes of IC cause the safe failure of the system, the IC node translates into NS state exponentially with the repair rate μ_{SR}; if the number of detected failure of parent nodes of IC is equal or larger than 1, the IC node translates into NS state exponentially with the repair rate μ_{TR}; otherwise, it maintains a DU state.

2.2.2 CPTs of CC Node in Proof Test Interval Phase

The transition relationships of CC node for KooM architecture are provided in Table 1. The CC node has five states: NS, SD, SU, DD, and DU; the failure rates have the relationship that $\lambda_C = \lambda_{SDC} + \lambda_{SUC} + \lambda_{DDC} + \lambda_{DUC}$. The rules are defined according to practical condition of engineering. When the current state of CC node is NS, it degrades to SD, SU, DD, and DU states exponentially. When the current state of CC node is SD or SU, it causes the safe failure of the system. Moreover, the system is translated into NS state exponentially with the repair rate μ_{SR}. When the current state of CC node is DD, it causes the dangerous detected failure of the system, and the CC node translates into to NS state exponentially with the repair rate μ_{TR}. When the current state of CC node is DU, it causes the dangerous undetected failure of the system because the self-diagnosis of the system cannot detect the dangerous undetected failure, and the CC node maintains a DU state. To calculate the SIL of a specified system, the safety-case authors should collect and obtain the rates of common mode failure of components using statistical approaches or from failure database directly first. Subsequently, the SIL can be calculated using the proposed MDBNs methodology.

Similar transition relationships of CC node for KooMD architecture are provided in Table 2. The self-diagnosis of the system can translate the dangerous detected failure into safe detected failure. Therefore, when the current state of CC node is DD, the CC node causes the safe failure of the system and translates into NS state exponentially with the repair rate μ_{SR}.

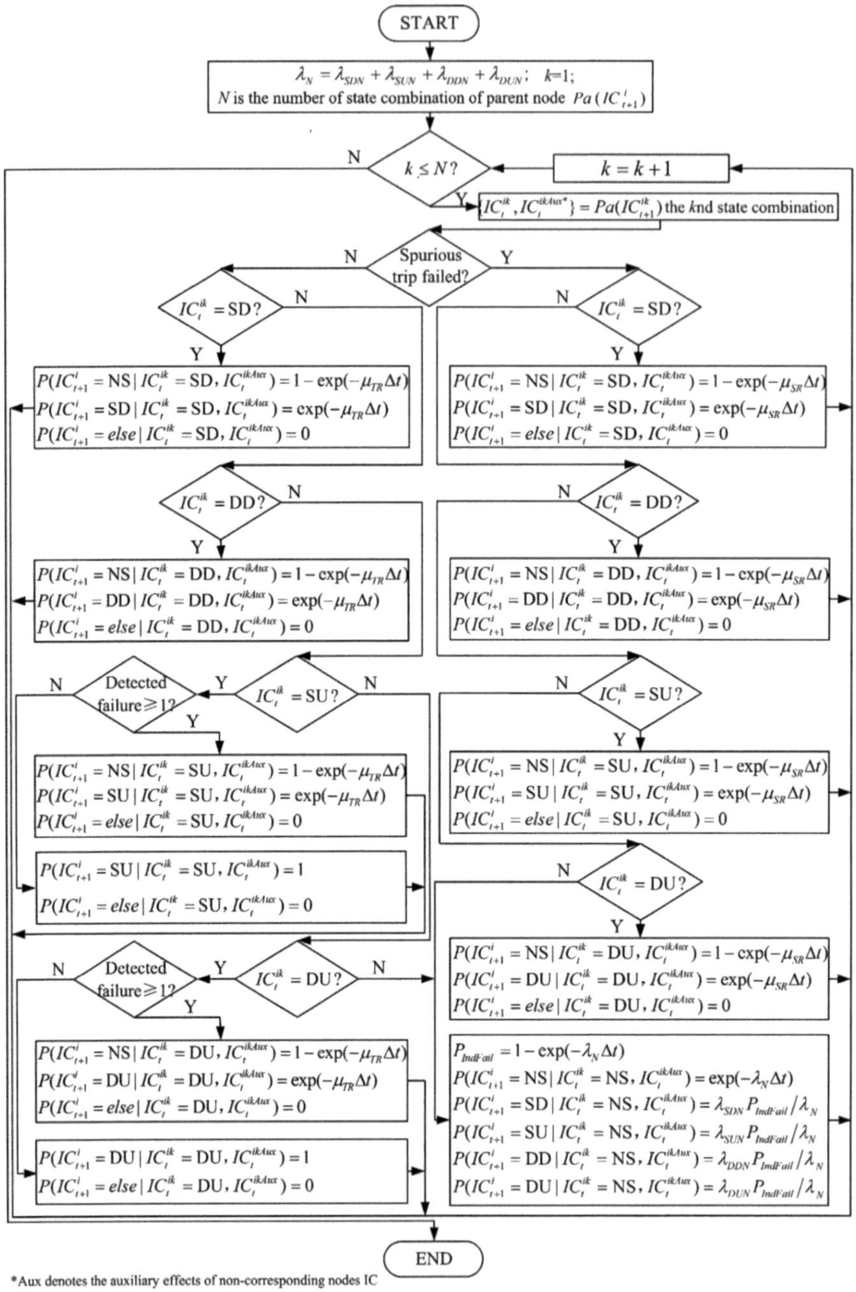

Fig. 4 Modeling flowchart of CPTs of IC node in proof test interval phase

Table 1 Transition relationships of CC node for KooM architecture

t	$t+\Delta t$				
	NS	SD	SU	DD	DU
NS	$e^{-\lambda_C \Delta t}$	$\frac{\lambda_{SDC}}{\lambda_C}\left(1-e^{-\lambda_C \Delta t}\right)$	$\frac{\lambda_{SUC}}{\lambda_C}\left(1-e^{-\lambda_C \Delta t}\right)$	$\frac{\lambda_{DDC}}{\lambda_C}\left(1-e^{-\lambda_C \Delta t}\right)$	$\frac{\lambda_{DUC}}{\lambda_C}\left(1-e^{-\lambda_C \Delta t}\right)$
SD	$1-e^{-\mu_{SR}\Delta t}$	$e^{-\mu_{SR}\Delta t}$	0	0	0
SU	$1-e^{-\mu_{SR}\Delta t}$	0	$e^{-\mu_{SR}\Delta t}$	0	0
DD	$1-e^{-\mu_{TR}\Delta t}$	0	0	$e^{-\mu_{TR}\Delta t}$	0
DU	0	0	0	0	1

Table 2 Transition relationships of CC node for KooMD architecture

t	$t+\Delta t$				
	NS	SD	SU	DD	DU
NS	$e^{-\lambda_C \Delta t}$	$\frac{\lambda_{SDC}}{\lambda_C}\left(1-e^{-\lambda_C \Delta t}\right)$	$\frac{\lambda_{SUC}}{\lambda_C}\left(1-e^{-\lambda_C \Delta t}\right)$	$\frac{\lambda_{DDC}}{\lambda_C}\left(1-e^{-\lambda_C \Delta t}\right)$	$\frac{\lambda_{DUC}}{\lambda_C}\left(1-e^{-\lambda_C \Delta t}\right)$
SD	$1-e^{-\mu_{SR}\Delta t}$	$e^{-\mu_{SR}\Delta t}$	0	0	0
SU	$1-e^{-\mu_{SR}\Delta t}$	0	$e^{-\mu_{SR}\Delta t}$	0	0
DD	$1-e^{-\mu_{SR}\Delta t}$	0	0	$e^{-\mu_{SR}\Delta t}$	0
DU	0	0	0	0	1

2.2.3 CPTs of CH Node in Proof Test Interval Phase

CH node has five states. NS is considered a normal state, and SD, SU, DD, and DU are considered abnormal states. For 1oo1 and 1oo1D architecture, the state of CH node is determined by IC node, which means that the architectures have the same probability. For KooM and KooMD redundancy architecture, the state of CH node is determined by IC and CC nodes. When both of the states of IC and CC nodes are NS, the probability that the state of CH node is NS is 1. When IC and CC nodes are both in abnormal and same states, the probability that CH node is in the same state of IC and CC nodes is 1. When IC and CC nodes are both in abnormal and different states, the probability that CH node is in the same state of CC node is w, and the probability that CH node is in the same state of IC node is $1-w$, where w is the common cause weight. The entire CPTs of CH node for KooM and KooMD redundancy architecture is given in Table 3.

2.2.4 CPTs of S Node in Proof Test Interval Phase

S node has four states: NS, SS, DD, and DU. The failure criterions of the system should be known for developing the CPTs of S node. The failure criterions for KooM and KooMD architectures proposed by Guo and Yang are used in the current work [40].

Table 3 CPTs of CH node for redundancy architecture

State of IC node	State of CC node	State of CH node				
		NS	SD	SU	DD	DU
NS	NS	1	0	0	0	0
SD	NS	0	1	0	0	0
SU	NS	0	0	1	0	0
DD	NS	0	0	0	1	0
DU	NS	0	0	0	0	1
NS	SD	0	1	0	0	0
SD	SD	0	1	0	0	0
SU	SD	0	w	$1-w$	0	0
DD	SD	0	w	0	$1-w$	0
DU	SD	0	w	0	0	$1-w$
NS	SU	0	0	1	0	0
SD	SU	0	$1-w$	w	0	0
SU	SU	0	0	1	0	0
DD	SU	0	0	w	$1-w$	0
DU	SU	0	0	w	0	$1-w$
NS	DD	0	0	0	1	0
SD	DD	0	$1-w$	0	w	0
SU	DD	0	0	$1-w$	w	0
DD	DD	0	0	0	1	0
DU	DD	0	0	0	w	$1-w$
NS	DU	0	0	0	0	1
SD	DU	0	$1-w$	0	0	w
SU	DU	0	0	$1-w$	0	w
DD	DU	0	0	0	$1-w$	w
DU	DU	0	0	0	0	1

n_{D}, n_{S}, n_{DD}, n_{SD}, n_{DU}, and n_{SU} represent the number of dangerous failures, safe failures, dangerous detected failures, safe detected failures, dangerous undetected failures, and safe undetected failures, respectively. For KooM architecture, the criterion of dangerous detected failure of the system is

$$M - n_{\mathrm{D}} < K \tag{1}$$

$$n_{\mathrm{DD}} + n_{\mathrm{SD}} > 0 \tag{2}$$

The criterion of dangerous undetected failure of the system is

$$M - n_{\mathrm{D}} < K \tag{3}$$

$$n_{\mathrm{DD}} + n_{\mathrm{SD}} = 0 \tag{4}$$

For KooMD architecture, the criterion of safe failure of the system is

$$n_{\mathrm{S}} + n_{\mathrm{DD}} \geq K \tag{5}$$

The criterion of dangerous detected failure of the system is

$$M - n_{\mathrm{DU}} < K \tag{6}$$

$$n_{DD} + n_{SD} > 0 \tag{7}$$

The criterion of dangerous undetected failure of the system is

$$M - n_{\mathrm{DU}} < K \tag{8}$$

$$n_{\mathrm{DD}} + n_{\mathrm{SD}} = 0 \tag{9}$$

A procedure of automatic creation of CPTs is developed. For a given KooM or KooMD architecture, the failure criterion is determined according to Eqs. (1)–(9) first. The state combinations of channels and the number of combinations, N_S, as well as list of all the combinations in the order of CPTs of S node are then determined. Finally, the CPTs are determined individually, and whether the number of state combinations that has been determined is equal or smaller than N_S. If so, the failure probability of each state is determined according to the failure criterion and the values of n_{DD}, n_{SD}, n_{DU}, and n_{SU}. If not, the automatic creation is ended.

2.2.5 CPTs of IC and CC Nodes in Proof Test Phase

The conditional probability relationships of IC and CC nodes in proof test phase are determined using Markov state transition diagram of five states, namely DU, DD, NS, SD, and SU, as shown in Fig. 5. Eight variables, namely ζ, δ, θ, σ, α, ε, μ, and γ, are used to describe the transition probabilities of states in each node. θ, μ, ζ, and δ are the transition rates from DU, DD, SD, and SU to NS, respectively. σ, ε, α, and γ are the transition rate factors from DU to DU, SU to SU, NS to SU, and NS to DU, respectively. Other transition rates can be calculated based on these transition rates and transition rate factors. Imperfect proof test and repair can be modeled using these variables, whose values are determined according to practical

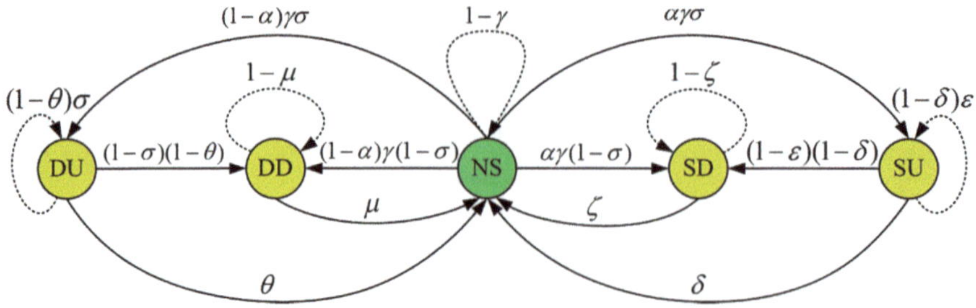

Fig. 5 State transition diagram of IC and CC nodes in proof test phase

Table 4 State transition CPTs of IC and CC nodes in proof test phase

Before proof test	After proof test				
	NS	SD	SU	DD	DU
NS	$1–\gamma$	$\alpha\gamma(1–\sigma)$	$\alpha\gamma\sigma$	$(1–\alpha)\gamma(1–\sigma)$	$(1–\alpha)\gamma\sigma$
SD	ζ	$1–\zeta$	0	0	0
SU	δ	$(1–\delta)(1–\varepsilon)$	$(1–\delta)\varepsilon$	0	0
DD	μ	0	0	$1–\mu$	0
DU	θ	0	0	$(1–\theta)(1–\varepsilon)$	$(1–\theta)\varepsilon$

condition of engineering. The CPTs of IC and CC nodes in proof test phase can be obtained from the transition diagram in Table 4.

2.3 *Calculation of Performance Parameters of SIS*

Probability of failure on demand (PFD), average probability of failure on demand (PFD_{avg}), probability of failing safely (PFS), and average probability of failing safely (PFS_{avg}) are main target failure measures of SISs operating in a low-demand mode. PFD is the probability that a SIS fails to perform its intended safety function during a potentially dangerous condition, which is called a dangerous failure. IEC 61508 focuses only on PFD and ignores the issues about system safe failure PFS [41]. The current work studies both dangerous failure and safe failure using the proposed MSBN method. PFD_{avg} and PFS_{avg} are the average values of PFD and PFS in a period of time, respectively, and the expressions are obtained by means of the compound trapezium rule as follows:

$$\text{PFD}_{\text{avg}} = \frac{1}{T}\int_0^T \text{PFD}(t)\text{d}t = \lim_{N_{\Delta t}\to\infty} \frac{1}{N_{\text{TI}}\cdot\text{TI}} \sum_{j=1}^{N_{\text{TI}}}\sum_{i=1}^{N_{\Delta t}} \frac{\text{PFD}\left(t_i^j\right) + \text{PFD}\left(t_{i+1}^j\right)}{2} \tag{10}$$

$$\mathrm{PFS}_{\mathrm{avg}} = \frac{1}{T}\int_{0}^{T} \mathrm{PFS}(t)\mathrm{d}t = \lim_{N_{\Delta t}\to\infty} \frac{1}{N_{\mathrm{TI}}\cdot \mathrm{TI}} \sum_{j=1}^{N_{\mathrm{TI}}}\sum_{i=1}^{N_{\Delta t}} \frac{\mathrm{PFS}\left(t_i^j\right) + \mathrm{PFS}\left(t_{i+1}^j\right)}{2} \tag{11}$$

where t_i^j is the time of the ith time slice in the jth proof test interval TI. N_{TI} is the combination of time interval TI in total time T, and T should be divisible by TI. $N_{\Delta t}$ is the number of small time intervals Δt in time interval TI, and TI should be divisible by Δt. Therefore, the number of time slices for MSBNs in a time interval TI should be $N_{\Delta t}$ plus one.

The most important measure of safety system performance, SIL, can be determined in terms of average probability of a dangerous failure on demand of the safety function [4]. Four discrete integrity levels are associated with SIL: SIL 1, SIL 2, SIL 3, and SIL 4. Higher SIL level means the associated safety level is also higher; consequently, the probability that a system will fail to perform properly is lower.

3 Effects of Parameters on SIS Performance

3.1 Effects of Time Interval of MDBNs on the Model Precision

To identify the effects of time interval of MDBNs Δt and number of total time slices N on the model precision, the following parameters are provided, including failure rate of a single channel, $\lambda = 2.0 \times 10^{-6}\ \mathrm{h}^{-1}$, total time, $T = 4038$ h, safe failure fraction (SFF), $R_{\mathrm{S}} = 0.5$, self-diagnostic coverage for safe (dangerous) failure, C_{S} (C_{D}) = 0.9, common cause weight factor, $w = 1$, undetected common cause failure fraction, $\beta = 0.02$, detected common cause failure fraction, $\beta_{\mathrm{D}} = 0.02$, mean time to repair, MTTR = 8 h, and mean time to system restoration, MTSR = 24 h. When the total time T is divided into 2^6, 2^7, 2^8, 2^9, 2^{10}, 2^{11}, 2^{12}, and 2^{13} small time intervals, the number of total time slices N will be the number of small time intervals plus one. PFD, PFS, $\mathrm{PFD}_{\mathrm{avg}}$, $\mathrm{PFS}_{\mathrm{avg}}$ for 2oo2D and 2oo3 architectures are plotted in Fig. 6. As can be seen from the three-dimensional figures, PFD and PFS decrease first with the decrease of time interval Δt and then converge to steady-state values. With increase of time t, PFD and PFS also increase. As can be seen from the two-dimensional figures, $\mathrm{PFD}_{\mathrm{avg}}$ and $\mathrm{PFS}_{\mathrm{avg}}$ decrease rapidly with the increase of number of total time slices N and then converge to steady-state values. Therefore, when N is sufficiently large, the calculated values of PFD, PFS, $\mathrm{PFD}_{\mathrm{avg}}$, and $\mathrm{PFS}_{\mathrm{avg}}$ can be considered true values.

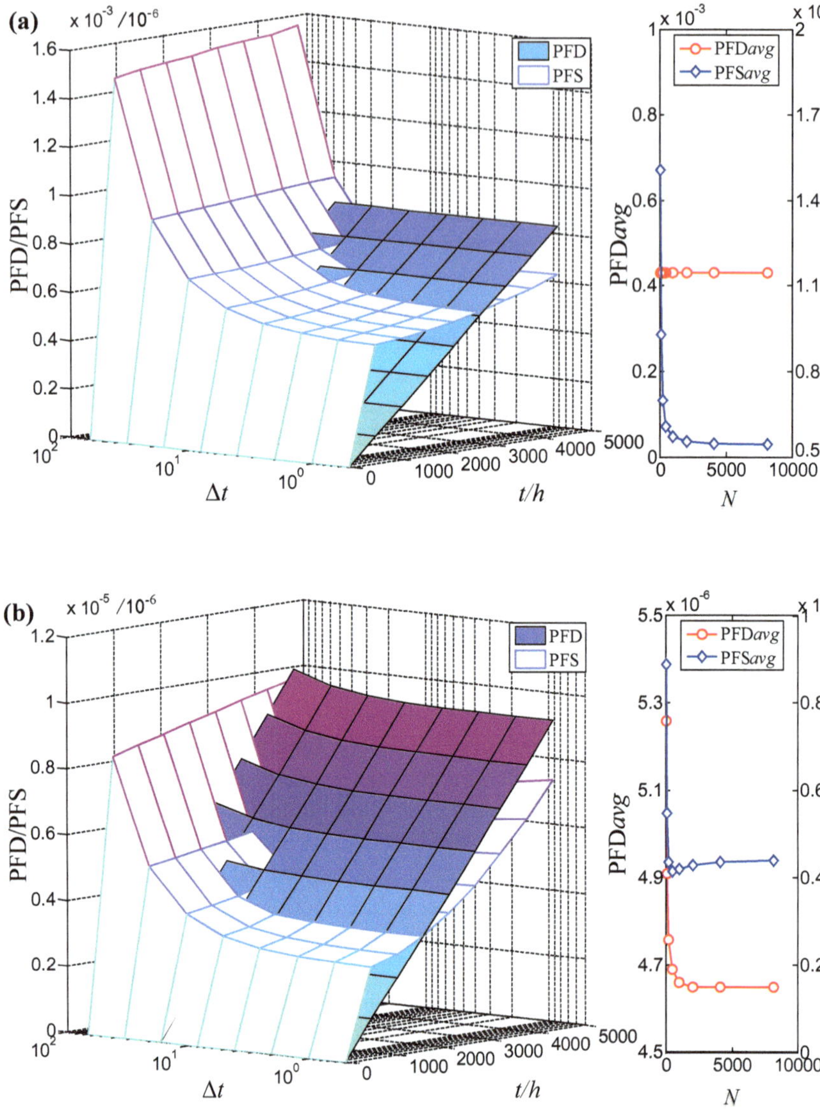

Fig. 6 Effects of time interval on the model precision **a** 2oo2D architecture and **b** 2oo3 architecture

3.2 Effects of Common Cause Weight on the Model Precision

To identify the effects of common cause weight w on the model precision, the following parameters are provided, including failure rate of a single channel, $\lambda = 2.0 \times 10^{-6}\ h^{-1}$, total time, $T = 4038$ h, safe failure fraction (SFF), $R_S = 0.5$, self-diagnostic coverage for safe (dangerous) failure, C_S (C_D) = 0.9, undetected common cause failure fraction, $\beta = 0.02$, detected common cause failure fraction, $\beta_D = 0.02$, mean time to repair, MTTR = 8 h, mean time to system restoration, MTSR = 24 h, and number of time slices, $N = 4097$. When the common cause weight w is 0, 0.2, 0.4, 0.6, 0.8, and 1, the PFD, PFS, PFD_{avg}, PFS_{avg} for 2oo2D and 2oo3 architectures are plotted, as provided in Fig. 7. As can be seen from the three-dimensional figures, PFD and PFS almost have no change with the increase of common cause weight w. As time t increases, PFD and PFS also increase. The two-dimensional figures show that PFD_{avg} and PFS_{avg} have little change despite the increase of common cause weight w. Above all, common cause weight has little effects on the four target failure measures.

3.3 Effects of Imperfect Proof Test and Repair on the Model Precision

To identify the effects of imperfect proof test and repair on the model precision, the following parameters are provided, including failure rate of a single channel, $\lambda = 2.0 \times 10^{-6}\ h^{-1}$, total time, $T = 8760$ h, safe failure fraction (SFF), $R_S = 0.5$, self-diagnostic coverage for safe (dangerous) failure, C_S (C_D) = 0.9, undetected common cause failure fraction, $\beta = 0.02$, detected common cause failure fraction, $\beta_D = 0.02$, mean time to repair, MTTR = 8 h, mean time to system restoration, MTSR = 24 h, and number of time slices, $N = 4097$. The values of eight variables, namely ζ, δ, θ, σ, α, ε, μ, and γ, reflect the degree of imperfect proof test and repair, and the difference in performance of SISs after the proof test. However, the eight variables have many combinations; thus, researching all of the combinations is rarely practical. Four typical combinations of variables, A, B, C, and D, are researched. The PFD, PFS, PFD_{avg}, PFS_{avg} for 2oo2D and 2oo3 architectures are plotted, as provided in Fig. 8. The variables in group A represent that the proof test coverage is 100%, and the repair in proof test phase is perfect. The variables in group B represent that the proof test coverage is 100%, and no repair is in proof test phase. The variables in group C represent that the proof test coverage is 100%, and the detected failure by proof test is repaired perfectly; nevertheless, the detected failure by self-diagnosis remains. The variables in group D represent that the proof test coverage is 0, and the proof test phase has no repair. Figure 8 shows that the curves of PFD, PFS, PFD_{avg}, and PFS_{avg} for group A are the best, whereas the curves for group D are the worst, for group B and C are in between best and worst. The results agree with the practical engineering situation.

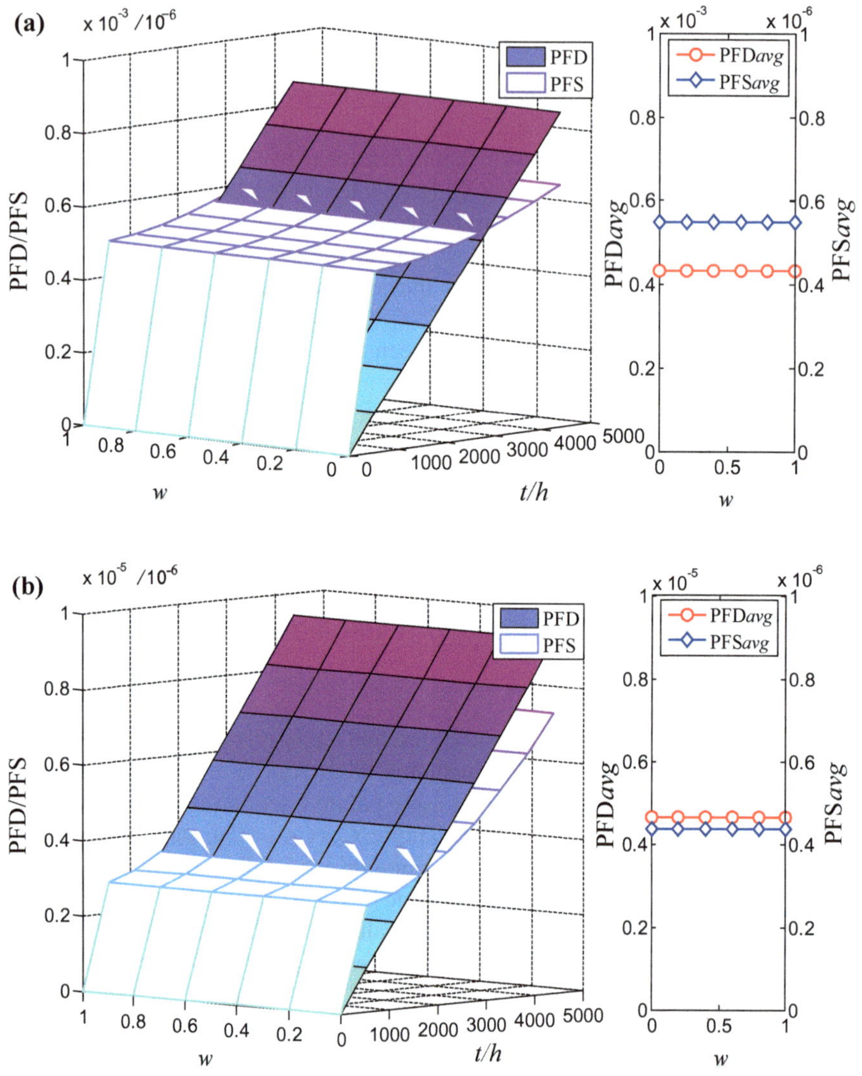

Fig. 7 Effects of common cause weight on the model precision **a** 2oo2D architecture and **b** 2oo3 architecture

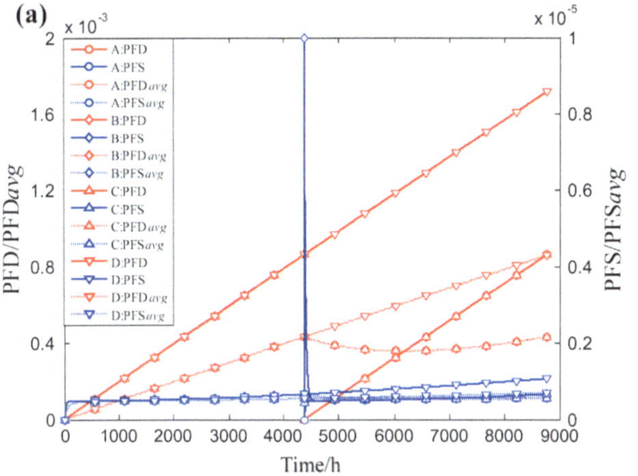

Variable	Value			
	A	B	C	D
γ	0	0	0	0
α	0	0	0	0
σ	0	0	0	0
ε	0	0	1	1
θ	1	0	1	0
μ	1	0	0	0
ζ	1	0	0	0
δ	1	0	1	0

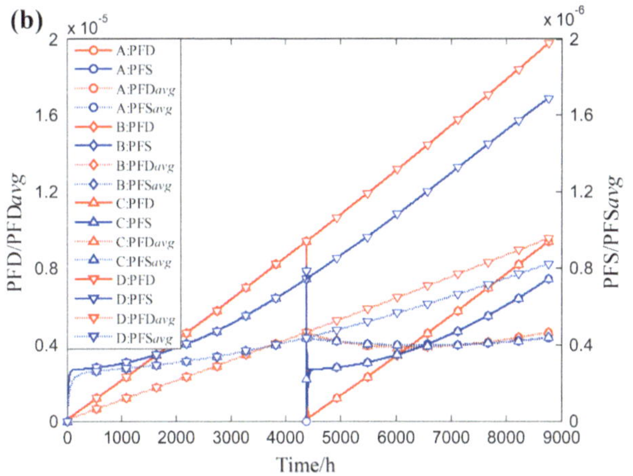

Variable	Value			
	A	B	C	D
γ	0	0	0	0
α	0	0	0	0
σ	0	0	0	0
ε	0	0	1	1
θ	1	0	1	0
μ	1	0	0	0
ζ	1	0	0	0
δ	1	0	1	0

Fig. 8 Effects of imperfect proof test and repair on the model precision **a** 2oo2D architecture and **b** 2oo3 architecture

4 SIL Software Development

The MDBN-based user-friendly SIL determination software is developed by using MATLAB graphical user interface (GUI) to assist engineers in determining SIL values without programming in the MATLAB environment. The SILs of all the general redundancy architectures can be determined, such as 1oo1, 1oo1D, 1oo2, 2oo2, 1oo2D, 2oo2D, 1oo3, 2oo3, and others, in the software. The MDBN models for these architectures are created automatically using the proposed method in Sect. 2. As shown in Fig. 9, the parameters, such as failure rates, time interval, and test parameters can be input in the top right part of the GUI. After calculation, the curves of PFD, PFD_{avg}, PFS, and PFS_{avg} are output in the top left part of the GUI, and the calculated results are displayed in the bottom. The 2oo3 architecture is taken

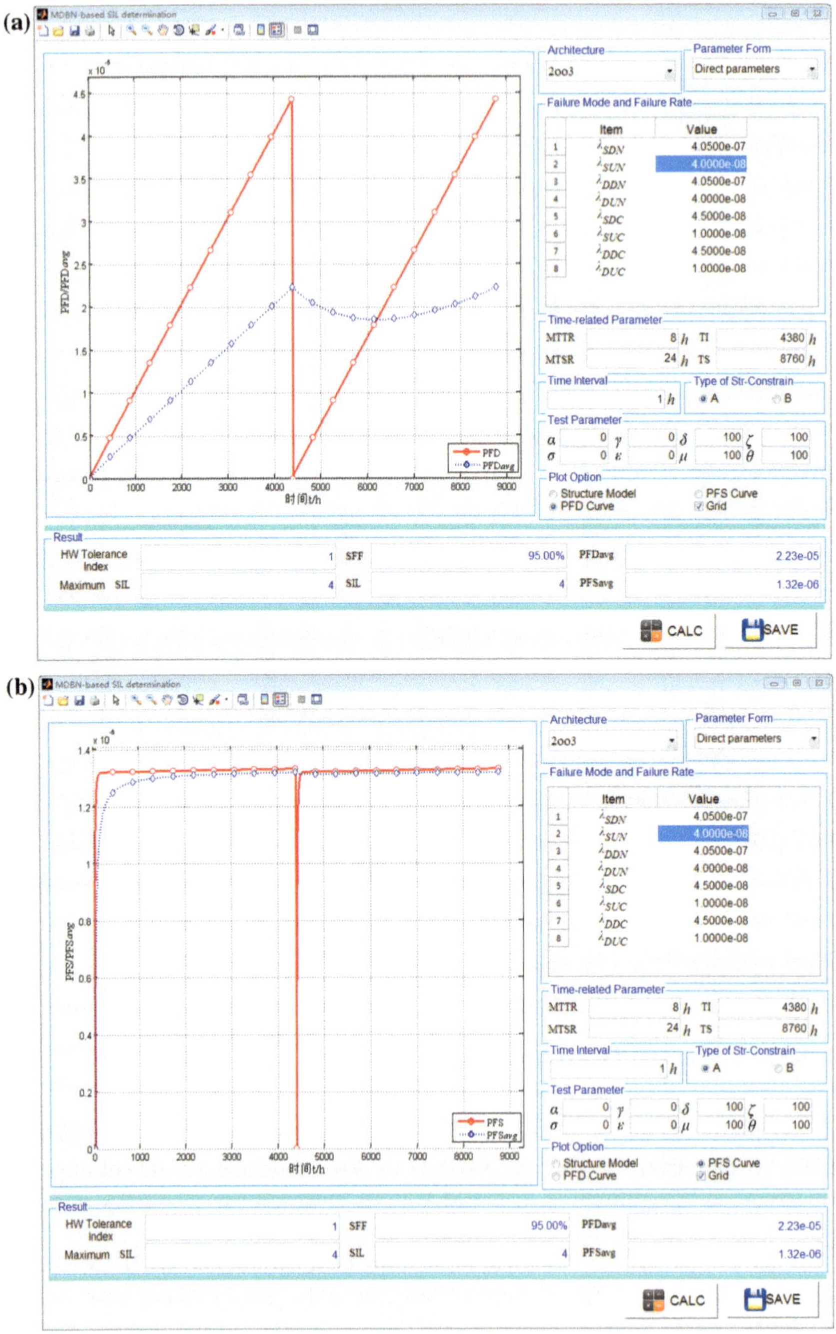

Fig. 9 MDBN-based SIL determination software **a** PFD/PFD_{avg} curves and **b** PFS/PFS_{avg} curves for 2oo3 architecture

as an example. When the parameters are input, the four curves are displayed in the interface, and the SIL determines that the level is 4. These results agree with the calculated results obtained by using Markov-based method proposed in IEC 61580, indicating that the MDBN-based SIL determination method is correct.

5 Conclusions

This study proposes a novel MDBN-based methodology for the determination of SILs of SISs operating in a low-demand mode. The unified models of MDBNs for KooM and KooMD architectures are constructed, and procedures of automatic creation of conditional probability tables are developed. The proposed MDBNs can be used to evaluate PFD, PFD_{avg}, PFS, and PFS_{avg} values and determine SIL values. The effects of time interval of MDBNs, common cause weight, imperfect proof test, and repair on model precision are researched. The results show that the calculated values of target failure measures can be considered true values when the number of total time slices is sufficiently large. Meanwhile, common cause weight has little effect on the target failure measures, and the effects of imperfect proof test and repair agree with practical engineering situations. The MDBN-based user-friendly SIL determination software is developed using MATLAB GUI to assist engineers in determining SIL values without programming in the MATLAB environment. With the aid of the software, PFD, PFD_{avg}, PFS, and PFS_{avg} curves can be calculated and displayed, and the SIL value can be determined.

Acknowledgements The authors wish to acknowledge the financial support of Hong Kong Scholars Program (No. XJ2014004), National Natural Science Foundation of China (No. 51309240), Specialized Research Fund for the Doctoral Program of Higher Education (No. 20130133120007), China Postdoctoral Science Foundation (No. 2015M570624), Applied Basic Research Programs of Qingdao (No. 14-2-4-68-jch), Science and Technology Project of Huangdao District (No. 2014-1-48), Fundamental Research Funds for the Central Universities (No. 14CX02197A), Theme-based Research Scheme of University Grants Council (No. T32-101/15-R), and Key Project of National Natural Science Foundation of China (No. 71532008).

References

1. B. Cai, Y. Liu, Z. Liu, F. Wang, X. Tian, Y. Zhang, Development of an automatic subsea blowout preventer stack control system using PLC based SCADA. ISA Trans. **51**(1), 198–207 (2012)
2. X. Iturbe, A. Ebrahim, K. Benkrid, C. Hong, T. Arslan, J. Perez, D. Keymeulen, M.D. Santambrogio, R3TOS-based autonomous fault-tolerant systems. IEEE Micro **34**(6), 20–30 (2014)
3. R. Hammett, Flight-critical distributed systems: Design considerations [avionics]. IEEE Aerosp. Electron. Syst. Mag. **18**(6), 30–36 (2003)
4. IEC 61508, Electric/Electronic/Programmable Electronic safety-related systems, parts 1–7. *Technical Report, International Electrotechnical Commission* (May 2010)
5. IEC 61511, Functional safety: Safety instrumented systems for the process industry sector, parts 1–3. *Technical Report, International Electrotechnical Commission* (Mar 2003)

6. IEC 61513, *Nuclear Power Plants: Instrumentation and Control Important to Safety, General Requirements for Systems* (Oct 2011)
7. IEC 62061, *Safety of Machinery: Functional Safety of Electrical, Electronic and Programmable Electronic Control Systems* (Jan 2005)
8. EN 50129, *Railway Applications: Communications, Signaling and Processing Systems, Safety Related Electronic Systems for Signaling* (May 2002)
9. ISO 26262, *Road Vehicles: Functional Safety* (Nov 2011)
10. M. Catelani, L. Ciani, V. Luongo, A simplified procedure for the analysis of safety instrumented systems in the process industry application. Microelectron. Reliab. **51**, 9–11 (2011)
11. I.W. Soro, M. Nourelfath, D. Aït-Kadi, Performance evaluation of multi-state degraded systems with minimal repairs and imperfect preventive maintenance. Reliab. Eng. Syst. Saf. **95**, 65–69 (2010)
12. O. Gemikonakli, E. Ever, A. Kocyigit, Approximate solution for two stage open networks with Markov-modulated queues minimizing the state space explosion problem. J. Comput. Appl. Math. **223**, 519–533 (2009)
13. S.K. Kim, Y.S. Kim, An evaluation approach using a HARA and FMEDA for the hardware SIL. J. Loss Prev. Process Ind. **26**, 1212–1220 (2013)
14. Y. Dutuit, F. Innal, A. Rauzy, J.P. Signoret, Probabilistic assessments in relationship with safety integrity levels by using fault trees. Reliab. Eng. Syst. Saf. **93**, 1867–1876 (2008)
15. K. Chang, S. Kim, D. Chang, J. Ahn, E. Zio, Uncertainty analysis for target SIL determination in the offshore industry. J. Loss Prev. Process Ind. **34**, 151–162 (2015)
16. M. Khalil, M.A. Abdou, M.S. Mansour, H.A. Farag, M.E. Ossman, A cascaded fuzzy-LOPA risk assessment model applied in natural gas industry. J. Loss Prev. Process Ind. **25**, 877–882 (2012)
17. W. Mechri, C. Simon, K. BenOthman, Switching Markov chains for a holistic modeling of SIS unavailability. Reliab. Eng. Syst. Saf. **133**, 212–222 (2015)
18. L. Ding, H. Wang, K. Kang, K. Wang, A novel method for SIL verification based on system degradation using reliability block diagram. Reliab. Eng. Syst. Saf. **132**, 36–45 (2014)
19. R. Nait-Said, F. Zidani, N. Ouzraoui, Modified risk graph method using fuzzy rule-based approach. J. Hazard. Mater. **164**, 651–658 (2009)
20. T. Daemi, A. Ebrahimi, M. Fotuhi-Firuzabad, Constructing the Bayesian network for components reliability importance ranking in composite power systems. Int. J. Electr. Power Energy Syst. **43**, 474–480 (2012)
21. Y.D. Shu, J.S. Zhao, A simplified Markov-based approach for safety integrity level verification. J. Loss Prev. Process Ind. **29**, 262–266 (2014)
22. M. Sallak, C. Simon, J.-F. Aubry, A fuzzy probabilistic approach for determining safety integrity level. IEEE Trans. Fuzzy Syst. **16**, 239–248 (2008)
23. H. Jahanian, Generalizing PFD formulas of IEC 61508 for KooN architectures. ISA Trans. **55**, 168–174 (2015)
24. R. Ouache, M.N. Kabir, A.A. Adham, A reliability model for safety instrumented system. Saf. Sci. **80**, 264–273 (2015)
25. F. Innal, Y. Dutuit, M. Chebila, Safety and operational integrity evaluation and design optimization of safety instrumented systems. Reliab. Eng. Syst. Saf. **134**, 32–50 (2015)
26. K. Tsilipanos, I. Neokosmidis, D. Varoutas, A system of systems framework for the reliability assessment of telecommunications networks. IEEE Syst. J. **7**, 114–124 (2013)
27. O. Doguc, R.-M.J. Emmanuel, An automated method for estimating reliability of grid systems using Bayesian networks. Reliab. Eng. Syst. Saf. **104**, 96–105 (2012)
28. Y. Jiang, H.H. Zhang, X.Y. Song, X. Jiao, W.N.N. Hung, M. Gu, J.G. Sun, Bayesian-network-based reliability analysis of PLC systems. IEEE Trans. Ind. Electron **60**, 5325–5336 (2013)
29. L.M. Zhang, X.G. Wu, M.J. Skibniewski, J.B. Zhong, Y.J. Lu, Bayesian-network-based safety risk analysis in construction projects. Reliab. Eng. Syst. Saf. **131**, 29–39 (2014)
30. T. Daemi, A. Ebrahimi, Detailed reliability assessment of composite power systems considering load variation and weather conditions using the Bayesian network. Int. Trans. Electr. Energy Syst. **24**, 305–317 (2014)

31. P. Baraldi, L. Podofillini, L. Mkrtchyan, E. Zio, V.N. Dang, Comparing the treatment of uncertainty in Bayesian networks and fuzzy expert systems used for a human reliability analysis application. Reliab. Eng. Syst. Saf. **138**, 176–193 (2015)
32. A. O'Connora, A. Moslehb, A general cause based methodology for analysis of common cause and dependent failures in system risk and reliability assessments. Reliab. Eng. Syst. Saf. **145**, 341–350 (2016)
33. B.P. Cai, Y.H. Liu, Z.K. Liu, X.J. Tian, X. Dong, S.L. Yu, Using Bayesian networks in reliability evaluation for subsea blowout preventer control system. Reliab. Eng. Syst. Saf. **108**, 32–41 (2012)
34. B. Cai, Y. Liu, Y. Ma, L. Huang, Z. Liu, A framework for the reliability evaluation of grid-connected photovoltaic systems in the presence of intermittent faults. Energy **93**, 1308–1320 (2015)
35. B. Cai, Y. Liu, Y. Ma, Z. Liu, Y. Zhou, J. Sun, Real-time reliability evaluation methodology based on dynamic Bayesian networks: A case study of a subsea pipe ram BOP system. ISA Trans. **58**, 595–604 (2015)
36. P.A.P. Ramírez, I.B. Utne, Use of dynamic Bayesian networks for life extension assessment of ageing systems. Reliab. Eng. Syst. Saf. **133**, 119–136 (2015)
37. F. Flammini, S. Marrone, N. Mazzocca, V. Vittorini, A new modeling approach to the safety evaluation of N-modular redundant computer systems in presence of imperfect maintenance. Reliab. Eng. Syst. Saf. **94**(9), 1422–1432 (2009)
38. F. Flammini, S. Marrone, N. Mazzocca, R. Nardone, V. Vittorini, Using Bayesian Networks to evaluate the trustworthiness of '2 out of 3' decision fusion mechanisms in multi-sensor applications. IFAC-Pap. OnLine **48**(21), 682–687 (2015)
39. P. Weber, C. Simon, D. Theilliol, Reconfiguration of over-actuated consecutive-k-out-of-n: F systems based on Bayesian network reliability model. ACD. **302** (2010)
40. H. Guo, X. Yang, Automatic creation of Markov models for reliability assessment of safety instrumented systems. Reliab. Eng. Syst. Saf. **93**, 807–815 (2008)
41. J. Jin, L. Pang, S. Zhao, B. Hu, Quantitative assessment of probability of failing safely for the safety instrumented system using reliability block diagram method. Ann. Nucl. Energy **77**, 30–34 (2015)

Availability-Based Engineering Resilience Metric and Its Corresponding Evaluation Methodology

Abstract Several resilience metrics have been proposed for engineering systems (e.g., mechanical engineering, civil engineering, critical infrastructure, etc.); however, they are different from one another. Their difference is determined by the performances of the objects of evaluation. This study proposes a new availability-based engineering resilience metric from the perspective of reliability engineering. Resilience is considered an intrinsic ability and an inherent attribute of an engineering system. Engineering system structure and maintenance resources are principal factors that affect resilience, which are integrated into the engineering resilience metric. A corresponding dynamic-Bayesian-network-based evaluation methodology is developed on the basis of the proposed resilience metric. The resilience value of an engineering system can be predicted using the proposed methodology, which provides implementation guidance for engineering planning, design, operation, construction, and management. Some examples for common systems (i.e., series, parallel, and voting systems) and an actual application example (i.e., a nine-bus power grid system) are used to demonstrate the application of the proposed resilience metric and its corresponding evaluation methodology.

Keywords Resilience · Availability · Metric · Engineering system

1 Introduction

Resilience is the capability of an entity to recover from an external disruptive event. To date, the concept of resilience has been spread from ecology [1, 2] to various fields, such as economics [3, 4], psychology [5, 6], and sociology [7, 8]. In comparison with the research in non-engineering contexts, only a small proportion of resilience-related research exists in the field of engineering [9–11]. For engineering systems, such as mechanical engineering, civil engineering, critical infrastructure, etc., different definitions are proposed depending on the objects of evaluation. The National Infrastructure Advisory Council defines critical infrastructure resilience as the capability to reduce the magnitude and/or duration of disruptive events. The effectiveness of a resilient infrastructure or enterprise depends upon its capability

B. Cai et al., *Bayesian Networks for Reliability Engineering*,
https://doi.org/10.1007/978-981-13-6516-4_11

to anticipate, absorb, adapt to, and/or rapidly recover from a potentially disruptive event [12]. The American Society of Mechanical Engineers defines resilience as the capability of a system to sustain external and internal disruptions without discontinuity of performing the system function or, if the function is disconnected, to fully recover the functions rapidly [13]. Haimes [45] defined resilience as the capability of the system to withstand a major disruption within acceptable degradation parameters and recover within an acceptable time and composite costs and risks. Many other researchers have also proposed their own engineering resilience definitions from different perspectives [14–17].

According to the definitions above, various resilience metrics and their corresponding evaluation methodologies have been developed [18–23]. For example, Dessavre et al. [18] defined a new model and visual tools that improve the capabilities to characterize the resilience behavior of complex systems by extending existing time-dependent resilience functions. Bruneau et al. [19] defined four dimensions of resilience, namely robustness, rapidity, resourcefulness, and redundancy, in the well-known resilience triangle model in civil infrastructure and proposed a deterministic static metric for measuring the resilience loss of a community to an earthquake. Henry et al. [20] proposed a time-dependent quantifiable resilience metric corresponding to a specific figure-of-metric, which was evaluated at a certain time period under disruptive events. Francis et al. [21] proposed a resilience metric that incorporates three resilience capabilities, including adaptive capacity, absorptive capacity, recoverability, and the time to recovery. Hosseini et al. [22] used static Bayesian networks to model infrastructure resilience and used a case of inland waterway port to demonstrate the proposed method.

Although various resilience metrics have been developed, quantifying the resilience for a specific engineering system remains a challenge because of internal and external factors involved in such metrics. From the above definitions and metrics, resilience overlaps with a number of existing concepts, such as adaptability [24], robustness [24, 25], redundancy [26], flexibility [27], survivability [27], recoverability [28, 29], rapidity [25, 30], and resourcefulness [30]. Here, we consider resilience as an intrinsic capability and an inherent attribute of an engineering system itself. It is composed of two properties, namely performance- and time-related properties. System structure determines performance-based properties, such as robustness, adaptability, redundancy, flexibility, and survivability, whereas maintenance resource determines time-related properties, such as reparability, recoverability, rapidity, and resourcefulness. Similar to the reliability in reliability engineering, external factors, such as disturbance, attack, and disaster events, are not intrinsic properties of resilience in engineering system and are thus not involved in the resilience metric (see Fig. 1). Therefore, when an engineering system is designed and maintenance resource is allocated, the resilience of this system is determined. Hence, the structure and maintenance resource in the engineering system form a unified whole, thereby determining the engineering resilience of the system. The promoted viewpoint may be different from the dominant ones [46, 47]; however, it can provide implementation guidance for engineering planning, design, operation, construction, and management.

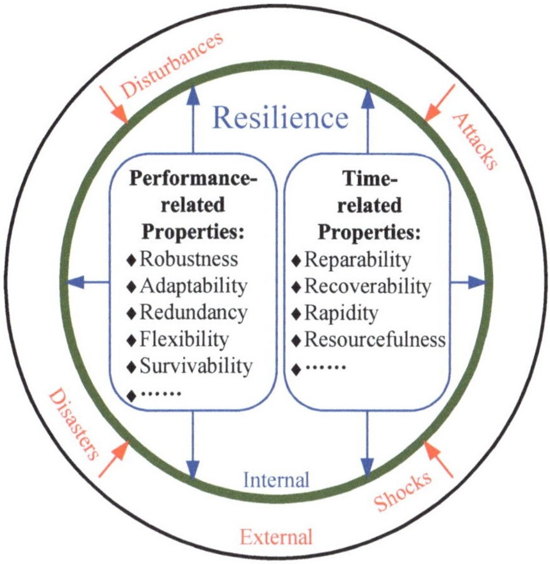

Fig. 1 Resilience-related properties of engineering system

In this study, we aim to develop a new availability-based engineering resilience metric from the essence and property of resilience in engineering system. From the perspective of reliability engineering, steady-state availability and steady-state time can be used to represent the performance- and time-related properties. Each engineering system has its own availability; thus, the resilience value can be obtained easily according to the steady-state availability and steady-state time. Therefore, the metric is suitable for every engineering system. The rest of this paper is organized as follows. Section 2 presents the proposed availability-based engineering resilience metric and its corresponding evaluation methodology. Section 3 adopts some examples for common systems to demonstrate the application of the proposed resilience metric and its evaluation methodology. Section 4 adopts an actual example for a nine-bus power grid system to demonstrate the application of the proposed method. Section 5 summarizes the contributions of this paper.

2 Resilience Metric and Evaluation Methodology

2.1 Availability-Based Engineering Resilience Metric

Each engineering system has its own availability, where an item is capable to be in a state of performing a required function under given conditions at a given time or time interval, assuming that the required external resources are provided. The availability of an engineering system decreases continuously to reach a steady-state availability

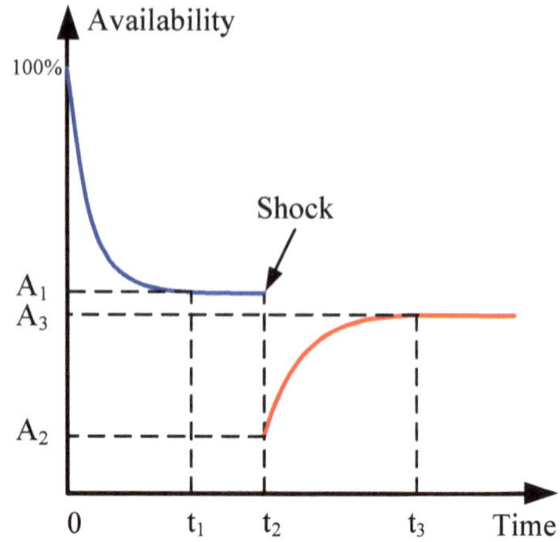

Fig. 2 Availability of a system subject to degradation and shock

A_1 at steady state t_1 from the initial time with the initial availability of 100%. This progress is caused by degradation of components and daily maintenance of the system. Suppose an external shock occurs at time t_2, the availability instantaneously decreases to a post-shock transient-state availability A_2 and then increases to a new equilibrium state A_3. This progress is also caused by emergency repair after shock, as well as degradation of components. The blue line in Fig. 2 represents the availability considering the degradation of components and daily system maintenance without external shocks, and the red line represents the availability considering the emergency repair after shock and degradation of components.

The steady-state availability A_1, post-shock transient-state availability A_2, post-shock steady-state availability A_3, steady-state time t_1, and post-shock steady-state time $(t_3 - t_2)$ are determined by the structure of the engineering system and maintenance resource, such as redundant structure, failure rate, and repair rate. High redundancy, low failure rate, or high repair rate results in high steady-state availability and short steady-state time before and after any shocks. This condition accords with the essence and property of resilience. Therefore, steady-state availability and low steady-state time are used to represent the performance- and time-related properties of engineering resilience. Thus, quantifying the resilience with an appropriate resilience metric is no longer a challenge given that steady-state availability and steady-state time are easy to obtain.

The proposed resilience metric aims to compare the resilience of different systems that achieve the same functions, thereby identifying the different internal factors that contribute to it. In this study, we develop a resilience metric that incorporates performance- and time-related properties using steady-state availability and steady-state time before and after external shocks. The value of resilience increases with the increase of availability A and the decrease of recovery time t. Thus, $A/\ln(t)$ is used to describe the degree of resilience. The natural logarithm function $\ln(x)$ is used to balance the level of effects between availability A and recovery time t. The resilience metric is considered to be the product of $A/\ln(t)$ before and after external shocks. Therefore, the final developed resilience metric is given as follows:

$$\rho = \frac{A_1}{n\ln(t_1)}\sum_{i=1}^{n}\frac{A_2^i A_3^i}{\ln\left(t_3^i - t_2^i\right)}, \tag{1}$$

where n is the number of shocks, and $i \in [1, n]$.

Given that the external factors are random and unpredictable, they are not factors of resilience and not involved in the resilience metric. The external factors only trigger a "bounce back," which is similar to the spring system, where an external force F can extend or compress a spring by some distance X and the spring can bounce back to the initial balance once the force F is removed. According to Hooke's law, the stiffness of the spring can be expressed as $k = F/X$. It is a constant factor characteristic of spring, which is determined by the spring itself and not by the force. The defined resilience ρ is similar with k; however, for the engineering system, the external factors and responses of this system are not directly proportional. Therefore, we determine a series of shocks on the engineering system, which result in the common cause failure of components. The prior probability of common cause failure for each component is defined as

$$p_i = \frac{i}{n+1}, \tag{2}$$

where $i \in [1, n]$. A shock can lead to a common cause failure of components with any prior probability. The proposed resilience metric is used to evaluate, optimize, compare, and design systems only if n is the same for each system. A larger value of n indicates more simulated shocks. When the number of shocks is more than 9, the resilience slightly changes. Therefore, we select $n = 9$ to evaluate system resilience.

For different shocks, the repair rates of components are completely different with fixed maintenance resources. When a shock is serious, the maintenance resources are dispersed, thereby causing low repair rates of components. That is, a larger prior probability of common cause failure indicates a smaller repair rate of components. To simplify, we define the repair rate of component under different shocks as

$$\mu_i = (1 - p_i)\mu, \tag{3}$$

where μ is the repair rate of each component under normal circumstances. Notably, other relationships between repair rate and prior probability of common cause failure, $\mu_i = f\,(p_i)$, can also be modeled and used to calculate the availability and subsequent resilience.

Based on the prior probabilities of common cause failure and corresponding repair rates of components, the steady-state availability of the engineering system can be obtained as follows:

$$A(\infty) = \lim_{t \to \infty} A(t), \tag{4}$$

Notably, a real steady-state availability does not exist. In practice, we therefore define steady-state availability as the availability when the difference within five continuous time point (hour) is equal to or less than 10^{-5}, and the time is termed as steady-state time.

2.2 *Engineering Resilience Evaluation Methodology*

The availability evaluation of engineering systems is very important for resilience calculation. Several approaches can be used to evaluate the steady-state availability, such as reliability block diagram [31], fault tree [32], Monte Carlo simulation [33], and Markov chain [34]. In this current work, a dynamic-Bayesian-network-based evaluation methodology is proposed to calculate the steady-state availability, steady-state time, and subsequent resilience of engineering systems. Bayesian network is a probabilistic graphical model that represents a set of random variables, including their conditional dependencies through directed acyclic graphs. It is considered to be one of the most useful models in the field of probabilistic knowledge representation and reasoning [35, 43, 44]. Dynamic Bayesian networks are a long-established extension to ordinary Bayesian networks and allow the explicit modeling of changes over time. In view of classical probabilistic temporal models, such as Markov chains, dynamic Bayesian networks are stochastic transition models factored over a number of random variables, over which a set of conditional dependency assumption is defined [36]. We adopt dynamic Bayesian networks to predict the future state of variables considering the current observation of variables. That is, we can predict the steady-state availability and steady-state time of an engineering system on the basis of the current state of components, such as when external shocks destroy some components.

Engineering resilience evaluation methodology with dynamic Bayesian networks consists of the following five procedures:

(1) Structural modeling of dynamic Bayesian networks is completed by using structural relationship methods, mapping algorithms, or structure learning methods.
(2) Expert elicitation with noisy models or parameter learning methods are used to model the parameters of dynamic Bayesian networks.
(3) Availability is evaluated by using exact or approximated inference algorithms.
(4) Resilience is evaluated using the resilience metric shown in Eqs. (1)–(4).
(5) Sensitivity analysis is conducted to research the influences of failure and repair actions on the resilience of engineering systems.

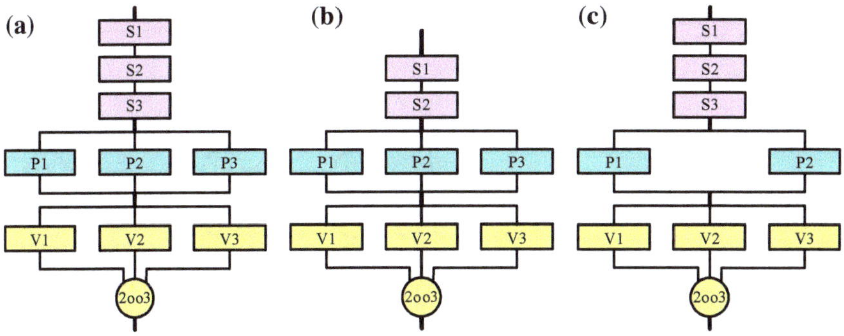

Fig. 3 Simple systems composed of series, parallel, and voting subsystems: **a** S3P3V3, **b** S2P3V3, and **c** S3P2V3

3 Examples for Common Systems

3.1 Series, Parallel, and Voting Systems

Many systems in practical engineering can be abstracted as series, parallel, or voting system. Taking subsea blowout preventer system as an example, the control stations, control pods, annular preventer, and ram preventer are redundantly configured; thus, three control stations, two control pods, several annular preventers, and several ram preventers are considered in parallel. The entire system can be considered a series of control stations, triple modular redundancy controllers, subsea control pods, annular preventers, lower marine riser package connector, ram preventer, and wellhead connector because the complete failure of each component category causes failure of the subsea blowout preventer system [34]. In the current work, we adopt some examples for the common systems to demonstrate the application of the availability-based resilience metric and its evaluation methodology.

Series, parallel, and voting systems are an abstract system composed of three series components, three parallel components, and three voting components, denoted by S3P3V3 (see Fig. 3a). The series subsystem works only when all of the three series components S1, S2, and S3 work; the parallel subsystem works when any of the three components P1, P2, or P3 works; the 2-out-of-3 (2oo3) voting system works when at least two components of V1, V2, and V3 work. The entire system works only when all of the three subsystems work, which is equivalent to a series system. Similarly, we use S2P3V3 to denote a series system composed of two series components, three parallel components, and three voting components (see Fig. 3b), and S3P2V3 to denote a series system composed of three series components, two parallel components, and three voting components (see Fig. 3c).

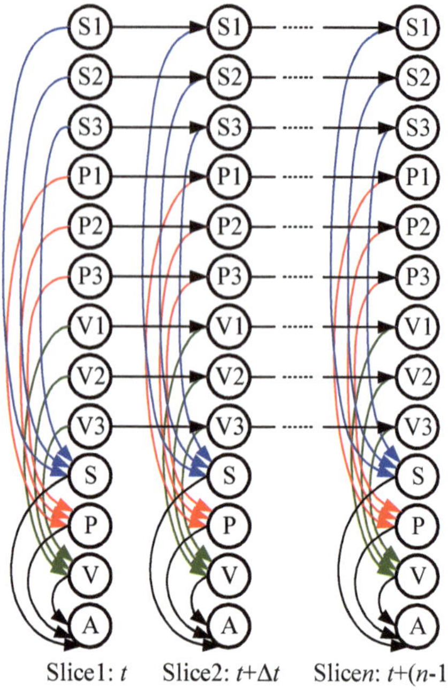

Fig. 4 Dynamic Bayesian networks of S3P3V3 system

3.2 *Structural Modeling of Dynamic Bayesian Networks*

For these series, parallel, and voting systems, the structural models of dynamic Bayesian networks are established by using the structural relationship of each component. Taking S3P3V3 system as an example, the components and their states are denoted by root nodes, including S1, S2, S3, P1, P2, P3, V1, V2, and V3 at a specific time slice of dynamic Bayesian networks (e.g., Slice1: *t*; see Fig. 4). A is the final leaf node of the network, which represents the state of the entire system. Each root node has two states (i.e., *work* and *fail*). For node A, the probabilities of work and fail indicate the transient availability and transient unavailability, respectively. We artificially added several intermediate nodes, including S, P, and V, to simplify the conditional probability table of related nodes. The causal relationship between the nodes is connected by intra-slice arcs.

Dynamic Bayesian networks are essentially replications of static Bayesian networks over n time slices between t and $t + (n - 1)\Delta t$. A set of inter-arcs between adjacent time slices t and $t + \Delta t$ connects the corresponding nodes of components, which represent the dynamic degradation process and daily maintenance or emergency repair of components. All the information required to predict a state at time $t + \Delta t$ is contained in the description at time t, and no information about earlier times is required; thus, the process possesses the Markov property.

Table 1 Failure and repair rates of components in series, parallel, and voting systems

System	Component	Failure rate	Repair rate
Series, parallel, and voting systems	S	0.833e−3	0.500
	P	2.083e−3	0.330
	V	1.389e−3	0.670

3.3 Parameter Modeling of Dynamic Bayesian Networks

The parameter model of dynamic Bayesian networks is composed of intra- and inter-slice parameter models. For the intra-slice parameter model, the marginal prior probabilities are assigned to them according to the resilience metric in Eq. (2), and the conditional probability tables are determined using the series, parallel, and voting relationship. For the inter-slice parameter model, we use Markov state transition relationship to determine the dynamic degradation process and daily maintenance or emergency repair of components.

In dynamic Bayesian networks, the inter-slice parameter model is the probability of nodes between time slices t and $t + \Delta t$. For the components of S3V3P3 system, we suppose that the failure and repair follow an exponential distribution; that is, all of the transition rates, including failure and repair rates, are constant. Given that the process possesses the Markov property, the probability is determined using a Markov state transition relationship. Hence, the transition relationships between consecutive nodes can be expressed as follows:

$$p(X_{t+\Delta t} = \text{work}|X_t = \text{work}) = \mathrm{e}^{-\lambda \Delta t}, \tag{5}$$

$$p(X_{t+\Delta t} = \text{fail}|X_t = \text{work}) = 1 - \mathrm{e}^{-\lambda \Delta t}, \tag{6}$$

$$p(X_{t+\Delta t} = \text{fail}|X_t = \text{fail}) = \mathrm{e}^{-\mu \Delta t}, \tag{7}$$

$$p(X_{t+\Delta t} = \text{work}|X_t = \text{fail}) = 1 - \mathrm{e}^{-\mu \Delta t}, \tag{8}$$

where X is the root node, λ is the failure rate of a component, and μ is the repair rate of a component. For the S3P3V3, S2P3V3, and S3P2V3 systems, we provide the same failure and repair rates for each component (see Table 1).

3.4 Resilience Evaluation

The goal of inference in a dynamic Bayesian network is to compute the marginal $p(X_{t+h}|y_{1:t})$ when $y_{1:t}$ is observation. $h = 0$, $h < 0$, and $h > 0$, indicate filtering, smoothing, and prediction, respectively. We use the following prediction of dynamic Bayesian networks to evaluate the resilience value of engineering systems. Junction

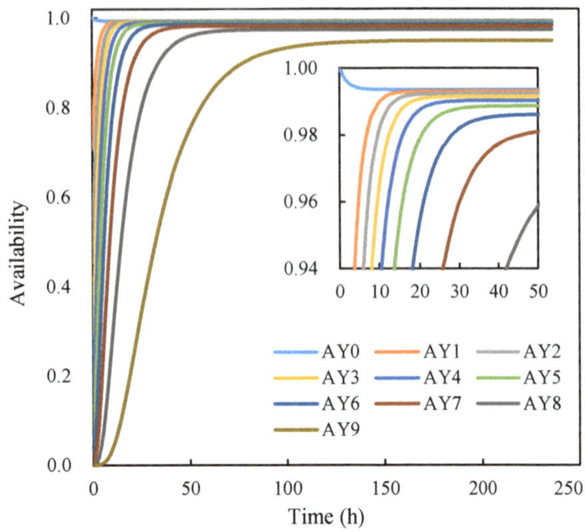

Fig. 5 Availability of S3P3V3 system subject to degradation and different shocks

tree algorithm for propagation analysis is conducted, where the joint probability for the model from the conditional probability structure of the dynamic Bayesian networks is calculated in a computationally efficient manner.

$$p\left(Y_{t+h} = h|y_{1:t}\right) = \sum_{x} p(Y_{t+h} = h|X_{t+h} = x)p\left(X_{t+h} = x|y_{1:t}\right) \tag{9}$$

The trends of availability without external shocks or with different external shocks are different (see Fig. 5). The curve AY0 indicates the availability without any shocks, and the curves AY1–AY9 indicate the availability with different shocks, that is, different prior probabilities ($p1$–$p9$) of common cause failure for each component. When no shocks occur, that is, during normal running of the S3P3V3 system, the availability decreases rapidly from 100% and reaches a stable level of 99.365% at the 15th hour. The steady-state availability and steady-state time are therefore 99.365% and 15, respectively. During emergency circumstances, the availabilities decrease to a minimum valve the moment shocks occur and then increase to reach different stable levels. Taking AY1 as an example, the prior probability of 10% of common cause failure is assigned to each component at the original time. The availability decreases to 70.790% immediately, increases rapidly with emergency repair, and reaches a stable level of 99.311% at the 27th hour. Hence, the post-shock transient-state availability, post-shock steady-state availability, and post-shock steady-state time are 70.790, 99.311, and 27%, respectively. With the increase in probability of shocks, the post-shock transient-state availability and the post-shock steady-state availability decrease, and the post-shock steady-state time increases. This finding agrees with the fact. Using all the characteristic values, we obtain the resilience value of the S3P3V3 system of 1.90% using the proposed availability-based resilience metric.

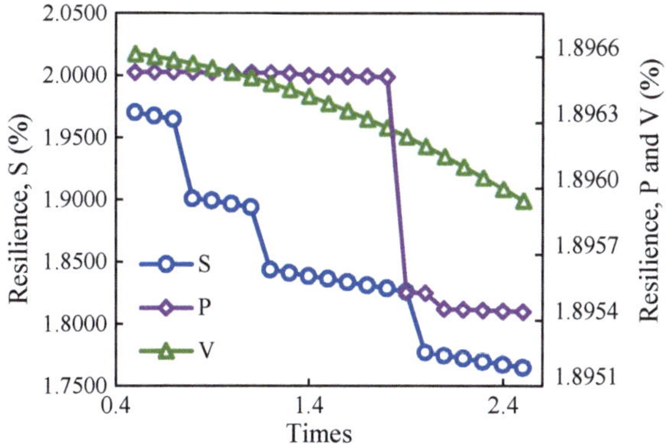

Fig. 6 Sensitivity of failure rates of components of S3P3V3 system

Resilience is determined by the engineering system itself and not by external shocks; thus, the factors that affect system performance are certain to affect the resilience value. System structures and failure and repair rates of components are main influencing factors.

3.5 *Sensitivity Analysis*

The sensitivity analysis of the failure rates of components S, P, and V are conducted by changing the failure rates of each component of the same category in multiples. The curves represent the resilience values with the changes of the failure rates from 0.5 times to 2.5 times of series, parallel, and voting components. The resilience of the S3P3V3 system decreases with the increase in time of failure rates (see Fig. 6). For components S and P, the resilience values present a ladder-form decrease with the increase of failure rates. For component V, the resilience values continuously decrease with the increase of failure rates. Under the same time variation of failure rates of components, the resilience value for S decreases fastest, that for V decreases slowest, and that for P is in between. Therefore, the resilience of the S3P3V3 system is the most sensitive to the failure rates of component S and is the least sensitive to the failure rates of component V.

The sensitivity analysis of the repair rates of components S, P, and V is conducted by changing the repair rates of each component of the same category in multiples. The curves represent the resilience values with the changes of the repair rates from 0.1 times to 3.0 times of series, parallel, and voting components. The resilience of the S3P3V3 system increases with the increase in time of repair rates (see Fig. 7). With the increase of repair rates, the resilience value for component S continuously decreases when time is minimal, whereas it presents a ladder-form increase when the time is considerable. For components P and V, the resilience values increase rapidly when the times is minimal and then reach stable levels with the increase of repair

Fig. 7 Sensitivity of repair rates of components of S3P3V3 system

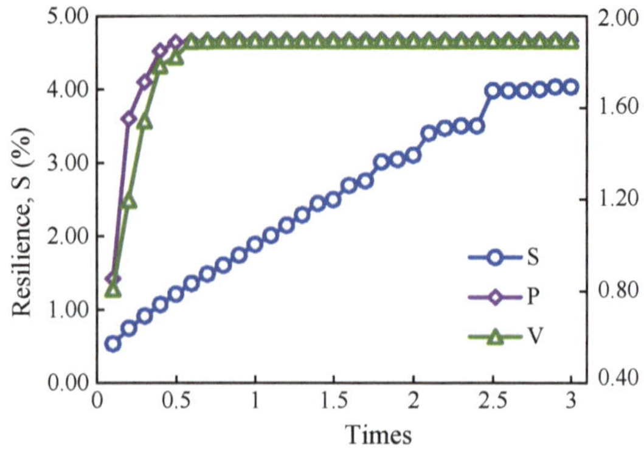

Fig. 8 Resilience values of S3P3V3, S2P3V3, and S3P2V3 systems

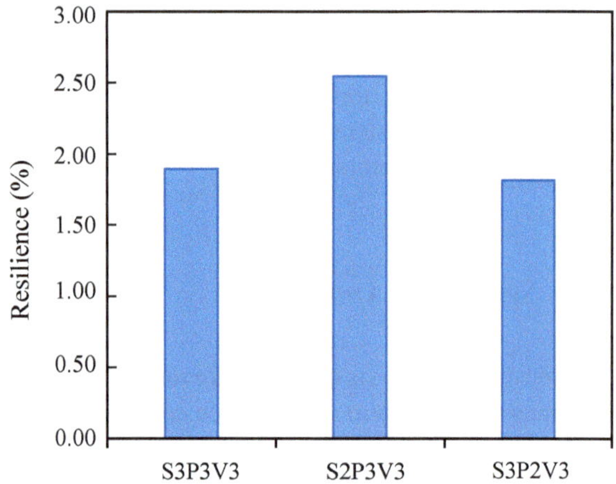

rates. Under the same time variation of repair rates of components, the resilience value for S increases fastest, and those for P and V increase slowest; the resilience value of V is slightly higher than that for P. Therefore, the resilience of the S3P3V3 system is the most sensitive to the repair rates of component S and is the least sensitive to the failure rates of components P and V.

We suppose that S3P3V3, S2P3V3, and S3P2V3 systems can achieve the same functions. Comparison of S2P3V3 and S3P3V3 systems reveal that less series components improve engineering resilience; comparison of S3P2V3 and S3P3V3 systems shows that less parallel components can reduce engineering resilience (see Fig. 8). Therefore, redundancy of engineering systems plays an important role in engineering resilience.

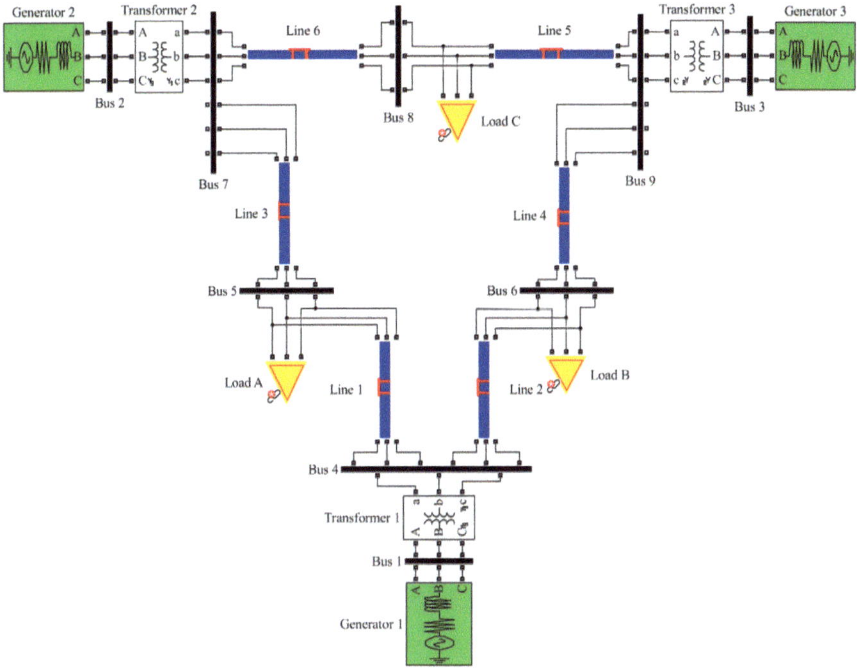

Fig. 9 Nine-bus power grid system

4 Actual Application Example for Nine-Bus Power Grid System

4.1 Resilience of Nine-Bus Power Grid System

Aside from abstracting the series, parallel, and voting systems, using a practical engineering system is necessary to demonstrate the application of the proposed availability-based resilience metric. Various resilience indexes or evaluation methods have been demonstrated using power systems [37, 38]. Here, we study the resilience of a nine-bus power grid system (see Fig. 9). The system consists of nine buses, three generators, three two-winding power transformers, six lines, and three loads. To simplify, we ignore the power energy volume and dynamic response and focus on the system structure. Hence, when loads A, B, and C are powered, the entire nine-bus power grid system works.

The dynamic Bayesian network model of the nine-bus power grid system is established similar to the S3P3V3 system, and the resilience is evaluated (see Fig. 10 and Table 2). Node B denotes the bus; G denotes the generator; T represents the transformer; L represents the line; and LA, LB, and LC are loads A, B, and C, respectively. Node S represents the state of the entire power grid system. Each root node has two

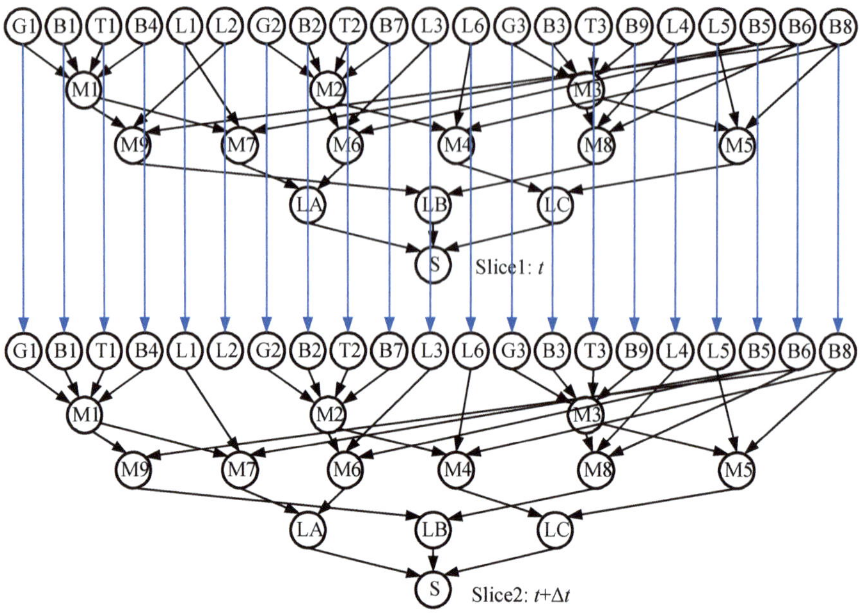

Fig. 10 Dynamic Bayesian networks of the nine-bus power grid system

Table 2 Failure and repair rates of components in the nine-bus power grid system

System	Component	Failure rate	Repair rate
Nine-bus power grid system	Generator	1.631e−5	0.050
	Transformer	1.903e−5	0.067
	Line	2.854e−5	0.083
	Bus	0.951e−5	0.250

states (i.e., *work* and *fail*). For node S, the probabilities of work and fail indicate the transient availability and transient unavailability of this power system, respectively. For the nine-bus power grid system, we review the literature [39] and determine the failure and repair rates of each components. Additional general distributions, such as Weibull distribution, can also be modeled using dynamic Bayesian networks [40]. The results show that the resilience value of the system is 0.54%. Although the resilience of the S3P3V3 system is larger than the nine-bus power grid system, comparing these two systems is illogical because they do not achieve the same functions. That is, the resilience of engineering systems that achieve the same functions is comparable. Therefore, for a specified engineering system, resilience provides implementation guidance for planning, design, operation, construction, and management.

4.2 Discussion of the Proposed Resilience Metric

(1) Resilience is an Intrinsic Capability and an Inherent Attribute

We consider resilience as an intrinsic capability and an inherent attribute of an engineering system itself. It is not influenced by external factors, such as disturbance, attack, and disaster events. This definition is different from other viewpoints [41, 42]. The developed availability-based resilience metric in this study involves the performance- and time-related properties of the engineering system but not the external factors. We adopt steady-state availability and steady-state time before and after several shocks to define the metric. Therefore, the resilience value of the system is determined when a system is developed and the repair resources are assigned and fixed in this system. Resilience value is helpful in planning, design, operation, construction, and management of an engineering system. When an emergency incident occurs (e.g., an earthquake destroys some components of a power grid system), although some emergency maintenance teams outside the system might be assigned to this system and hence shorten the repair time, the primary system does not possess this intrinsic capability. Therefore, the repair actions of the emergency maintenance teams from other systems cannot improve the resilience of this system. Another example is two similar systems in earthquake-prone area and non-earthquake area; these systems should have the same resilience values when the same maintenance teams are involved because the systems have the same capability to suffer from an earthquake and recover from the destruction. To improve the capability against earthquakes, one should improve the performance of the system or increase the maintenance teams of the system itself, but not to obtain help from other systems.

(2) Comparison of Resilience-Based Engineering Systems

The proposed availability-based resilience metric aims to compare the resilience values of different systems that achieve the same functions, thereby identifying different internal factors that contribute to it. For systems with the same functions, a larger resilience value indicates that the system is more resilient. Notably, the objects for comparison should be systems that achieve the same functions. In this study, the series, parallel, and voting systems S3P3V3, S2P3V3, and S3P2V3 aim to connect two terminations, and the nine-bus power grid system aims to supply power for loads A, B, and C. Comparing the resilience values of S3P3V3, S2P3V3, and S3P2V3 systems seems logical, whereas comparing the S3P3V3 system with the nine-bus power grid system is irrational. Similarly, if we develop a new power grid system to supply power for loads A, B, and C, then comparing it with the nine-bus power grid system and selecting a more resilient system are important.

(3) Optimization of Resilience-Based Engineering System

When an engineering system is planned and designed, identifying the weak components that affect the resilience significantly is important. Resilience is influenced by the internal factors of an engineering system; thus, the sensitivity of these factors (e.g., redundancy and failure and repair rates) to resilience can be quantified. We can change the system structure and failure or repair rates in multiples to evaluate the resilience values and analyze the sensitivity. We should pay more attention to the structure or component that is most sensitive to resilience (e.g., increase the redundancy of that structure to decrease the failure rate or increase the repair rate of that component).

(4) Design of Resilience-Based Engineering System

From the perspective of reliability engineering, several reliability-based engineering system design methods are available. For example, we can design an automation control system of subsea blowout preventers with the reliability of 99.9999%. Similarly, a resilience-based engineering system design method might be useful because it involves the characteristics of recovery after shocks. In practical guidance documents, the recommended resilience value of engineering systems should be specified. For example, suppose that the resilience value for a power grid system in Qingdao City is specified to be 0.60%; a new power grid system should be designed to meet this requirement. Notably, the specified resilience values are determined by practical engineering environment even for the same system. For example, the resilience value for a power grid system in a large city might be 0.80%, but 0.50% in a small village, 0.90% in an industrial park, and 0.60% in a residential area. Such values are determined by practical guidance documents produced by experts.

5 Conclusion

A new availability-based engineering resilience metric is proposed from the perspective of reliability engineering. The corresponding dynamic-Bayesian-network-based evaluation methodology is developed based on the proposed metric. Series, parallel, and voting systems and a nine-bus power grid system are used to demonstrate the application of the metric and its corresponding evaluation methodology. The results show that the engineering resilience metric is reasonable and that the corresponding evaluation methodology is precise. System structures and failure and repair rates of components are the main influencing factors of engineering resilience. Redundancy of components plays an important role in increasing resilience. The proposed metric can be used for engineering comparison, optimization, and design. Therefore, we can use the metric to compare the resilience of different systems that achieve the same functions, thereby identifying different internal factors that contribute to it. Moreover, we can change the system structure and failure or repair rates in multiples to evaluate resilience values and analyze the sensitivity. Furthermore, we can conduct resilience-based design for engineering systems based on the required resilience values determined by practical guidance documents.

Notably, only the series, parallel, voting, and nine-bus power grid systems with binary state are used in this study to demonstrate the proposed availability-based resilience metric and its corresponding evaluation methodology. Bayesian networks are a powerful tool in modeling any complex system, such as multistate system, linear consecutively connected systems, and general networks with sources and sinks. Therefore, the proposed metric and methodology are general for any complex systems. Future scopes of work can be directed toward resilience evaluation of a complex system (not limited to the binary state systems) and further system comparison, system optimization, and system design by using the proposed metric and methodology.

References

1. L.E. Cole, S.A. Bhagwat, K.J. Willis, Recovery and resilience of tropical forests after disturbance. Nat. Commun. **5**, 3906 (2014)
2. T.H. Oliver, N.J. Isaac, T.A. August, B.A. Woodcock, D.B. Roy, J.M. Bullock, Declining resilience of ecosystem functions under biodiversity loss. Nat. Commun. **6**, 10122 (2015)
3. R. Martin, Regional economic resilience, hysteresis and recessionary shocks. J. Econ. Geogr. **12**, 1–32 (2011)
4. A. Rose, S.Y. Liao, Modeling regional economic resilience to disasters: a computable general equilibrium analysis of water service disruptions. J. Reg. Sci. **45**(1), 75–112 (2005)
5. S.M. Southwick, D.S. Charney, The science of resilience: implications for the prevention and treatment of depression. Science **338**, 79–82 (2012)
6. J.Y. Pan, C.L.W. Chan, Resilience: a new research area in positive psychology. Psychologia **50**(3), 164–176 (2007)
7. L. Olsson, A. Jerneck, H. Thoren, J. Persson, D. O'Byrne, Why resilience is unappealing to social science: theoretical and empirical investigations of the scientific use of resilience. Sci. Adv. **1**(4), e1400217 (2015)
8. M. Keck, P. Sakdapolrak, What is social resilience? Lessons learned and ways forward. Erdkunde **67**(1), 5–19 (2013)
9. S. Hosseini, K. Barker, J.E. Ramirez-Marquez, A review of definitions and measures of system resilience. Reliab. Eng. Syst. Saf. **145**, 47–61 (2016)
10. J. Gao, B. Barzel, A.L. Barabási, Universal resilience patterns in complex networks. Nature **530**, 307–312 (2016)
11. Y.P. Fang, N. Pedroni, E. Zio, Resilience-based component importance measures for critical infrastructure network systems. IEEE Trans. Reliab. **65**(2), 502–512 (2016)
12. National Infrastructure Advisory Council (US), Critical infrastructure resilience: Final report and recommendations. National Infrastructure Advisory Council (2009)
13. American Society of Mechanical Engineers (US), All-hazards risk and resilience: prioritizing critical infrastructures using the RAMCAP Plus^{SM} approach. Am. Soc. Mech. Eng. (2009). ISBN: 978-0-7918-0287-8
14. B.D. Youn, C. Hu, P. Wang, Resilience-driven system design of complex engineered systems. J. Mech. Des. **133**(10), 101011 (2011)
15. M. Ouyang, Z. Wang, Resilience assessment of interdependent infrastructure systems: with a focus on joint restoration modeling and analysis. Reliab. Eng. Syst. Saf. **141**, 74–82 (2015)
16. N. Yodo, P. Wang, Resilience modeling and quantification for engineered systems using Bayesian networks. J. Mech. Des. **138**(3), 031404 (2016)
17. P.E. Roege, Z.A. Collier, J. Mancillas, J.A. McDonagh, I. Linkov, Metrics for energy resilience. Energ. Policy **72**, 249–256 (2014)
18. D.G. Dessavre, J.E. Ramirez-Marquez, K. Barker, Multidimensional approach to complex system resilience analysis. Reliab. Eng. Syst. Saf. **149**, 34–43 (2016)

19. M. Bruneau, S.E. Chang, R.T. Eguchi, G.C. Lee, T.D. O'Rourke, A.M. Reinhorn, D. von Winterfeldt, A framework to quantitatively assess and enhance the seismic resilience of communities. Earthq. Spectra **19**(4), 733–752 (2003)
20. D. Henry, J.E. Ramirez-Marquez, Generic metrics and quantitative approaches for system resilience as a function of time. Reliab. Eng. Syst. Saf. **99**, 114–122 (2012)
21. R. Francis, B. Bekera, A metric and frameworks for resilience analysis of engineered and infrastructure systems. Reliab. Eng. Syst. Saf. **121**, 90–103 (2014)
22. S. Hosseini, K. Barker, Modeling infrastructure resilience using Bayesian networks: a case study of inland waterway ports. Comput. Ind. Eng. **93**, 252–266 (2016)
23. S. Zhao, X. Liu, Y. Zhuo, Hybrid hidden Markov models for resilience metrics in a dynamic infrastructure system. Reliab. Eng. Syst. Saf. **164**, 84–97 (2017)
24. D.D. Woods, Four concepts for resilience and the implications for the future of resilience engineering. Reliab. Eng. Syst. Saf. **141**, 5–9 (2015)
25. T. McDaniels, S. Chang, D. Cole, J. Mikawoz, H. Longstaff, Fostering resilience to extreme events within infrastructure systems: characterizing decision contexts for mitigation and adaptation. Glob. Environ. Change **18**(2), 310–318 (2008)
26. L. Molyneaux, C. Brown, L. Wagner, J. Foster, Measuring resilience in energy systems: insights from a range of disciplines. Renew. Sustain. Energy Rev. **59**, 1068–1079 (2016)
27. R. Arghandeh, A. von Meier, L. Mehrmanesh, L. Mili, On the definition of cyber-physical resilience in power systems. Renew. Sustain. Energy Rev. **58**, 1060–1069 (2016)
28. R. Filippini, A. Silva, A modeling framework for the resilience analysis of networked systems-of-systems based on functional dependencies. Reliab. Eng. Syst. Saf. **125**, 82–91 (2014)
29. J. Lundberg, B.J. Johansson, Systemic resilience model. Reliab. Eng. Syst. Saf. **141**, 22–32 (2015)
30. C.W. Zobel, L. Khansa, Characterizing multi-event disaster resilience. Comput. Oper. Res. **42**, 83–94 (2014)
31. K. Bourouni, Availability assessment of a reverse osmosis plant: comparison between reliability block diagram and fault tree analysis methods. Desalination **313**, 66–76 (2013)
32. I.H. Choi, D. Chang, Reliability and availability assessment of seabed storage tanks using fault tree analysis. Ocean Eng. **120**, 1–14 (2016)
33. M. Naseri, P. Baraldi, M. Compare, E. Zio, Availability assessment of oil and gas processing plants operating under dynamic arctic weather conditions. Reliab. Eng. Syst. Saf. **152**, 66–82 (2016)
34. B. Cai, Y. Liu, Z. Liu, X. Tian, Y. Zhang, J. Liu, Performance evaluation of subsea blowout preventer systems with common-cause failures. J. Petrol. Sci. Eng. **90**, 18–25 (2012)
35. B. Cai, Y. Zhao, H. Liu, M. Xie, A data-driven fault diagnosis methodology in three-phase inverters for PMSM drive systems. IEEE Trans. Power Electron. **32**(7), 5590–5600 (2017)
36. B. Cai, L. Huang, M. Xie, Bayesian networks in fault diagnosis. IEEE Trans. Industr. Inf. **13**(5), 2227–2240 (2017)
37. Y. Fang, G. Sansavini, Optimizing power system investments and resilience against attacks. Reliab. Eng. Syst. Saf. **159**, 161–173 (2017)
38. H. Fotouhi, S. Moryadee, E. Miller-Hooks, Quantifying the resilience of an urban traffic-electric power coupled system. Reliab. Eng. Syst. Saf. **163**, 79–94 (2017)
39. B. Yssaad, M. Khiat, A. Chaker, Reliability centered maintenance optimization for power distribution systems. Int. J. Electr. Power Energy Syst. **55**, 108–115 (2014)
40. A. Zaidi, B. Ould Bouamama, M. Tagina, Bayesian reliability models of Weibull systems: state of the art. Int. J. Appl. Math. Comput. Sci. **22**(3), 585–600 (2012)
41. M. Panteli, P. Mancarella, Modeling and evaluating the resilience of critical electrical power infrastructure to extreme weather events. IEEE Syst. J. (2015). https://doi.org/10.1109/jsyst.2015.2389272
42. C. Ji, Y. Wei, H. Mei, J. Calzada, M. Carey, S. Church, J. White, Large-scale data analysis of power grid resilience across multiple US service regions. Nat. Energy **1**, 16052 (2016)
43. B. Cai, H. Liu, M. Xie, A real-time fault diagnosis methodology of complex systems using object-oriented Bayesian networks. Mech. Syst. Sig. Process. **80**, 31–44 (2016)

44. B. Cai, Y. Liu, Q. Fan, Y. Zhang, Z. Liu, S. Yu, R. Ji, Multi-source information fusion based fault diagnosis of ground-source heat pump using Bayesian network. Appl. Energy **114**(2), 1–9 (2014)
45. Y.Y. Haimes, On the definition of resilience in systems. Risk Anal. **29**(4), 498–501 (2009)
46. E. Hollnagel, D. Woods, N. Leveson, *Resilience Engineering, Concepts and Precepts* (Ashgate Publishing, 2006)
47. N.O. Attoh-Okine, *Resilience Engineering, Models and Analysis* (Cambridge University Press, 2016)

MIX
Papier aus verantwortungsvollen Quellen
Paper from responsible sources
FSC® C105338

If you have any concerns about our products,
you can contact us on
ProductSafety@springernature.com

In case Publisher is established outside the EU,
the EU authorized representative is:
Springer Nature Customer Service Center GmbH
Europaplatz 3, 69115 Heidelberg, Germany

Printed by Libri Plureos GmbH
in Hamburg, Germany